TOPICS IN
STEREOCHEMISTRY

VOLUME 5

A WILEY-INTERSCIENCE SERIES

ADVISORY BOARD

STEPHEN J. ANGYAL, *University of New South Wales, Sydney, Australia*
JOHN C. BAILAR, Jr., *University of Illinois, Urbana, Illinois*
OTTO BASTIANSEN, *University of Oslo, Oslo, Norway*
GIANCARLO BERTI, *University of Pisa, Pisa, Italy*
DAVID GINSBURG, *Technion, Israel Institute of Technology, Haifa, Israel*
WILLIAM KLYNE, *Westfield College, University of London, London, England*
KURT MISLOW, *Princeton University, Princeton, New Jersey*
SAN-ICHIRO MIZUSHIMA, *Japan Academy, Tokyo, Japan*
GUY OURISSON, *University of Strasbourg, Strasbourg, France*
GERHARD QUINKERT, *Technische Hochschule Braunschweig, Braunschweig, Germany*
VLADO PRELOG, *Eidgenössische Technische Hochschule, Zurich, Switzerland*
JIRI SICHER, *University of Lausanne, Lausanne, Switzerland*
HANS WYNBERG, *University of Groningen, Groningen, The Netherlands*

TOPICS IN STEREOCHEMISTRY

EDITORS

ERNEST L. ELIEL

Professor of Chemistry
University of Notre Dame
Notre Dame, Indiana

NORMAN L. ALLINGER

Professor of Chemistry
University of Georgia
Athens, Georgia

VOLUME 5

WILEY-INTERSCIENCE

A DIVISION OF JOHN WILEY & SONS

New York · London · Sydney · Toronto

Copyright © 1970, by John Wiley & Sons, Inc.

All rights reserved. No part of this book may be reproduced by any means, nor transmitted, nor translated into a machine language without the written permission of the publisher.

Library of Congress Catalogue Card Number: 67-13943
ISBN 0-471-23750-7

Printed in the United States of America

10 9 8 7 6 5 4 3 2 1

To the 1969 Nobel Laureates
in Chemistry
DEREK H. R. BARTON
and
ODD HASSEL

INTRODUCTION TO THE SERIES

During the last decade several texts in the areas of stereochemistry and conformational analysis have been published, including *Stereochemistry of Carbon Compounds* (Eliel, McGraw-Hill, 1962) and *Conformational Analysis* (Eliel, Allinger, Angyal, and Morrison, Interscience, 1965). While the writing of these books was stimulated by the high level of research activity in the area of stereochemistry, it has, in turn, spurred further activity. As a result, many of the details found in these texts are already inadequate or out of date, although the student of stereochemistry and conformational analysis may still learn the basic concepts of the subject from them.

For both human and economic reasons, standard textbooks can be revised only at infrequent intervals. Yet the spate of periodical publications in the field of stereochemistry is such that it is an almost hopeless task for anyone to update himself by reading all the original literature. The present series is designed to bridge the resulting gap.

If that were its only purpose, this series would have been called "Advances (or "Recent Advances") in Stereochemistry." It must be remembered, however, that the above-mentioned texts were themselves not treatises and did not aim at an exhaustive treatment of the field. Thus the present series has a second purpose, namely to deal in greater detail with some of the topics summarized in the standard texts. It is for this reason that we have selected the title *Topics in Stereochemistry*.

The series is intended for the advanced student, the teacher, and the active researcher. A background of the basic knowledge in the field of stereochemistry is assumed. Each chapter is written by an expert in the field and, hopefully, covers its subject in depth. We have tried to choose topics of fundamental import, aimed primarily at an audience of organic chemists but involved frequently with fundamental principles of physical chemistry and molecular physics, and dealing also with certain stereochemical aspects of inorganic chemistry and biochemistry.

It is our intention to bring out future volumes at approximately annual intervals. The Editors will welcome suggestions as to suitable topics.

We are fortunate in having been able to secure the help of an international board of Editorial Advisors who have been of great assistance by suggesting topics and authors for several articles and by helping us avoid duplication of topics appearing in other, related monograph series. We are grateful to the

Editorial Advisors for this assistance, but the Editors and Authors alone must assume the responsibility for any shortcomings of *Topics in Stereochemistry*.

N. L. Allinger
E. L. Eliel

January 1967

PREFACE

Volume 5 continues our annual publication schedule of *Topics* in *Stereochemistry*. There appears to be no shortage of topics to be discussed or of competent authors willing to discuss them. The increased number of chapters in Volume 5—six, as compared to four in each of the previous volumes—is in part a reflection of this fact.

This volume, for the first time, presents a trend which will, undoubtedly become more important in future volumes: some of the subjects represent, in some measure, elaborations of certain aspects of chapters published earlier. Thus, a very brief account of the stereochemistry of the Wittig reaction was included in the chapter on stereochemical aspects of phosphorus chemistry by Gallagher and Jenkins in Volume 3. In the intervening two years, the subject developed to the point where it merits treatment of its own at greater length. Dr. Manfred Schlosser, in the first chapter, presents a carefully organized and quite detailed summary of the stereochemistry of the Wittig reaction. His chapter should be of interest not only to those interested in the mechanism of what has become one of the major reactions in organic synthesis, but also to those chemists who wish to exploit the synthesis in a practical way to prepare pure *cis* or *trans* olefins.

The second chapter, by G. Krow, deals with a rather classical subject: the stereochemistry of compounds which owe their chirality to the presence of chiral axes or chiral planes rather than chiral centers. Although the *concepts* of chiral axes and chiral planes were only recently defined, the chiral axis was first recognized by van't Hoff himself in allenes, and spiranes and alkylidenecycloalkanes were resolved shortly after the turn of the century. The chiral plane seems to have made its first appearance in Lüttringhaus' ansa compounds in 1940. By now, a very large body of experimental material has accumulated and a review seems timely, especially since a number of absolute configurations of compounds of this type have been elucidated recently. It might be pointed out that Schlögl's chapter in Volume 1 deals with a particular class of compounds (the metallocenes) having a chiral plane.

Increasingly, stereochemical concepts are of interest in molecules of biochemical import. The third chapter, by M. Goodman, A. S. Verdini, and N. S. Choi, deals with the stereochemistry and conformation of polypeptides. In this chapter are brought together the results obtained by numerous physical techniques: ultraviolet, infrared, and nuclear magnetic resonance spectroscopy, optical rotatory dispersion, and circular dichroism—as well as by theoretical approaches, based on both statistical mechanics and semi-empirical calculation of conformation. This chapter should be of interest to those working in the fields of statistical mechanics, as well as to polymer chemists and to physical biochemists.

Volume 3 contained a chapter by Binsch on measurement of energy barriers by nmr. The topic of energy barriers and stable ground-state conformations is elaborated in two chapters in the present volume: the fourth chapter by G. J. Karabatsos and D. J. Fenoglio and the fifth chapter by E. Wyn-Jones and R. A. Pethrick. The former chapter deals mainly with the stable conformations of relatively small molecules about single bonds adjacent to double bonds. The latter chapter, in contrast, is oriented methodologically toward the determination of energy barriers by ultrasonic relaxation and infrared spectral methods, thus complementing the earlier chapter dealing with nmr. It might be mentioned here that a compilation of barriers and stable conformations of saturated molecules (by J. P. Lowe) has appeared in Volume 6 of the Streitwieser-Taft series *Progress in Physical Organic Chemistry*.

The sixth and last chapter by J. McKenna deals with the interpretation of quaternization rates in conformationally mobile amines. This is an area in which there has been much activity recently and also a certain amount of confusion in the interpretation of the data. Dr. McKenna has organized the experimental material carefully, considered it critically, and stated what conclusions may and may not be drawn. Hopefully, this chapter contains a lesson useful in other mechanistic studies.

We are dedicating this volume to Derek H. R. Barton and Odd Hassel who, on December 10, 1969, received the Nobel Prize for their pioneering work in the field of conformational analysis. The impact of conformational analysis on the progress of stereochemical thinking has been enormous; without it there would probably be no need for this Series. Suffice it to say that two chapters (the fourth and fifth) in the present volume and four in the four previous volumes are more or less directly concerned with conformational ideas and two additional ones in this volume (the third and sixth chapters) bear a strong relation, so our indebtedness to the 1969 Nobel Laureates is very evident.

Norman L. Allinger
Ernest L. Eliel

January 1970

CONTENTS

THE STEREOCHEMISTRY OF THE WITTIG REACTION
    by Manfred Schlosser, Organisch-Chemisches Institut der Universität and Institute für experimentelle Krebsforschung, Heidelberg, Germany 1

THE DETERMINATION OF ABSOLUTE CONFIGURATION OF PLANAR AND AXIALLY DISSYMMETRIC MOLECULES
    by Grant Krow, Department of Chemistry, Temple University, Philadelphia, Pennsylvania. 31

POLYPEPTIDE STEREOCHEMISTRY
    by Murray Goodman, Antonio S. Verdini, Nam S. Choi, and Yukio Masuda, Department of Chemistry, Polytechnic Institute of Brooklyn, New York 69

ROTATIONAL ISOMERISM ABOUT sp^2-sp^3 CARBON–CARBON SINGLE BONDS
    by Gerasimos J. Karabatsos and David J. Fenoglio, Department of Chemistry, Michigan State University, East Lansing, Michigan . 167

THE USE OF ULTRASONIC ABSORPTION AND VIBRATIONAL SPECTROSCOPY TO DETERMINE THE ENERGIES ASSOCIATED WITH CONFORMATIONAL CHANGES
    by E. Wyn-Jones and R. A. Pethrick, Department of Chemistry and Applied Chemistry, University of Salford, Salford, England . . 205

THE STEREOCHEMISTRY OF THE QUATERNIZATION OF PIPERIDINES
    by James McKenna, Chemistry Department, The University of Sheffield, Sheffield, England 275

Author Index 309

Subject Index 323

Cumulative Index, Volumes 1–5 337

TOPICS IN
STEREOCHEMISTRY

VOLUME 5

A WILEY-INTERSCIENCE SERIES

The Stereochemistry of the Wittig Reaction

MANFRED SCHLOSSER*

Organisch-Chemisches Institut der Universität and Institut für experimentelle Krebsforschung, Heidelberg, Germany

	Abstract	1
I.	Introduction	2
II.	Stereoselectivity in Reactions of Stable Ylids	4
III.	Stereoselectivity in Reactions of Reactive Ylids	13
IV.	Application of Stereoselective Carbonyl Olefination Procedures	20
V.	Discussion	25
	References	27

ABSTRACT

Stereochemical control in Wittig olefin syntheses may be accomplished in three different ways:

1. In salt-free solution the normal tendency of ylids is to combine with aldehydes to give betainlike intermediates which are very largely in the *erythro* configuration. If betaine formation can be made irreversible, high amounts of *cis* olefins will thus be obtained.

2. Several types of olefinic compounds, such as stilbenes and α,β-unsaturated ketones and esters, are significantly more stable as *trans* isomers than as *cis* isomers. Wittig reactions will afford such products *trans*-stereoselectively if equilibration of the intermediate betaines through reversible decomposition to the reactants is rapid.

3. In the presence of lithium salts, the adducts from triphenylphosphonium alkylids and aldehydes are thermodynamically much more stable in the *threo* configuration. Betaine equilibration is conveniently achieved by α-metallation followed by reprotonation of the resultant β-oxido phosphorus ylids. After completion of the reaction sequence, almost pure *trans* olefins can be isolated, provided that subsequent epimerizations are excluded.

*Present address: Organisch-Chemisches Institut der Universität, Heidelberg, Germany, Tiergartenstrasse.

Stereochemical implications as well as preparative applications of these procedures will be pointed out. Reaction rates, reversibility of betaine formation, and stereoselectivity are affected by the "stationary" ligands bound to the phosphorus atom, by the nature of the solvent, and by special additives, such as carboxylic acids or inorganic salts. These effects on the reaction will be discussed under mechanistic and practical aspects.

I. INTRODUCTION

Only a few years after its discovery, the Wittig carbonyl olefination reaction by means of phosphorus ylids has become a favorite tool in preparative organic chemistry (1–3). One of the main virtues of this synthetic method is its complete structural specificity. While, for instance, the addition of a Grignard reagent to a carbonyl compound followed by dehydration normally leads to a mixture of positionally isomeric olefins, the new carbon–carbon double bond created in the Wittig reaction appears exclusively at the site of the former carbonyl function (4) (eq. (1)).

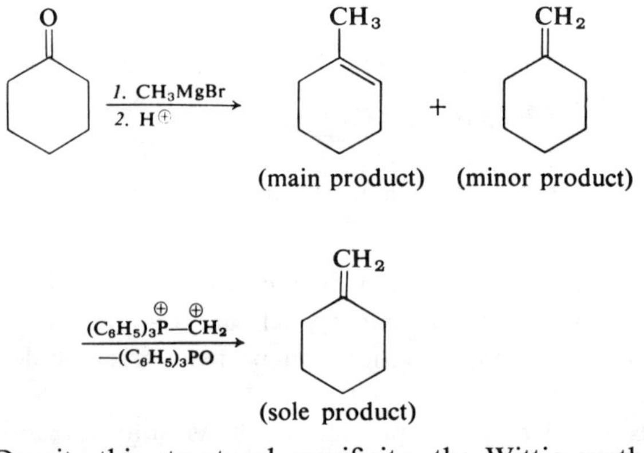

Despite this structural specificity, the Wittig synthesis may yet afford more than one olefin if the reaction product exhibits *cis–trans* isomerism. Indeed, the classical paper of Wittig and Schöllkopf (4) already mentions that the reaction between α-monosubstituted phosphorus methylids and aldehydes produces *cis* and *trans* olefins in about equal amounts. This finding was in agreement with expectation and so at that time the carbonyl olefination reaction seemed to offer no stereochemical problems. But soon Bohlmann et al. (5) recognized that phosphorus alkylids do not necessarily react nonstereoselectively, but may in fact preferentially yield the thermodynamically less stable *cis* olefins in some instances. Many other investigators were able to confirm this puzzling observation (6–10) and special credit must be given to Shemyakin and co-workers (11–13), who were the first to study the *cis* selectivity systematically and to point out its preparative value.

The first case of the opposite stereoselectivity was revealed by House and Rasmusson (14). They demonstrated that the condensation of triphenylphosphonium carbomethoxy-ethylid with acetaldehyde led to a mixture of isomeric α,β-unsaturated methyl esters, $CH_3CH=C(CH_3)COOCH_3$, in which the thermodynamically more stable tiglic acid derivative (*E* configuration) prevailed in a ratio of 96:4 over the less stable angelic acid ester, which bears the carbomethoxy residue on the same side of the double bond as the vicinal methyl group, i.e., has the *Z* configuration. Later, *trans*-selectivity was found to be a general feature of the Wittig reaction wherever so-called "stable" ylides are involved, that is to say, ylids bearing a powerful electron-attracting substituent, such as carbonyl or carboxyl functions, in the α-position to the phosphorus atom. Moreover, the *trans*-selectivity is preserved when the phenyl groups linked to the phosphorus are exchanged for other "stationary" ligands, even if one switches from ylids derived from phosphonium salts to the α-carbanions of phosphine oxides or phosphonic acid esters (15).

Any attempt to rationalize the complex stereochemistry of olefin formation will have to be concerned with the mechanism of the Wittig reaction. Unfortunately many mechanistic details of this reaction have not yet been satisfactorily elucidated. Thus, it is even uncertain whether the reacting system (ylid plus carbonyl compound) passes, on its way to phosphine oxide and olefin, through an open-chain zwitterion, i.e., a betaine, through a cyclic oxaphosphetane (16), or through both of them consecutively. The reaction sequence depicted in Figure 1 has no other merit than that of being believed to be the most probable one.

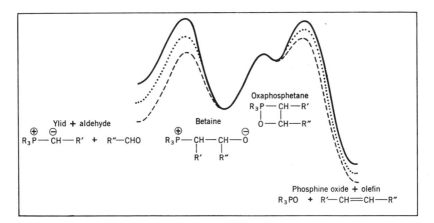

Fig. 1. (Assumed) energy profile of the Wittig carbonyl olefination reaction, effected by (———) "reactive", (· · · ·) "moderated" or (----) "stable" phosphonium ylids. (In the following figures the hypothetical oxaphosphetane intermediate will be omitted, since it has no further bearing on the discussion.)

Closely related to the olefin-forming process are cyclopropane (17–19) and epoxide (20) forming ylid reactions. Nevertheless, this chapter will be restricted to stereochemical effects governing the *cis–trans* ratios of products obtained in olefin syntheses. Accordingly, the configurational changes (21–24) at the phosphorus atom in the course of the Wittig reaction will also be disregarded.

II. STEREOSELECTIVITY IN REACTIONS OF STABLE YLIDS

So-called stable ylids are characterized by extensive delocalization of the negative charge through participation of resonance structures, e.g.,

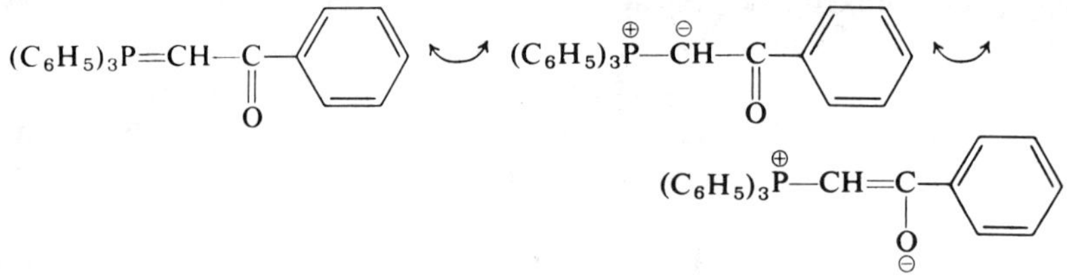

$$(C_6H_5)_3P=CH-\underset{O}{\overset{\|}{C}}-\underset{}{\bigcirc} \rightleftharpoons (C_6H_5)_3\overset{\oplus}{P}-\overset{\ominus}{CH}-\underset{O}{\overset{\|}{C}}-\underset{}{\bigcirc} \rightleftharpoons$$

$$(C_6H_5)_3\overset{\oplus}{P}-CH=\underset{\underset{\ominus}{O}}{C}-\underset{}{\bigcirc}$$

Because of the relatively low basicity of resonance-stabilized ylids, their addition to carbonyl compounds is an endergonic process. Thus, as is frequently observed in aldol-type reactions, no reaction intermediates can be

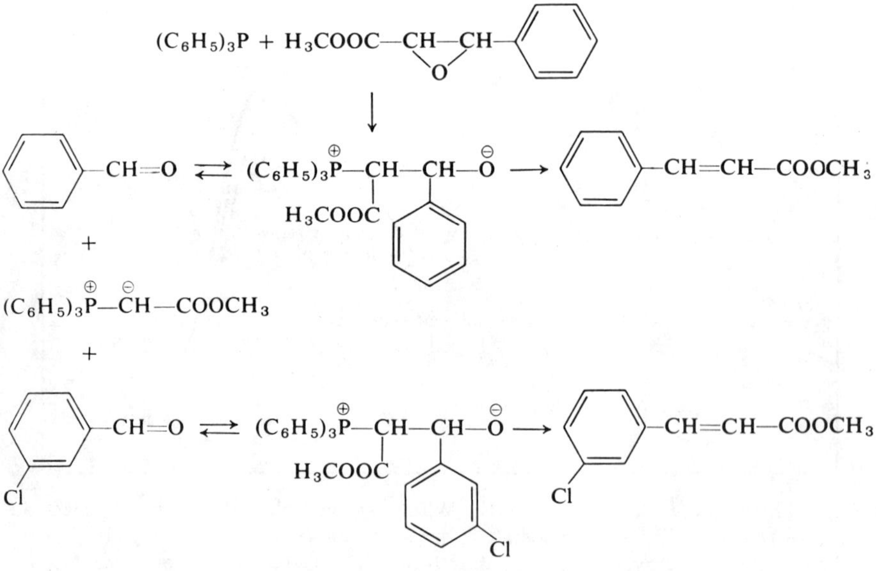

Scheme 1

captured. Yet, the hypothetical zwitterionic intermediates turned out to be accessible by an independent route, thus offering an elegant means for demonstrating the reversibility of betaine formation involving stable ylids. The addition of triphenylphosphine to phenylglycidic ester afforded a betaine which was cleaved rapidly and reversibly to yield benzaldehyde and triphenylphosphonium carbomethoxymethylid which could be trapped by added *m*-chlorobenzaldehyde (25) (Scheme 1).

The *trans* stereoselectivity of olefin syntheses effected with stable ylids may now be explained as follows (14). The reactants combine to give a betaine with either an *erythro* or *threo* configuration, the two species being in equilibrium. This intermediate eliminates triphenylphosphine oxide completely irreversibly, and in a *cis* manner (21), so that the *erythro* and *threo*-betaine must give *cis* and *trans* olefin, respectively. Because of conjugation, the *trans* isomer is approximately 4 kcal/mol more stable and it is reasonable to assume that the corresponding transition state is also lowered in energy to some extent. As a consequence, the *threo* epimer is preferentially consumed, but is continuously replenished by the mobile equilibrium through which both betaine diastereoisomers can be rapidly interconverted. Under these circumstances an overall preference of the *threo* → *trans* route by one or two powers of ten is what one might expect (see Fig. 2).

Resonance-stabilized phosphorus ylids need not, however, always yield the thermodynamically more stable olefins. If the carbonyl compound is highly reactive and the nucleophilicity of the ylid is not too low, the reversible decomposition of the reaction intermediates may be outrun by the triphenylphosphine oxide elimination, in which case product formation will be kinetically controlled. In this way it may be understood why phthalic anhydride affords the *trans* olefin by action of triphenylphosphonium carbamidomethylid and carbalkoxymethylids, but the *cis* olefin by action of triphenylphosphonium acetylmethylid and a 1:4 mixture of both by action of triphenylphosphonium phenacetylid (26) (eq. (2)).

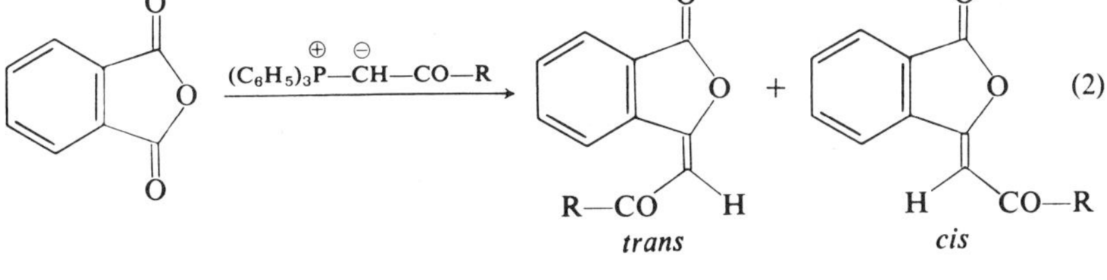

R	ratio
R = NH$_2$	> 10 : 1
R = OCH$_3$	> 10 : 1
R = C$_6$H$_5$	1 : 4
R = CH$_3$	< 1 : 10

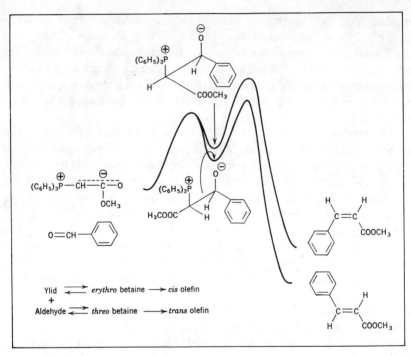

Fig. 2. Energy profile for the reaction between a stable phosphorus yield and an aldehyde. The eclipsed betaine conformations are not expected to be the most highly populated; they are depicted in this manner only for the sake of clarity.

Similarly, the less stable *trans* isomer (or *E* isomer) results from the reaction of triphenylphosphonium carbethoxymethylid with *N*-methylphthalimide (27).

Still somewhat obscure is the puzzling catalytic effect of weakly acidic additives on the reaction rates of stable ylids. In a typical case, the formation of cinnamic ester from triphenylphosphonium carbomethoxymethylid and benzaldehyde in benzene solution is accelerated by a factor of about 20 upon addition of one equivalent of benzoic acid. Replacement of dimethylformamide by methanol as the solvent leads to a rate enhancement by a factor of over 100 (28). Obviously, all the species involved in this carbonyl olefination reaction, including the phosphine oxide (29), can act as hydrogen-bond acceptors and, therefore, their energy levels will be lowered by proton-donor substances. But the weakly polar oxaphosphetane and the transition state will, of course, be less stabilized by hydrogen bonding than the ylid and betaine zwitterions, so that one might have predicted an overall rate decrease. Since the contrary is true, an additional effect must be operative. One possible mode of operation might be hydrogen bonding to the ester residue at the oxaphosphetane stage. In this manner electron demand increases at the position neighboring the phosphorus atom, an effect which is known to assist the elimination of triphenylphosphine oxide from the intermediate. Some further ideas in this respect may be derived from the report (30) that the cleavage of oxiranes yielding carbonyl compounds and olefins is accelerated by acid catalysis.

If the height of the second energy barrier is really reduced relative to the first barrier upon hydrogen bonding (see Fig. 3), this change should lead, at least in some cases, to a shift in isomer distribution. Indeed, in the reaction between phosphorus carbalkoxymethylids and aliphatic (31) or aromatic (32) aldehydes, the amount of the *cis* isomer in the product mixture increases significantly if an aprotic solvent is replaced by a protic one or if soluble lithium salts are added. For instance, triphenylphosphonium carbethoxymethylid and benzaldehyde afford *cis*- and *trans*-cinnamic esters in the ratio 2:98 if benzene serves as the solvent (28), but in the ratio 15:85 if the reaction is carried out in ethanol (32).

On the other hand, the *trans*-stereoselectivity is improved if the "stationary" phenyl ligands are replaced by alkyl groups (Table I). Alkyl substitution destabilizes the phosphorus ylid (33), but causes a considerable stabilization of phosphine oxides (29) and by extrapolation, since the phosphorus atom is then even more electron demanding, particularly of betaines. The olefin-generating transition state, however, will be lowered in energy content to a much lesser extent. This can easily be seen if one makes the reasonable assumption that the transition state resembles the oxaphosphetane in which the phosphorus atom bears only a small charge, if any. As a consequence,

TABLE I

cis–trans Ratios for Ethyl Cinnamate Resulting from the Interaction Between Different Phosphorus Carbethoxymethylids and Benzaldehyde in Ethanol at 25°C

R in $R_3\overset{\oplus}{P}-\overset{\ominus}{C}H-COOC_2H_5$	 Ph—CH=CH—$COOC_2H_5$ *cis–trans* ratio
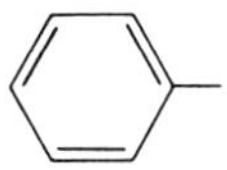 Ph	15:85
n-C_4H_9	5:95
n-$C_{10}H_{21}$	4:96
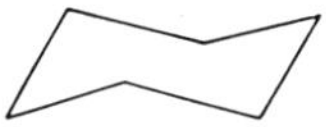 cyclohexyl	1:99

the energy barrier, which must be overcome in going from the betaine to the olefin, rises and the *erythro–threo* equilibration through reversible betaine decomposition becomes more successful (Fig. 3).

Wittig reactions effected with "moderated" ylids, such as triphenylphosphonium benzylid, are usually devoid of marked stereoselectivity (31–31b), although the principles governing the stereochemical course of these reactions are the same as in the case of olefination reactions through "stable" ylids. The reason for the difference is that aryl, alkenyl, and alkinyl residues at the position α to the phosphorus atom cause less effective resonance stabilization than, say, acyl, carbalkoxy, or cyano groups. As a consequence, betaine formation is only weakly endergonic and the rates of forward and backward decomposition of the intermediate are frequently of the same order of magnitude. Accordingly, electron-donating substituents which enhance the nucleophilicity of the ylid reduce the *trans*-stereoselectivity, while electron-withdrawing ligands increase it (34,35) (Scheme 2). Conversely, the *cis*-stereoselectivity of triphenylphosphonium propinylid (36) indicates that it is less effectively resonance stabilized than triphenylphosphonium allylid (4,37), which usually yields *cis–trans* isomer mixtures with a slight preponderance of the *trans* olefin.

A delicate balance of rate-retarding and rate-accelerating effects seems to be responsible for the changes in stereochemistry observed when moderated ylids are allowed to react in the presence of protic solvents or soluble lithium salts. A careful study of stilbene formation from methyldiphenylphosphonium benzylid and benzaldehyde has revealed (38) that reversible

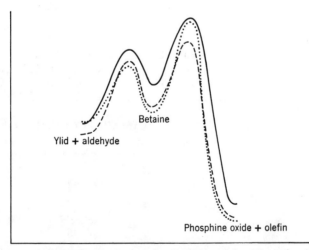

Fig. 3. Energy profile of cinnamic ester formation by action of tri*phenyl*phosphonium carbethoxymethylid (———) in benzene and (- - - -) in ethanol and (· · · ·) by tricy*clohexyl*phosphonium carbethoxymethylid in ethanol.

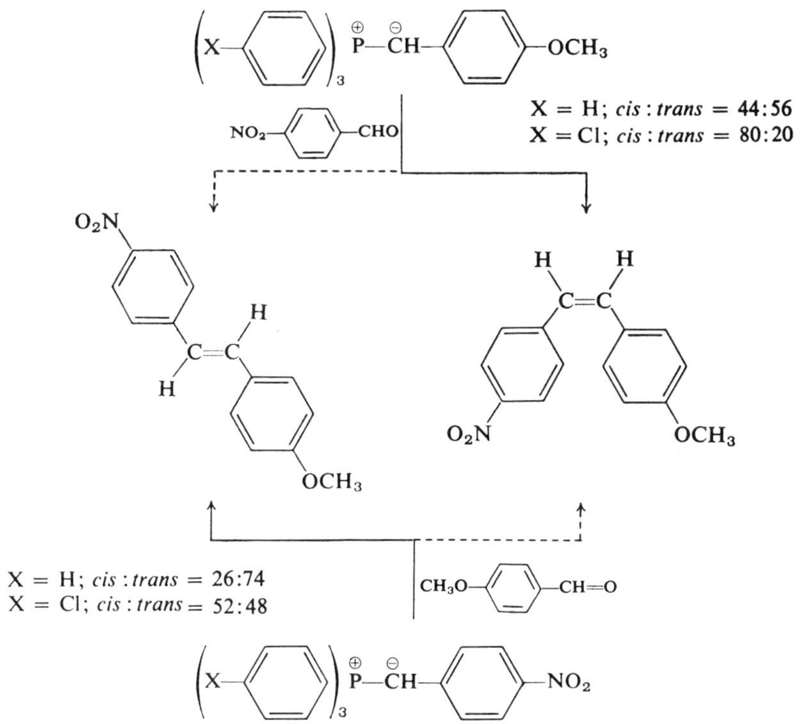

THE STEREOCHEMISTRY OF THE WITTIG REACTION

Scheme 2

X = H; cis : trans = 44 : 56
X = Cl; cis : trans = 80 : 20

X = H; cis : trans = 26 : 74
X = Cl; cis : trans = 52 : 48

betaine decomposition is slow compared to triphenylphosphine oxide elimination in aprotic solution. The "reversibility factor" which may be defined as the ratio $k(erythro$ betaine $\to$ reactants$)/k(erythro$ betaine $\to cis$ olefin) ranges from 0.1 in dimethyl sulfoxide to 3.3 in methanol. Correspondingly, the cis–trans ratio in the two solvents decreases from 47:53 to 22:78. In tetrahydrofuran in the presence of lithium bromide, an even higher proportion of cis-stilbene (cis : trans = 53 : 47) is obtained (39) (Table II).

A comparison of the distribution of stilbene isomers obtained from ethanol solutions clearly demonstrates that the more alkyl groups replace phenyl groups at the phosphorus atom, and the bigger these groups are, the more trans olefin results (Table II). As already stated above, the electron-rich alkyl residues lower the energy level of the betaine intermediate relative to the olefin-generating transition state and thus enhance the rate of betaine equilibration through reversible decomposition.

In principle, the same change is observed on passing from the "P$^\oplus$ ylids," derived from triphenylphosphonium salts, to the "PO ylids," as we may call the carbanions of phosphine oxides and phosphonic esters. In

TABLE II

Relative Yields of cis- and trans-Stilbene from the Reaction of Different Phosphorus Benzylids $RR'R''\overset{+}{P}-\overset{-}{C}HC_6H_5$ with Benzaldehyde at Room Temperature

Solvent	R, R', R''  = C$_6$H$_5$ (ref. 39, 53)	R = CH$_3$, R', R''  = C$_6$H$_5$ (ref. 38)	R, R', R'' = n-C$_4$H$_9$ (ref. 39)	R, R', R'' 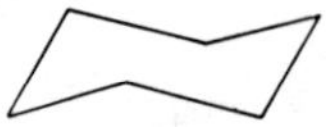 = C$_6$H$_{11}$ (ref. 40)
Tetrahydrofuran/LiBr	67:33	53:47	—	—
Dimethyl sulfoxide	—	47:53	—	—
t-Butanol	—	32:68	—	—
Ethanol	58:42	28:72	9:91	5:95
Methanol	47:53	22:78	—	—
Benzene/toluene	44:56	—	—	—

contrast to the rapid reaction of PO ylids with aldehydes in the addition step of the olefination sequence, the second, product-forming step can, in general, be brought about only if the intermediate is activated at the position α to phosphorus. The corresponding betaines containing a phosphinyl group and the final phosphorus acid salts are so stable that the transition state in between becomes very unfavorable. Therefore, the "PO modification" is limited mainly to the preparation of diarylethylenes and α,β-unsaturated carbonyl compounds (3,41) (eq. (3)).

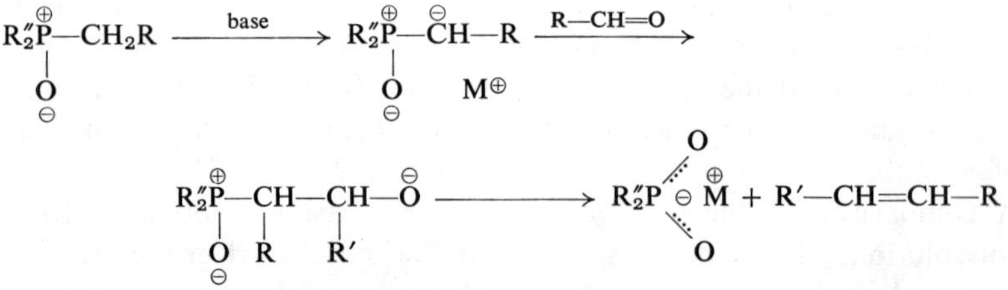

$$R''_2\overset{\oplus}{P}-CH_2R \xrightarrow{\text{base}} R''_2\overset{\oplus}{P}-\overset{\ominus}{C}H-R \xrightarrow{R-CH=O}$$
$$\underset{\ominus}{O} \qquad \underset{\ominus}{O} \quad M^{\oplus}$$

$$R''_2\overset{\oplus}{P}-CH-CH-\overset{\ominus}{O} \longrightarrow R''_2\overset{\oplus}{P}\overset{O}{\underset{O}{\cdots}}M^{\oplus} + R'-CH=CH-R \quad (3)$$
$$\underset{\ominus}{O} \quad R \quad R'$$

$$R'' = C_6H_5, C_2H_5O \ldots$$
$$R = C_6H_5, COCH_3, COOC_2H_5, CN \ldots$$

As exemplified above, a low barrier between the intermediate adduct and the reactants and a high barrier separating the intermediate from the products constitute excellent conditions for diastereoisomeric equilibration. As a consequence, the trans isomers, which are considerably more stable than the cis isomers in the series of stilbenes or α,β-unsaturated carbonyl compounds,

(4)

(5)

80–90% isomerically pure

should be formed preferentially or almost exclusively. Indeed, in all PO-activated carbonyl olefinations so far investigated, the proportion of the *trans* isomer exceeded 90% (in most cases 96%) (42).

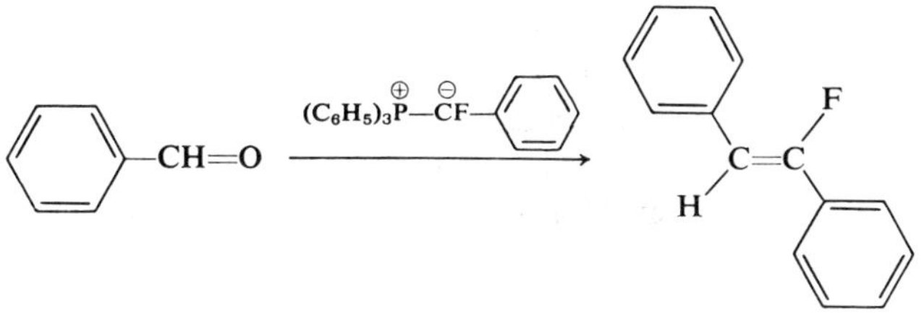

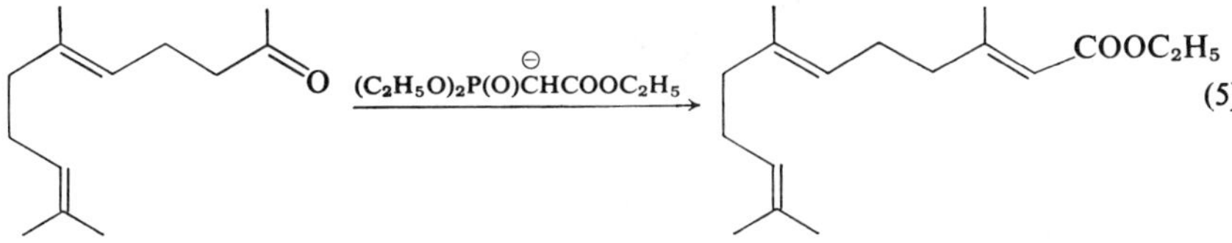

If a trisubstituted ethylene is produced in the reaction between an α,α-disubstituted phosphonate and an aldehyde, the aldehydic residue will normally show up at the olefinic double bond in the position *trans* to that ylid ligand which has the higher "mesomeric potential" in order to minimize steric hindrance to resonance (43–45) (eq. (4)). By the same token, the principal isomer resulting from the interaction of an unsymmetrical ketone and an unbranched PO ylid contains the larger ketone residue and the former ylid side chain on opposite sides of the double bonds (46,47) (eq. (5)).

trans (*E*) *cis* (*Z*)

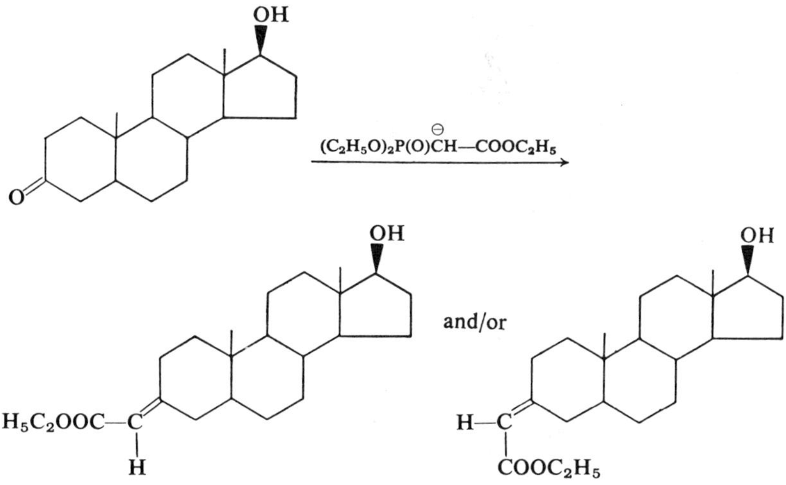

In view of the consistency observed throughout, it remains somewhat mysterious why the stereochemical outcome should depend on the nature of the base utilized for phosphonate generation to the extent that is reported for the condensation between diethyl carbethoxymethylphosphonate and dihydrotestosterol. In the presence of sodium hydride the steroid derivative

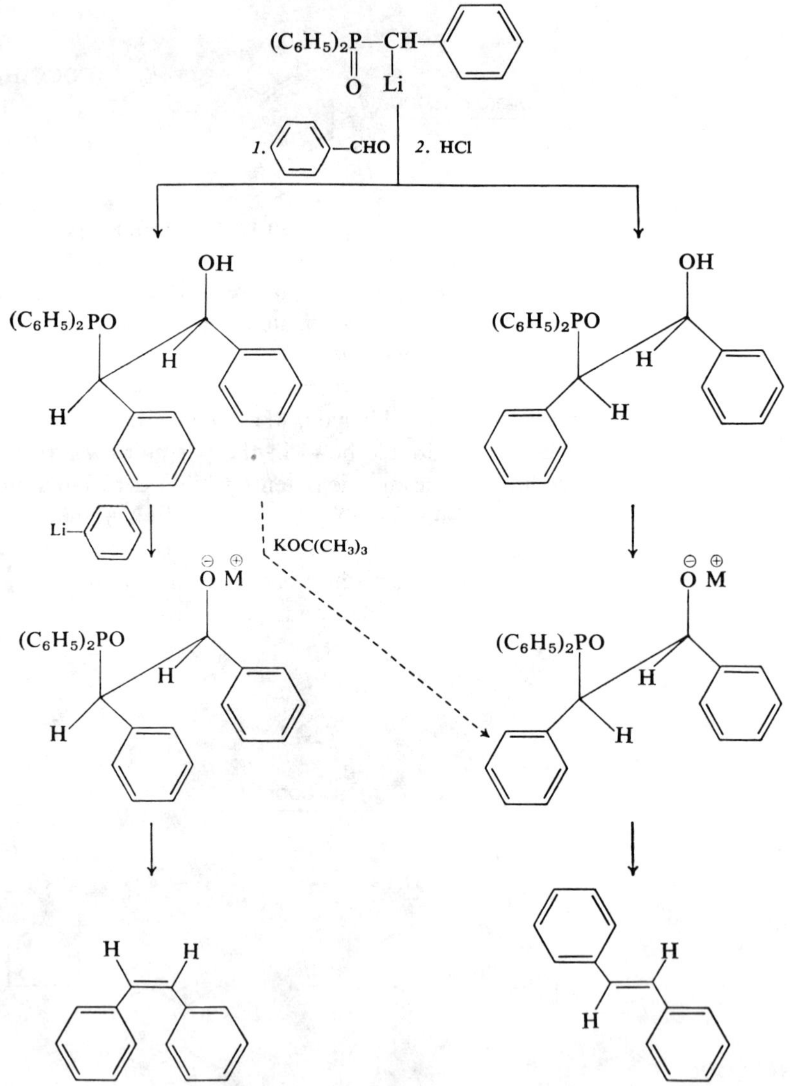

M = Li, K

Scheme 3

with the ester group placed in the position *trans* to ring B was found to be the sole product, while the *cis* product seemed to predominate when the reaction was carried out by means of potassium *t*-butoxide (48) (eq. (6)).

It became possible to study, in detail, the ability of the zwitterionic intermediate to equilibrate after accomplishment of successful isolation and separation of the *erythro–threo* mixture of (β-hydroxy-α,β-diphenylethyl)-diphenylphosphine oxide, which had been prepared by action of α-lithium benzyldiphenylphosphine oxide on benzaldehyde followed by hydrolysis. Upon treatment with exactly one equivalent of phenyllithium and heating, the *erythro* component afforded pure *cis*-stilbene. However, when potassium *t*-butoxide served as the base, only *trans*-stilbene could be isolated (49) (Scheme 3).

III. STEREOSELECTIVITY IN REACTIONS OF REACTIVE YLIDS

The addition step of most Wittig reactions brought about by means of stable or moderated ylids has been found to occur endergonically. But even if a highly nucleophilic PO ylid, such as α-lithium benzyldiphenylphosphine oxide, is involved, a rapid equilibrium between the zwitterionic reaction intermediate and the reactants will be established on account of the very great height of the second activation barrier. This situation changes profoundly as soon as we pass on to the so-called "reactive" phosphorus ylids, that is to say, ylids with a saturated aliphatic side chain. The interaction of such a reactive ylid with a carbonyl group is usually an exergonic process. In addition, the olefin-generating transition state lies low enough to ensure that the betaine intermediates are converted predominantly to the products before returning to the reactants. (Fig. 4).

A more detailed study has revealed that the reaction between triphenylphosphonium alkylids and primary *aliphatic* aldehydes is irreversible within the limits of experimental precision. Some reversible decomposition is observed if *α,β-unsaturated* or *aromatic* aldehydes serve as carbonyl components. Thus, addition of excess *m*-chlorobenzaldehyde to the preformed adduct from triphenylphosphonium methylid and benzaldehyde at $-50°C$ followed by warming to about $-25°C$ resulted in the formation of 6% *m*-chlorostyrene in addition to 70% styrene itself. Since the introduction of alkyl groups at the position α to the phosphorus atom slows down the elimination of triphenylphosphine oxide by one order of magnitude, reversible return is enhanced in these compounds. Thus, upon decomposition in the presence of *m*-chlorobenzaldehyde, the adduct from triphenylphosphonium ethylid and benzaldehyde yields 39% β-methylstyrene and 42% *m*-chlorophenylpropene (50).

If both forward steps of the Wittig reaction sequence are completely

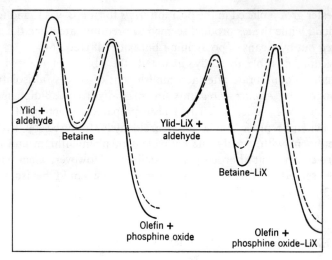

Fig. 4. Energy profiles of the reaction between triphenylphosphonium ethylid and propanal (left) in the absence and (right) in the presence of soluble lithium salts. *erythro* → *cis* course (----); *threo* → *trans* course (——).

irreversible, the olefin *cis–trans* ratio strictly reflects the *erythro–threo* composition of the intermediate betaine. Even in the case of partial reversibility the isomeric composition of the final reaction products still allows a fair estimate on the primary distribution of betaine diastereoisomers. Interestingly, all the stereochemical results observed lead to the conclusion that, particularly in salt-free solution, the formation of the *erythro* betaine is greatly favored over the formation of the *threo* betaine. *cis–trans* Ratios measured for the reaction of triphenylphosphonium ethylid with aliphatic and aromatic aldehydes are always greater than 85:15 provided the aldehyde is not highly electrophilic. Homologous unbranched triphenylphosphonium alkylids actually yield an average *cis–trans* distribution of 95:5 (Table III) (50–52) (eq. (7)).

$$(C_6H_5)_3\overset{\oplus}{P}-\overset{\ominus}{C}H-CH_2R + R'-CH=O \longrightarrow$$

$$\underset{\text{Main product}}{\overset{H}{\underset{R'}{>}}C=C\overset{H}{\underset{CH_2R}{<}}} + \underset{\text{Minor product}}{\overset{R'}{\underset{H}{>}}C=C\overset{H}{\underset{CH_2R}{<}}} \quad (7)$$

R, R': residues to be chosen arbitrarily

It has been recommended (53) that *cis*-stereoselective carbonyl olefinations be carried out in salt-free benzene, toluene, or tetrahydrofuran solutions at 0°C. In another convenient procedure a mixture of the phosphonium salt

Table III

cis–trans Ratios of Olefins Prepared from Triphenylphosphonium Alkylids and Aldehydes in Benzene, Toluene, or Tetrahydrofuran Solution at 0°C

R in R—CH=O	cis:trans Ratio		
	$^{\oplus}P(C_6H_5)_3$ $^{\ominus}CH$—CH_3	$^{\oplus}P(C_6H_5)_3$ $^{\ominus}CH$—C_2H_5	$^{\oplus}P(C_6H_5)_3$ $^{\ominus}CH$—n-C_3H_7
C_2H_5	—	97:3	—
n-C_3H_7	—	95:5	95:5
n-C_5H_{11}	91:9	96:4	—
$C_6H_5CH=CH-$	87:13	—	—
$C_6H_5C\equiv C-$	79:21	—	—
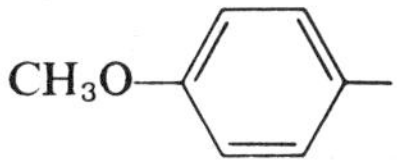 $CH_3O-C_6H_4-$	90:10	92:8	90:10
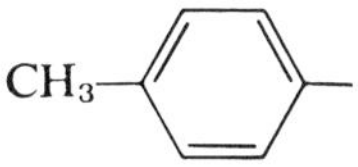 $CH_3-C_6H_4-$	89:11	95:5	92:8
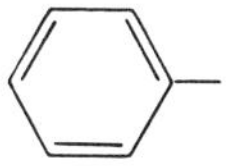 C_6H_5-	87:13	96:4	94:6
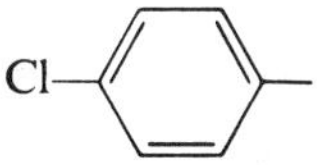 $Cl-C_6H_4-$	82:18	93:7	92:8
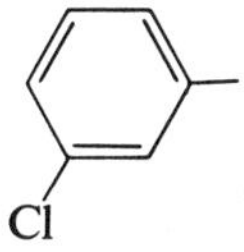 m-Cl-C_6H_4-	74:26	—	—
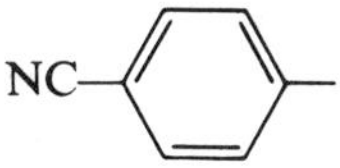 $NC-C_6H_4-$	74:26	—	—
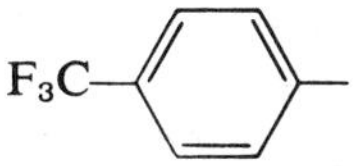 $F_3C-C_6H_4-$	69:31	—	—

and the aldehyde is treated with sodium *t*-pentoxide dissolved in benzene containing trace amounts of dimethyl sulfoxide (53a). Olefins are also formed *cis*-stereoselectively in some special types of Wittig reaction, as for instance if the reacting ylid is formed intermediately by a Michael-type addition of triphenylphosphine to acrylonitrile (54) (eq. (8)).

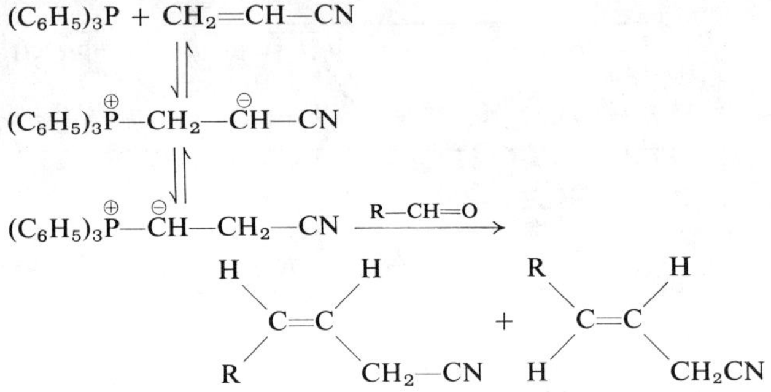

$$(C_6H_5)_3P + CH_2=CH-CN$$

$$\updownarrow$$

$$(C_6H_5)_3\overset{\oplus}{P}-CH_2-\overset{\ominus}{C}H-CN$$

$$\updownarrow$$

$$(C_6H_5)_3\overset{\oplus}{P}-\overset{\ominus}{C}H-CH_2-CN \xrightarrow{R-CH=O}$$

$$\overset{H}{\underset{R}{>}}C=C\overset{H}{\underset{CH_2-CN}{<}} + \overset{R}{\underset{H}{>}}C=C\overset{H}{\underset{CH_2CN}{<}} \quad (8)$$

R = alkyl

cis-Stereoselectivity is, however, lost if a secondary or tertiary alkyl group is linked to the position α to the phosphorus atom (55) (eq. (9)). One obvious reason may be sought in the fact that bulky groups cause an increase of steric hindrance mainly in the intermediate *erythro* diastereoisomers. It must, however, be noted that the rigid cyclopropene moiety (56,57) seems to exert nearly the same effect as the more bulky cyclohexyl group (55).

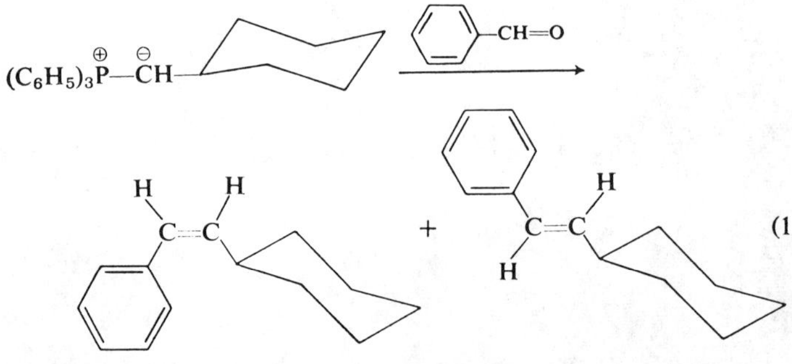

$$(C_6H_5)_3\overset{\oplus}{P}-\overset{\ominus}{C}H-\text{Cy} \xrightarrow{\text{Ph-CH=O}}$$

$$\overset{H}{\underset{Ph}{>}}C=C\overset{H}{\underset{Cy}{<}} + \overset{Ph}{\underset{H}{>}}C=C\overset{H}{\underset{Cy}{<}} \quad (1:1) \quad (9)$$

If the ylid is prepared from a phosphonium salt by action of *n*-butyllithium or phenyllithium, lithium salts are inevitably generated. Only rarely do such inorganic salts precipitate and thus not interfere with the reactants. More frequently, they alter the stereochemistry of the Wittig reaction at two different stages. At the beginning the lithium salts associate with the phosphorus ylid and later, after the addition of the phosphorus ylid to the carbonyl compound, they associate with the betaine. As a consequence, the free energy level of both species (ylid and betaine) is significantly lowered. The olefin-generating transition state, however, is stabilized only to a small extent, if at all. Thus, the reversibility factor must increase and, as a logical consequence, more of the kinetically favored *erythro* betaine will be converted to the *threo* diastereoisomer in the course of the reaction (53,58) (Fig. 3).

Even in those cases where reversible return seems to be unimportant, as with olefination reactions involving aliphatic aldehydes, lithium salts may affect the reaction course in another manner. For reasons which are not yet well understood, they shift the position of the *erythro–threo* equilibrium of the betaine epimers dramatically to the side of the *threo* compound. While there is good reason to assume that the equilibrium composition of the salt-free diastereoisomeric triphenylphosphonium ethylid/benzaldehyde adduct consists of *erythro* and *threo* components in the approximate ratio of 35:65 (53), a ratio of 5:95 has been found experimentally in the presence of excess lithium bromide (58). The increase in thermodynamic stability by complex formation counterbalances the kinetic disadvantage of the *threo* epimer. As expected, the advantage of the *erythro* betaine formation diminishes the more, the less nucleophilic the anion of the lithium salt (Table IV) (53).

TABLE IV

Ratios of *erythro* and *threo* Betaines Immediately after Mixing Triphenylphosphonium Ethylid or *n*-Propylid with Benzaldehyde at 0°C

	erythro/threo Ratio	
Additive	$(C_6H_5)_3\overset{\oplus}{P}-\overset{\ominus}{C}H-CH_3$	$(C_6H_5)_3\overset{\oplus}{P}-\overset{\ominus}{C}H-C_2H_5$
Salt-free	90:10[a]	96:4
LiCl	81:19	90:10
LiBr	61:39	86:14
LiI	58:42	83:17
LiB(C$_6$H$_5$)$_4$	50:50	52:48

[a] determined after addition at −78°C.

The remarkable shifting of the epimeric equilibrium to the side of the *threo* diastereoisomer in the presence of lithium halides opened a convenient route for *trans*-stereoselective synthesis of simple olefins. The essential point was to find a reliable method for rapid equilibration of diastereoisomeric betaines, since equilibration via reversible betaine decomposition had turned out to be too slow for practical reasons (59). A much superior method makes use of α-metallation of the betaines, which creates a center of rapid configurational inversion in the position neighboring the phosphorus atom (60). The diastereoisomeric β-oxido phosphorus ylids ("betaine ylids") are interconverted, even at temperatures as low as −78°C. After the completion of

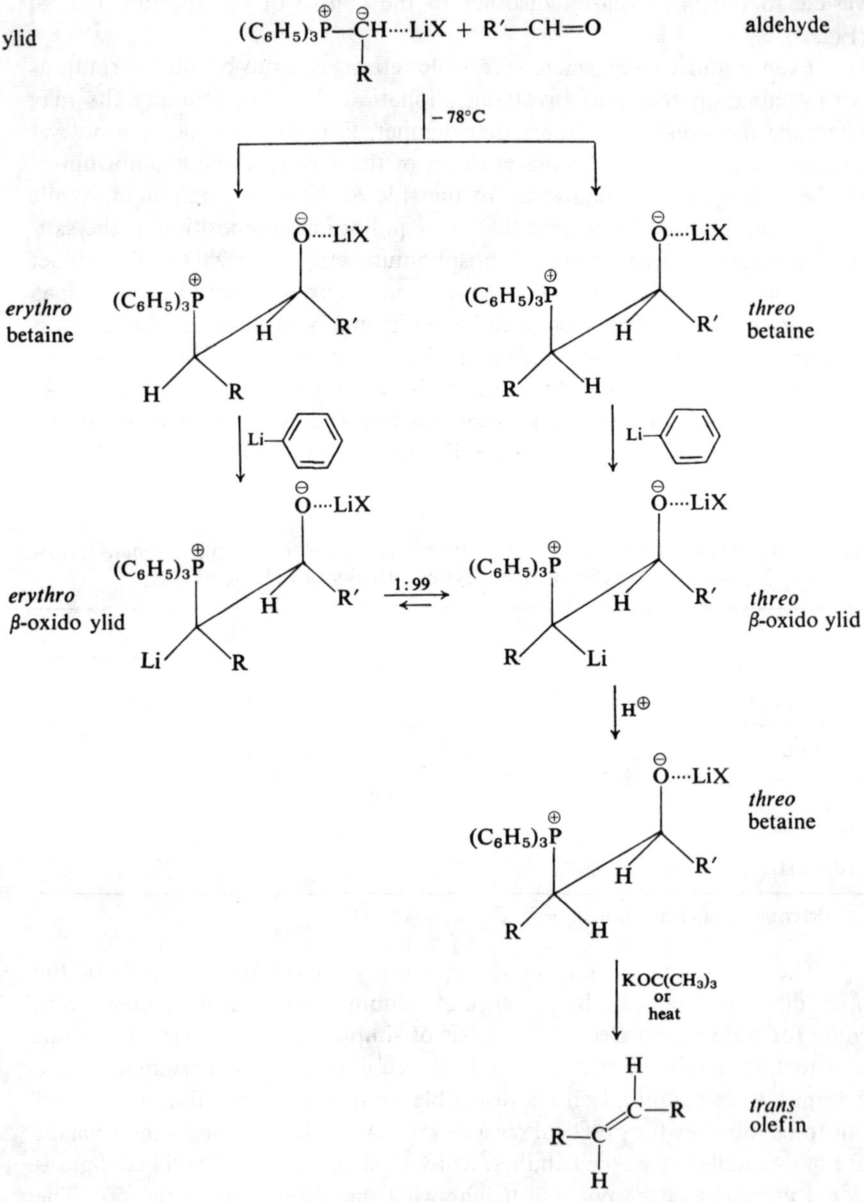
ylid
(C6H5)3P—CH····LiX + R′—CH=O
R
aldehyde
− 78°C
erythro
betaine
threo
betaine
O····LiX
(C6H5)3P
H
R′
Li
erythro
β-oxido ylid
1:99
threo
β-oxido ylid
H⊕
KOC(CH3)3
or
heat
trans
olefin

Scheme 4

THE STEREOCHEMISTRY OF THE WITTIG REACTION

equilibration (within a few minutes) the β-oxido ylid can be converted back to the betaine stage by addition of one equivalent of a proton donor (hydrogen chloride, *t*-butanol, or water). Through this procedure the betaine assumes the *threo* configuration to the extent of about 99% and may be converted to nearly pure *trans* olefin at higher temperatures with the aid of an activator (61) or a polar solvent (10,53) (Scheme 4).

trans-Stereoselective Wittig reactions via β-oxido ylids may be carried out using triphenylphosphonium alkylids in combination with a great variety of aliphatic, unsaturated, and aromatic aldehydes. *trans*-Stereoselectivity is observed even when an unsymmetrical ketone serves as the carbonyl reagent. Thus, triphenylphosphonium ethylid and acetophenone afford *trans*- and *cis*-2-phenyl-2-butene in a ratio of 40:60 in the conventional manner, but in a ratio of 11:89 through the β-oxido ylid (53) (Scheme 5).

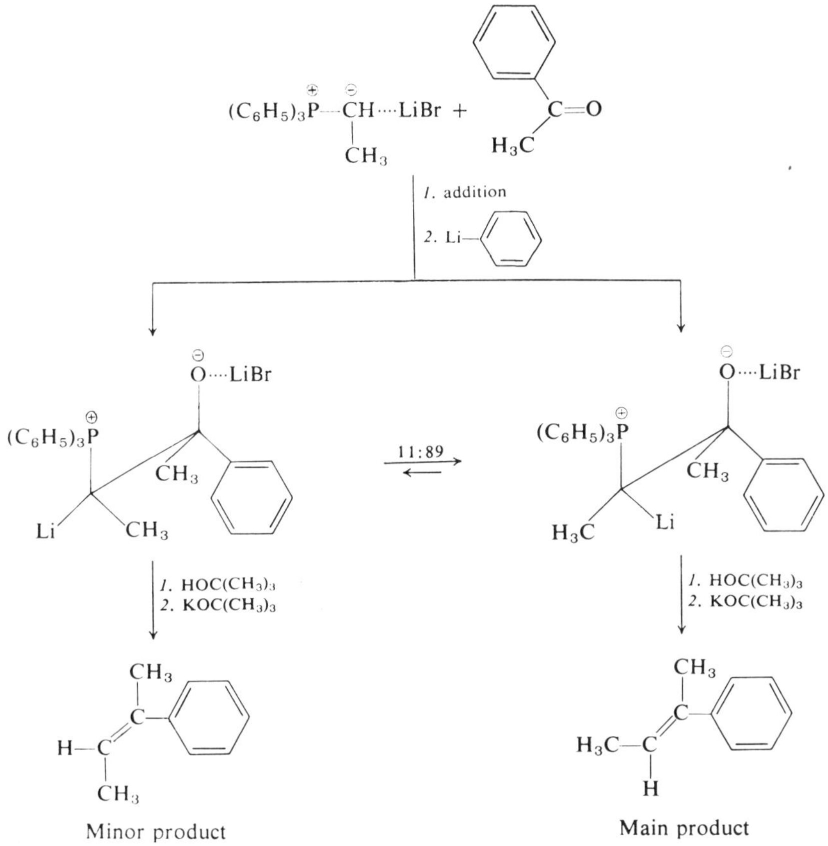

Investigations concerning PO ylid counterparts of reactive phosphonium ylids are rare. Only recently it was shown that phosphonic acid amides are easily deprotonated by *n*-butyllithium and that the resultant α-metallated derivatives combine readily with carbonyl compounds. The betainelike adducts of lithium alcoholate structure proved to be stable to heating, but it was noticed that the corresponding β-hydroxyphosphonamides liberated upon hydrolysis could be converted to phosphinic amides and olefins at reflux temperature in toluene solution (62) (Scheme 6).

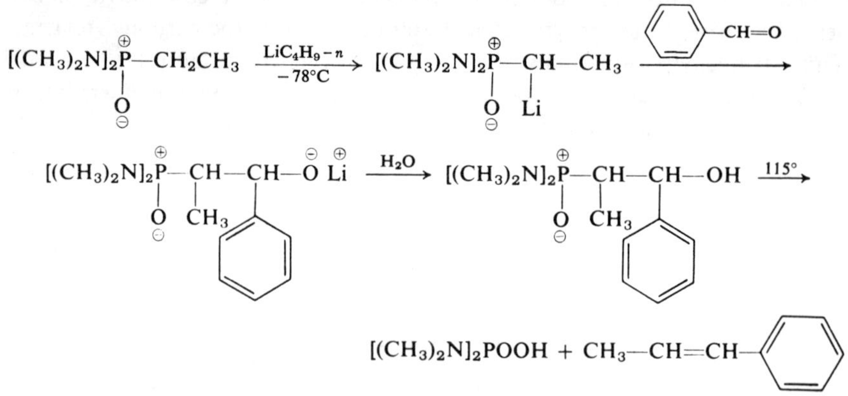

$$[(CH_3)_2N]_2\overset{\oplus}{P}-CH_2CH_3 \xrightarrow[-78°C]{LiC_4H_{9-n}} [(CH_3)_2N]_2\overset{\oplus}{P}-CH-CH_3 \xrightarrow{C_6H_5-CH=O}$$

$$[(CH_3)_2N]_2\overset{\oplus}{P}-CH-CH-\overset{\ominus}{O}\overset{\oplus}{Li} \xrightarrow{H_2O} [(CH_3)_2N]_2\overset{\oplus}{P}-CH-CH-OH \xrightarrow{115°}$$

$$[(CH_3)_2N]_2POOH + CH_3-CH=CH-C_6H_5$$

Scheme 6

The stereochemical outcome of this procedure is very similar to that found with reactive P⊕ ylids. Again it is the *erythro* diastereoisomer which is formed perferentially; thus, in Scheme 6 the *erythro–threo* ratio is 7:2. As in the P⊕ series, the *erythro* component may be separated by fractional crystallization and stereospecifically converted to the corresponding *cis* olefin. Finally, almost pure *trans* olefin can be obtained through equilibration of the epimeric mixture of intermediate adducts. For preparative purposes, however, oxidation of the reaction intermediates to β-ketophosphonamides, e.g., $[(CH_3)_2N]_2P(O)-CH(CH_3)-CO-C_6H_5$, followed by reconversion to β-hydroxyphosphonamides through reduction with sodium borohydride (which stereoselectively yields *threo*-diastereoisomers) appears to be the more successful procedure (62).

IV. APPLICATION OF STEREOSELECTIVE CARBONYL OLEFINATION PROCEDURES

cis-Stereoselective Wittig reactions offer a convenient and versatile route to many natural products. The method was first applied to the preparation of long-chain unsaturated fatty acids (12,63). Later, the isolation of *cis*-con-

THE STEREOCHEMISTRY OF THE WITTIG REACTION

figurated carotinoids by the Wittig reaction was reported (7,64). Salt-free triphenylphosphonium alkylids were found to be very suitable for the synthesis of naturally occurring pyrethrins and rethrones, such as *cis*-pyrethrolone (65) (eq. (10)) and *cis*-jasmone (66) (eq. (11)). Alternative procedures, such as Lindlar hydrogenation, proved to be less successful.

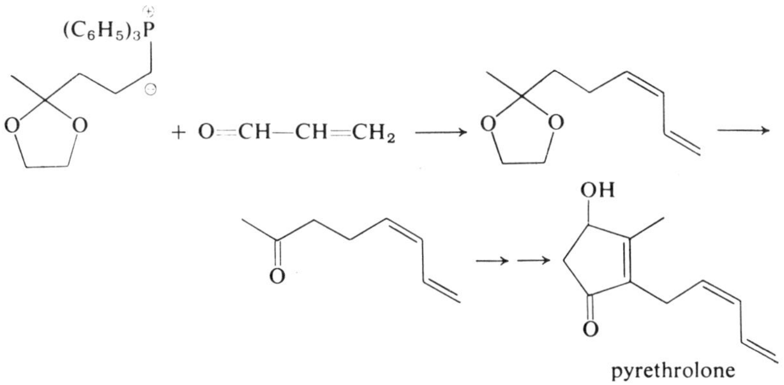

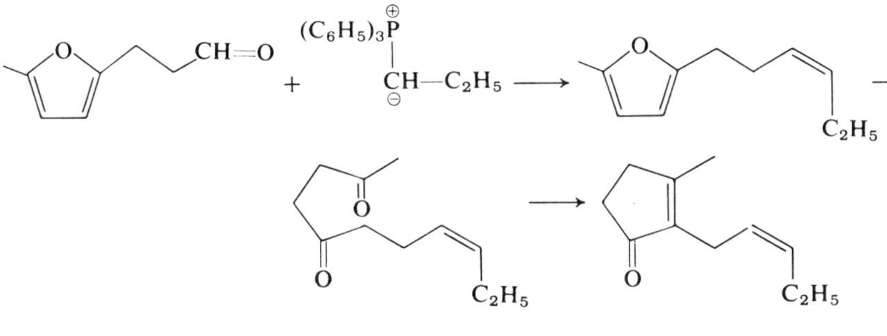

cis-Stereoselective carbonyl olefinations also appear to be acquiring practical importance in the field of steroid chemistry. In a series of ylid reactions with steroidal 17-ketones, the side chain derived from the ylid always preferred the sterically more hindered position at the olefinic double bond, though it has not yet been possible to determine the exact *cis–trans* ratios (67–69) (eqs. (12)–(14)).

From a preparative point of view, *trans*-stereoselective syntheses of α,β-unsaturated ketones and esters or stilbenes possess only limited value, since *cis–trans* mixtures of such compounds can normally be isomerized without difficulty to yield the more stable *trans*-configurated products anyway. The *trans*-stereoselective Wittig reaction via β-oxido ylids does close a preparative gap, however, since it yields almost pure *trans* olefins even when they are aliphatic and therefore thermodynamically favored over the corresponding cis isomers only to a small extent. One interesting field of application should lie in the stereoselective syntheses of geraniol, farnesol, and squalene derivatives.

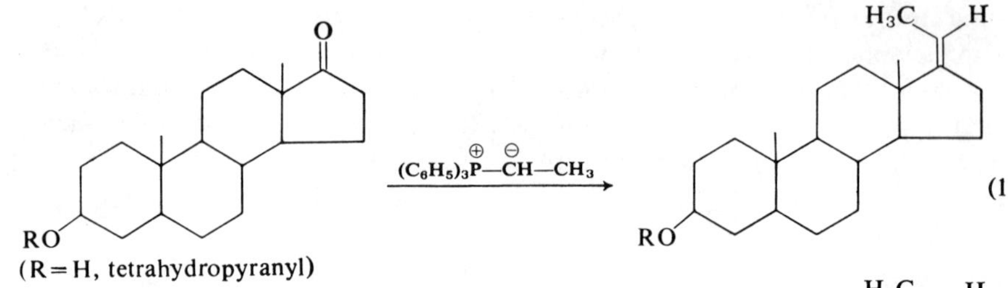

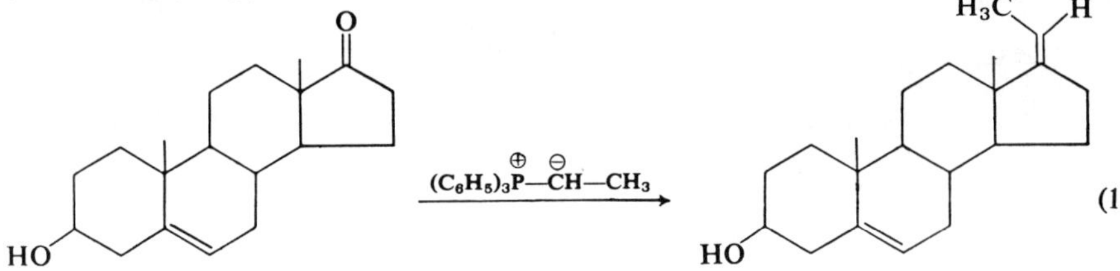

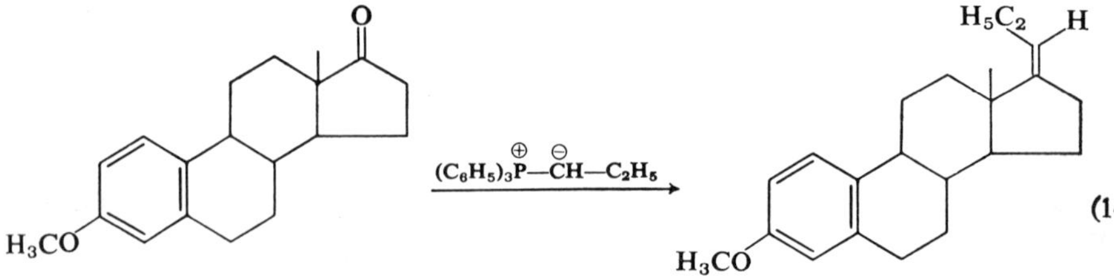

The synthesis of *trans* olefins from β-oxido phosphorus ylids through equilibration and acid quenching of the intermediate "betaine ylids" does not exhaust the synthetic potential of these β-oxido ylids. As is common with ylids (70), β-oxido ylids are susceptible to addition of a variety of other electrophilic agents. Treatment with a donor of deuterium, halogen or alkyl group, with carbonyl compounds and with many other electrophilic reagents leads to the correspondingly substituted betaine, which may be converted to the olefinic product substituted at a vinylic position (71,72). The total reaction sequence may thus be referred to as "*α-s*ubstitution plus *c*arbonyl *o*lefination via β-*o*xido *p*hosphorus *y*lids (*scoopy*)" and can be formulated in a most general fashion as shown in Scheme 7. (Instead of X–X', unsaturated compounds of type Y=Y' or Z≡Z' may serve as electrophilic agents as well.)

Since rapidly inverting β-oxido ylids are once again involved in the reaction sequence, the residues derived from the ylid side chain and the aldehyde

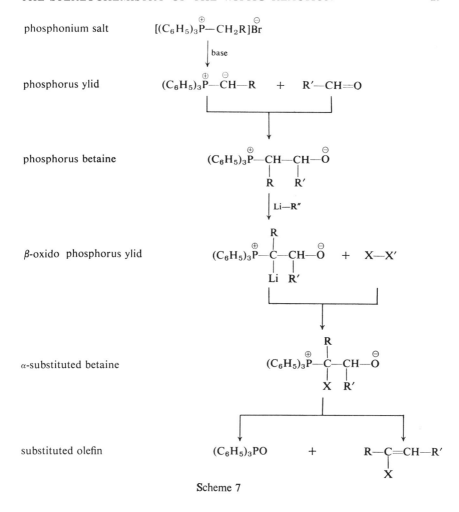

Scheme 7

continue to occupy *trans* positions in the resultant unsaturated product almost exclusively, provided that the preparations are carried out in the presence of soluble lithium salts [*e.g.*, eq. (15)]. The virtue of stereoselectivity is lost only when the unsubstituted triphenylphosphonium methylid is utilized as the ylid component. The reason is that now two ligands of similar small size, hydrogen and lithium, compete sterically at the β-oxido ylid stage so that no significant discrimination can result (eq. (16)).

Alkylation of β-oxido phosphorus ylids affords a stereoselective approach to either *cis*- or *trans*-configurated trisubstituted ethylenes. The method seems attractive enough to be explored in syntheses of polyisoprenoid olefins (eq. (17)).

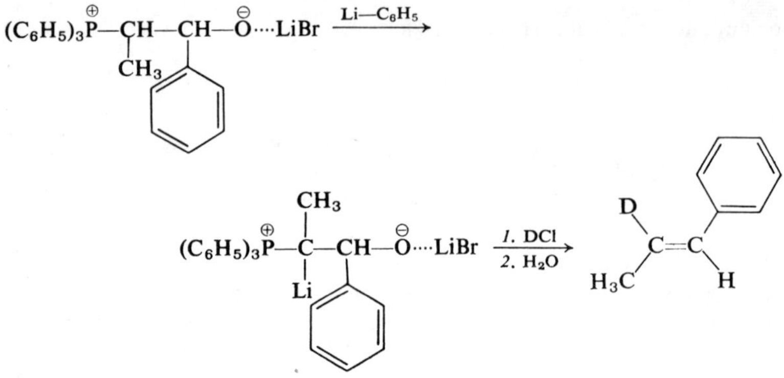

$$(C_6H_5)_3\overset{\oplus}{P}-CH-CH-\overset{\ominus}{O}\cdots LiBr \xrightarrow{Li-C_6H_5}$$
$$\phantom{(C_6H_5)_3\overset{\oplus}{P}-CH}\underset{|}{CH_3}\phantom{-CH-\overset{\ominus}{O}\cdots LiBr}$$

$$(C_6H_5)_3\overset{\oplus}{P}-\underset{\underset{Li}{|}}{\overset{\overset{CH_3}{|}}{C}}-CH-\overset{\ominus}{O}\cdots LiBr \xrightarrow[2.\ H_2O]{1.\ DCl} \underset{H_3C}{\overset{D}{\diagdown}}C=C\underset{H}{\overset{C_6H_5}{\diagup}} \quad (15)$$

$$(C_6H_5)_3\overset{\oplus}{P}-CH_2-CH-\overset{\ominus}{O}\cdots LiBr \xrightarrow{Li-C_6H_5}$$

$$(C_6H_5)_3\overset{\oplus}{P}-\underset{}{\overset{\overset{Li}{|}}{CH}}-CH-\overset{\ominus}{O}\cdots LiBr \xrightarrow[2.\ H_2O]{1.\ DCl} \underset{H}{\overset{D}{\diagdown}}C=C\underset{H}{\overset{C_6H_5}{\diagup}} + \underset{D}{\overset{H}{\diagdown}}C=C\underset{H}{\overset{C_6H_5}{\diagup}} \quad (16)$$

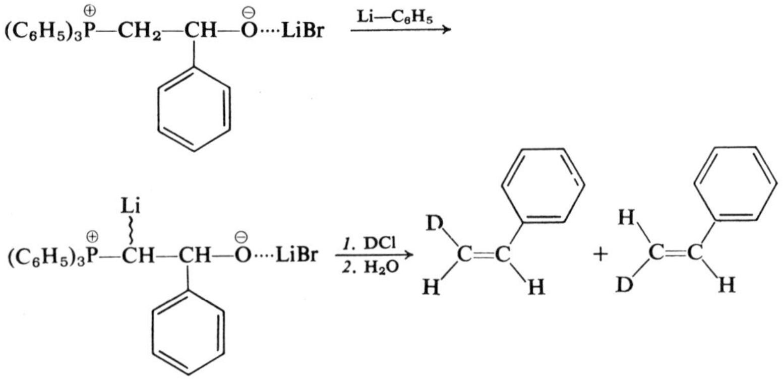

While yields are good to excellent in deuteration and alkylation and fair to good in halogenation, only small amounts, if any, of the desired products could be identified when the β-oxido ylid was treated with donors of acyl groups, such as methyl chloroformate and benzoyl chloride, or metal halides, such as trimethylchlorosilane and phenylmercuric bromide. Presumably the olefin-forming elimination of triphenylphosphine oxide cannot keep abreast of the rapid decomposition of the betaine to aldehyde and a novel ylid stabilized by the electron-attracting residue which has entered into the α-position (eq. (18)).

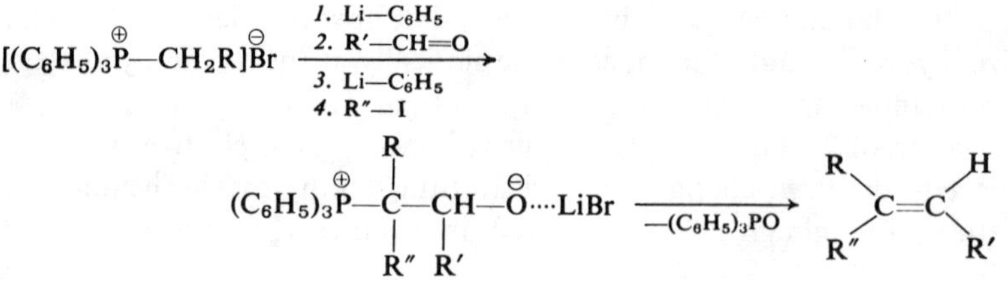

$$[(C_6H_5)_3\overset{\oplus}{P}-CH_2R]\overset{\ominus}{Br} \xrightarrow[\substack{1.\ Li-C_6H_5 \\ 2.\ R'-CH=O \\ 3.\ Li-C_6H_5 \\ 4.\ R''-I}]{}$$

$$(C_6H_5)_3\overset{\oplus}{P}-\underset{\underset{R''}{|}}{\overset{\overset{R}{|}}{C}}-\underset{\underset{R'}{|}}{CH}-\overset{\ominus}{O}\cdots LiBr \xrightarrow{-(C_6H_5)_3PO} \underset{R''}{\overset{R}{\diagdown}}C=C\underset{R'}{\overset{H}{\diagup}} \quad (17)$$

(18)

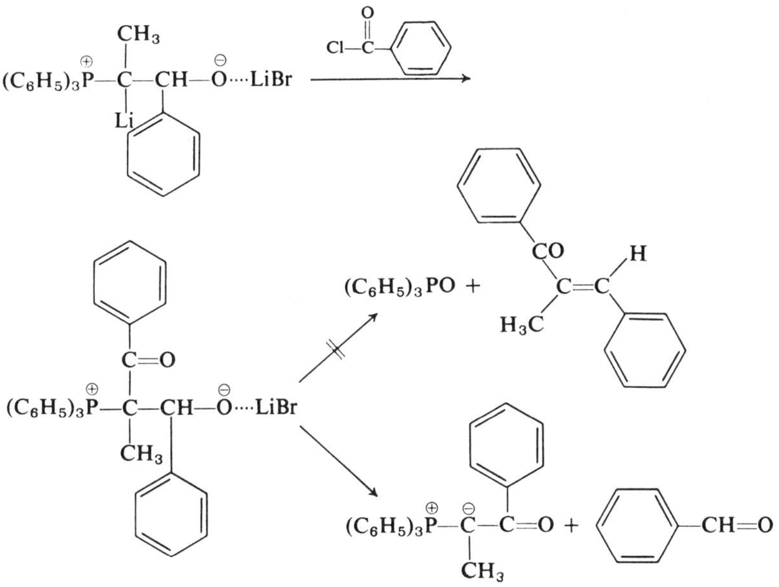

V. DISCUSSION

Novel procedures for the synthesis of either *cis* or *trans* olefins have been based on the finding that the Wittig reaction, utilizing reactive triphenylphosphonium alkylids, may follow one or the other of two possible stereochemical avenues:

1. Under conditions of kinetic control, ylids and aldehydes combine predominantly to yield *erythro* betaines.

2. Under equilibrium conditions, *threo* betaines are the favored products, provided that lithium salts are present.

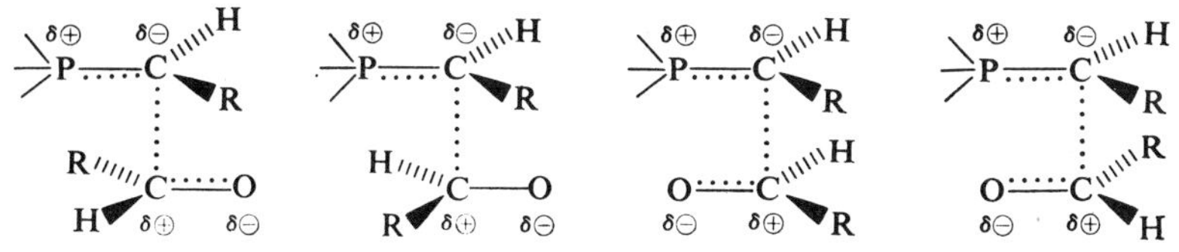

Explanations for these empirically discovered rules have been looked for intensively, but until now only tentative hypotheses can be put forward. Any discussion of the *cis*-stereoselectivity phenomenon must be related to the geometry attributed to the transition state of the addition step. If the phosphorus and oxygen atoms are considered to be in the *anti* orientation, the

erythro-anti *threo-anti* *erythro-syn* *threo-syn*

erythro configuration of the intermediate should be favored for steric reasons. The difference between nonbonded interaction in the alternative configurations cannot, however, be large enough to account for the strikingly high *erythro–threo* ratios observed in kinetically controlled betaine formation.

Moreover, at least in nonpolar solution, the *syn* arrangement of the heteroatoms would seem to be the preferred one, since attractive dipole interactions should be strong enough to more than outweigh opposed eclipsing effects. Without any doubt steric strain in the *erythro–syn* array must be considerably higher than in the *threo–syn* array. However, it appears to be a fundamental (though rather obscure) principle in organic chemistry that carbon–carbon linking usually occurs in such a manner as to yield the thermodynamically less stable isomer preferentially. Well-known examples are presented by the Diels-Alder reaction (73–75), the cyclodimerization reaction leading to four-membered ring systems (76,77), and formation of cyclopropanes through carbene intermediates (78–80), as well as by the addition of enolates or other organometallic derivatives to carbon–oxygen (81–83) and carbon–carbon double bonds (84–86).

Therefore, the *cis*-stereoselectivity of the Wittig reaction is believed to be a consequence of a more general *syn*-stereoselectivity of carbon–carbon linking reactions as a whole (87). Compatible with this view is the fact that dominating *erythro* betaine formation is normal with all types of Wittig reactions regardless of the special nature of the ylid and aldehyde involved. *threo* Betaine formation becomes important under conditions of kinetic control only if the ylid side chain bears bulky ligands such as phenyl (88) or cyclohexyl.

The second phenomenon, the pronounced shift of the equilibrium position far to the side of *threo* betaines upon complexation with lithium halides, seems to present less of a problem at first sight. Dramatic changes in conformational equilibria effected by a variation of participating metal complexes have been reported in the cyclohexanol series (89), and simple replacement of a nonpolar solvent such as benzene by a polar solvent such as dimethyl sulfoxide was found to alter the free energy differences between α-haloacetaldehyde rotamers by as much as 1.2 kcal/mol (90).

Nevertheless, an unspecified salt effect on conformational distribution obviously cannot satisfactorily explain the 95:5 preference of the betaine *threo* epimer and the 99:1 preference of the β-oxido ylid *threo* epimer. A better understanding may be reached by a discussion in terms of betaine-lithium halide adduct structures. While no reliable data are available, some speculation regarding the constitution and geometry of these addition products may be permitted (91). It is evident that optimum charge compensation may be achieved in a six-membered ring array containing the lithium salt cation and anion in addition to the four characteristic centers of the betaine chain.

Inspection of molecular models (see Fig. 5) indicates that this betaine–lithium salt complex should exist in a twist-boat conformation rather than in a chairlike conformation. At the expense of semieclipsing of exocyclic bonds at the α- and β-carbon atom, this twist-boat structure avoids more unfavorable eclipsing between the α-carbon-to-β-carbon bond and one phosphorus–phenyl linkage and, moreover, secures perfect dipole–dipole interactions.

On the basis of this model, the strong preference of the *threo* configuration becomes intelligible. Nonbonded interactions of vicinal ligands in the *trans* position at the α- and β-carbon atoms are unimportant compared to corresponding strain interactions in the *cis* positions. Therefore, a rapid *erythro* to *threo* conversion will take place whenever configurational inversion is possible.

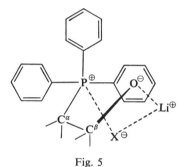

Fig. 5

REFERENCES

1. A. W. Johnson, *Ylid Chemistry*, Academic Press, New York, 1966.
2. A. Maercker, *Org. Reactions*, **14**, 270 (1965).
3. L. Horner, *Fortschr. Chem. Forsch.*, **7**, 1 (1966).
4. G. Wittig and U. Schöllkopf, *Chem. Ber.*, **87**, 1318 (1954).
5. F. Bohlmann, E. Inhoffen, and P. Herbst, *Chem. Ber.*, **90**, 1661 (1957).
6. P. C. Wailes, *Chem. Ind. (London)*, **1958**, 1086.
7. H. Pommer, *Angew. Chem.*, **72**, 817 (1960).
8. A. Butenandt, E. Hecker, M. Hopp, and W. Koch, *Ann. Chem.*, **658**, 42 (1962).
9. E. Truscheit and K. Eiter, *Ann. Chem.*, **658**, 65 (1962).
10. C. F. Hauser, T. W. Brooks, M. L. Miles, M. A. Raymond, and C. B. Butler, *J. Org. Chem.*, **28**, 372 (1963).
11. L. D. Bergelson and M. M. Shemyakin, *Tetrahedron*, **19**, 149 (1963).
12. L. D. Bergelson and M. M. Shemyakin, *Angew. Chem.*, **76**, 113 (1964); *Angew. Chem. Intern. Ed. Engl.*, **3**, 250 (1964).
13. L. D. Bergelson, V. A. Vaver, L. I. Barsukov, and M. M. Shemyakin, *Tetrahedron Letters*, **1965**, 2357.
14. H. O. House and G. R. Rasmusson, *J. Org. Chem.*, **26**, 4278 (1961).
15. L. Horner, *Fortschr. Chem. Forsch.*, **7**, 35 (1966).

16. Several members of this novel heterocyclic family have been isolated and characterized recently:
G. H. Birum and C. N. Matthews, *J. Org. Chem.*, **32**, 3554 (1967); F. Ramirez, *Accounts Chem. Res.*, **1**, 168 (1968); F. Ramirez, C. P. Smith, and J. F. Pilot, *J. Amer. Chem. Soc.*, **90**, 6726 (1968).
17. R. Mechoulam and F. Sondheimer, *J. Amer. Chem. Soc.*, **80**, 4386 (1958). The two propyl groups in the spiro compound derived from fluorenone occupy *trans* positions (R. Mechoulam, private communication to the author).
18. D. B. Denney and M. J. Boskin, *J. Amer. Chem. Soc.*, **81**, 6330 (1959); D. B. Denney, J. J. Vill, and M. J. Boskin, *J. Amer. Chem. Soc.*, **84**, 3944 (1962).
19. W. E. McEwen and A. P. Wolf, *J. Amer. Chem. Soc.*, **84**, 676 (1962).
20. V. Marks, *J. Amer. Chem. Soc.*, **85**, 1884 (1963); V. Marks, *Tetrahedron Letters*, **1964**, 3139; R. Burgada, *Ann. Chim. (Paris)*, **1966**, 15; F. Ramirez, A. S. Gulati, and C. P. Smith, *J. Org. Chem.*, **33**, 13 (1968).
21. A. Bladé-Font, C. A. Van der Werf, and W. E. McEwen, *J. Amer. Chem. Soc.*, **82**, 2396 (1960).
22. L. Horner and H. Winkler, *Tetrahedron Letters*, **1964**, 3265.
23. See also: L. Horner, *Helv. Chim. Acta*, **49**, 93 (1967). For a review, see M. J. Gallagher and I. D. Jenkins, *Topics in Stereochemistry*, Vol. 3, E. L. Eliel and N. L. Allinger, Eds., Interscience, New York, 1968.
24. I. Tömösközi, *Tetrahedron*, **19**, 1969 (1963).
25. A. J. Speziale and D. E. Bissing, *J. Amer. Chem. Soc.*, **85**, 3878 (1963).
26. P. A. Chopard, R. F. Hudson, and R. J. G. Searle, *Tetrahedron Letters*, **1965**, 2357
27. W. Flitsch and H. Peters, *Tetrahedron Letters*, **1969**, 1161.
28. C. Rüchardt, P. Panse, and S. Eichler, *Chem. Ber.*, **100**, 1144 (1967).
29. S. B. Hartley, W. S. Holmes, J. K. Jacques, M. F. Mole, and J. C. McCoubrey, *Quart. Rev.*, **17**, 204 (1963).
30. F. Nerdel and P. Weyerstahl, *Angew. Chem.*, **71**, 339 (1959).
31. H. O. House, V. K. Jones, and G. A. Frank, *J. Org. Chem.*, **29**, 3327 (1964).
31a. G. Drehfahl, D. Lorenz, and G. Schnitt, *J. Prakt. Chem.*, **23**[4], 143 (1964).
31b. In this connection attention should be called to the interesting results obtained by Huisman et al. in a series of bifunctional Wittig reactions (H. Heitman, U. K. Pandit, and H. O. Huisman, *Tetrahedron Letters*, **1963**, 915; I. H. Anthonissen, *Akademisch Proefschrift*, Amsterdam, 1968).
32. D. E. Bissing, *J. Org. Chem.*, **30**, 1296 (1965).
33. A. W. Johnson and R. B. LaCount, *Tetrahedron*, **9**, 130 (1960).
34. R. Ketcham, D. Jambotkar, and L. Martinelli, *J. Org. Chem.*, **27**, 4666 (1962).
35. A. W. Johnson and V. L. Kyllingstad, *J. Org. Chem.*, **31**, 334 (1966).
36. K. Eiter and H. Oedinger, *Ann. Chem.*, **682**, 62 (1965).
37. L. Crombie, P. Hemesley, and G. Pattenden, *J. Chem. Soc.*, C, **1969**, 1024.
38. M. E. Jones and S. Trippett, *J. Chem. Soc.*, C, **1966**, 1090.
39. M. Schlosser, K. F. Christmann, and G. Müller, unpublished results.
40. H. J. Bestmann and O. Kratzer, *Chem. Ber.*, **95**, 1894 (1962).
41. L. Horner, H. Hoffmann, H. Wippel, and G. Klahre, *Chem. Ber.*, **92**, 2499 (1958).
42. D. H. Wadsworth, O. E. Schupp, E. J. Seus, and J. A. Ford, *J. Org. Chem.*, **30**, 680 (1965).
43. E. D. Bergmann, I. Shahak, and J. Appelbaum, *Israel J. Chem.*, **6**, 73 (1968).
44. The opposite stereoselectivity has been reported for reactions with diethyl α-fluorocarbethoxymethanephosphonate; see ref. 45.
45. H. Machleidt and R. Wessendorf, *Ann. Chem.*, **674**, 1 (1964).

46. Y. Butsugan, M. Murai, and T. Matsuura, *J. Chem. Soc. Japan, Pure Chem. Sect.*, **89**, 431 (1968).
47. G. Jones and R. F. Maisey, *Chem. Commun.*, **1968**, 543.
48. H. Kaneko and M. Okazaki, *Tetrahedron Letters*, **1966**, 219.
49. L. Horner and W. Klink, *Tetrahedron Letters*, **1964**, 2467.
50. A. Piskala and M. Schlosser, to be published.
51. M. Schlosser, G. Müller, and K. F. Christmann, *Angew. Chem.*, **78**, 677 (1966); *Angew. Chem. Intern. Ed. Engl.*, **5**, 667 (1966).
52. M. Schlosser and G. Heinz, unpublished results.
53. M. Schlosser and K. F. Christmann, *Ann. Chem.*, **708**, 1 (1967).
53a M. Fétizon, personal communication.
54. J. D. McClure, *Tetrahedron Letters*, **1967**, 2401.
55. G. Heublein, private communication.
56. H. J. Bestmann and T. Denzel, *Tetrahedron Letters*, **1966**, 3591.
57. E. E. Schweitzer, J. G. Thompson, and T. A. Ulrich, *J. Org. Chem.*, **33**, 3082 (1968).
58. M. Schlosser and K. F. Christmann, *Angew. Chem.*, **77**, 682 (1965); *Angew. Chem. Intern. Ed. Engl.*, **4**, 689 (1965).
59. M. Schlosser and K. F. Christmann, unpublished observations.
60. M. Schlosser and K. F. Christmann, *Angew. Chem.*, **78**, 115 (1966); *Angew. Chem. Intern. Ed. Engl.*, **5**, 126 (1966).
61. M. Schlosser and K. F. Christmann, *Angew. Chem.*, **76**, 683 (1964); *Angew. Chem. Intern. Ed. Engl.*, **3**, 636 (1964).
62. E. J. Corey and G. T. Kwiatkowski, *J. Amer. Chem. Soc.*, **88**, 5653, 5654 (1966); **90**, 6816 (1968).
63. G. Pattenden and B. C. L. Weedon, *J. Chem. Soc.*, C, **1968**, 1984.
64. C. F. Garbers, D. F. Schneider, and J. P. van der Merwe, *J. Chem. Soc.*, C, **1968**, 1982.
65. L. Crombie, P. Hemesley, and G. Pattenden, *Tetrahedron Letters*, **1968**, 3021; *J. Chem. Soc.*, C **1969**, 1016.
66. L. Crombie, P. Hemesley, and G. Pattenden, *J. Chem. Soc.*, C, **1969**, 1024.
67. G. Drehfahl, K. Ponsold, and H. Schick, *Chem. Ber.*, **98**, 604 (1965).
68. A. M. Krubiner and E. P. Oliveto, *J. Org. Chem.*, **31**, 24 (1966).
69. A. M. Krubiner, N. Gottfried, and E. P. Oliveto, *J. Org. Chem.*, **33**, 1715 (1968).
70. For a thorough review see: H. J. Bestmann, *Angew. Chem.*, **77**, 609, 651, 850 (1965); *Angew. Chem. Intern. Ed. Engl.*, **4**, 583, 645, 830 (1965).
71. M. Schlosser, *Symp. Cancerologie*, **2**, 148 (1968); see also M. Schlosser and K. F. Christmann, *Liebigs Ann. Chem.*, **708**, 20 (1967) (footnote 72).
72. M. Schlosser and K. F. Christmann, *Synthesis*, **1969**, 38.
73. J. G. Martin and R. K. Hill, *Chem. Rev.*, **61**, 537 (1961).
74. J. Sauer, *Angew. Chem.*, **79**, 76 (1967); *Angew. Chem. Intern. Ed. Engl.*, **6**, 16 (1967).
75. R. Criegee, *Angew. Chem.*, **70**, 607 (1958); M. Avram, I. G. Dinulescu, E. Marica, G. Maateescu, E. Sliam, and C. D. Nenitzescu, *Chem. Ber.*, **97**, 382 (1964); L. Watts, J. D. Fitzpatrick, and R. Pettit, *J. Amer. Chem. Soc.*, **87**, 3253 (1965); **88**, 623 (1966); G. Wittig and J. Weinlich, *Chem. Ber.*, **98**, 471 (1965).
76. E. Vogel, *Ann. Chem.*, **615**, 1 (1958); G. Schröder and W. Martin, *Angew. Chem.*, **78**, 117 (1966); *Angew. Chem. Intern. Ed. Engl.*, **5**, 130 (1966).
77. T. Nagase, *Bull. Chem. Soc. Japan*, **34**, 139 (1961); P. D. Landor and S. R. Landor, *J. Chem. Soc.*, **1963**, 2707; H. A. Staab and H. A. Kurmeier, *Chem. Ber.* **101**, 2697 (1968); E. V. Dehmlow, *ibid.*, **100**, 3260 (1967); id., *Tetrahedron Letters* **1969**, 4283.

78. G. L. Closs, R. Moss, and J. J. Coyle, *J. Amer. Chem. Soc.*, **84**, 4985 (1962).
79. G. L. Closs, *Topics in Stereochemistry*, Vol. 3, E. L. Eliel and N. L. Allinger, Eds., Interscience, New York, 1968.
80. M. Schlosser and G. Heinz, *Angew. Chem.*, **80**, 849 (1968); *Angew. Chem. Intern. Ed. Engl.*, **7**, 820 (1968).
81. J. Canceill, J. Gabard, and J. Jacques, *Bull. Soc. Chim. France*, **1968**, 231.
82. J. Seyden-Penne and A. Roux, *Compt. Rend.*, **267** C, 1057 (1968); and private communication of Dr. Seyden-Penne.
83. J. E. Dubois and M. Dubois, *Chem. Commun.*, **1968**, 1567.
84. L. L. McCoy, *J. Amer. Chem. Soc.*, **80**, 6569; **84**, 2246 (1960).
85. See also: M. Mousseron, R. Fraisse, R. Jacquier, and G. Bonavent, *Compt. Rend.*, **248**, 1465 (1959).
86. A. Risaliti, M. Forchiassin, and E. Valentin, *Tetrahedron Letters*, **1966**, 6331.
87. Quite another approach has been discussed by *W. P. Schneider* (*Chem. Commun.*, **1969**, 785), who argues on the basis of an assumed initial attack of the carbonyl oxygen atom on the ylid phosphorus atom.
88. However, the *threo-erythro* ratios reported for the addition of stable and moderated ylids (38) and of α-lithium benzyldiphenylphosphine oxide (49) are felt to be astonishingly high. For a comment see also Ref. 50.
89. E. L. Eliel and M. N. Rerick, *J. Amer. Chem. Soc.*, **82**, 1367 (1960); E. L. Eliel and S. H. Schroeter, *J. Amer. Chem. Soc.*, **87**, 5031 (1965).
90. G. J. Karabatsos, D. J. Fenoglio, and S. S. Lande, *J. Amer. Chem. Soc.*, **91**, 3572 (1969); G. J. Karabatsos and D. J. Fenoglio, *J. Amer. Chem. Soc.*, **91**, 3577 (1969). See also the chapter by the same authors in this volume.
91. M. Schlosser, K. F. Christmann, and A. Piskala, *Chem. Ber.*, in press (1970).

The Determination of Absolute Configuration of Planar and Axially Dissymmetric Molecules

GRANT KROW

Department of Chemistry, Temple University, Philadelphia, Pennsylvania

I.	Introduction.	31
II.	Scope.	32
III.	Methods of Correlation of Absolute Configuration	33
	A. Chemical Methods	33
	1. Self-Immolative Asymmetric Syntheses	33
	2. Conservative Asymmetric Syntheses	44
	3. Direct Chemical Correlation	49
	B. Physical Methods.	52
	1. Thermal Analysis.	52
	2. Optical Methods.	52
	C. Theoretical Methods.	53
	1. Biaryls.	53
	2. Allenes.	55
	3. Alkylidenecycloalkanes	56
	4. Spiranes	56
	5. *trans*-Cycloalkenes	57
	6. Phenanthrenes.	57
	7. Hexahelicene	58
	D. X-ray Analysis.	58
IV.	Appendix: Specification of Molecular Chirality.	59
	References	65

I. INTRODUCTION

Three-dimensional space can be occupied asymmetrically about a chiral center, chiral axis, or chiral plane (1). Although an enormous number of correlations of absolute configuration of molecules with chiral centers has accumulated throughout the last hundred years, it is only during the last ten years that the problem of determining the absolute configuration of planar and axially dissymmetric molecules has been solved. The enormous challenge to stereochemists of the lack of obvious methods for correlation of chiral centers with dissymmetric molecules lacking chiral centers has stimulated a variety of novel and ingenious solutions to this problem.

II. SCOPE

According to the *factorization rule* (1) for configurational nomenclature overall chirality can be factorized into three elements, which can be treated in the order of chiral centers, chiral axes, and chiral planes whenever necessary. Some molecules (e.g., hexahelicene) may be viewed also in terms of their helicity.

The present paper discusses the determination of absolute configuration of molecules which by necessity cannot be treated in terms of a chiral center. In addition, for historical reasons (2,3), certain alkylidenecycloalkanes (1) and spiranes (2) which are conventionally treated in terms of chiral centers (1) are included. Specifically, configurational correlations of chiral paracyclophanes (3), biaryls (4), allenes (5), alkylidenecycloalkanes (6), *trans*-cycloalkenes (7), phenanthrenes (8), and hexahelicene (9) will be discussed.

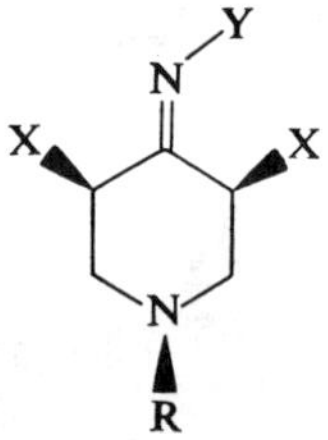

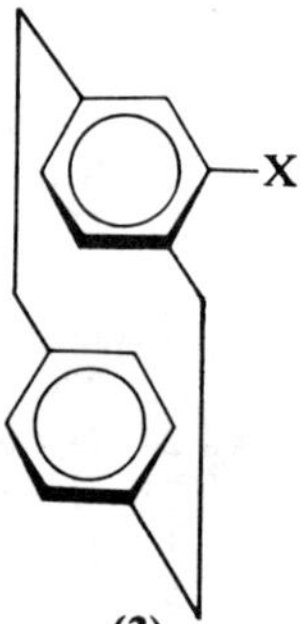

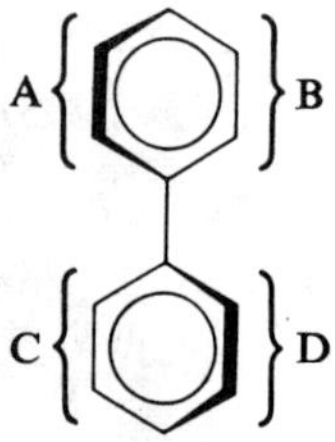

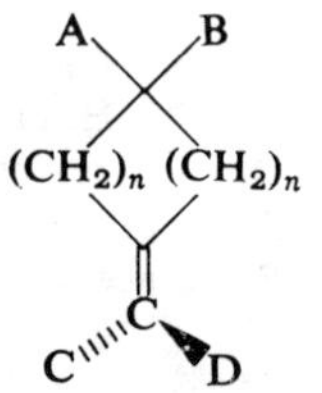

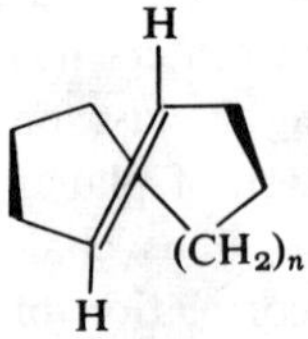

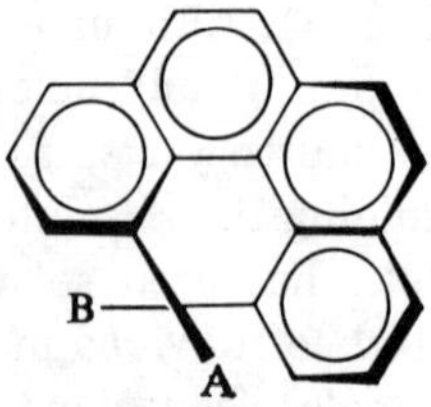

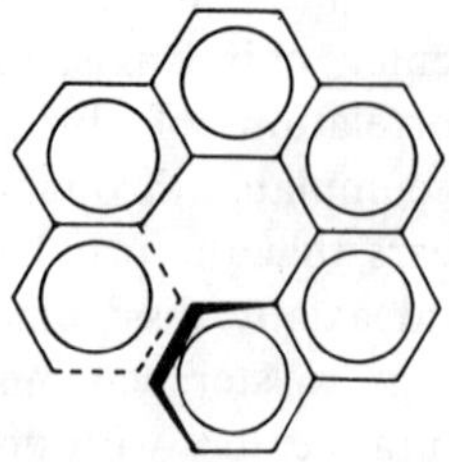

Although optical rotatory dispersion and circular dichroism have been used to correlate a variety of skewed styrenes (6–8) containing chiral (i.e., asymmetric) centers, no direct correlations of styrenes of type **10** have been made. Among other dissymmetric molecules whose absolute configurations have not been determined are axially dissymmetric spiranes of type **11**, adamantanes of type **12**, *meta*-bridged diphenic acids (**13**), aromatic Schiff bases (9) (**14**), and cyclooctatetraenes (10) (**15**). More extensive lists can be found elsewhere (1–5).

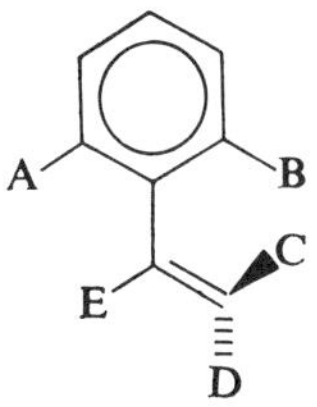

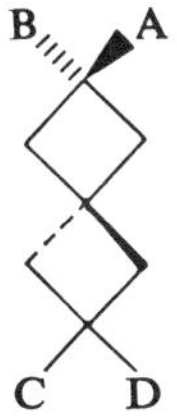

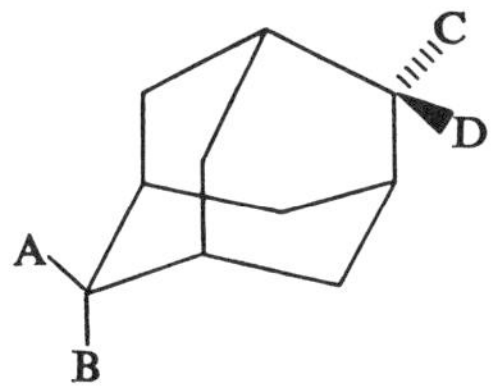

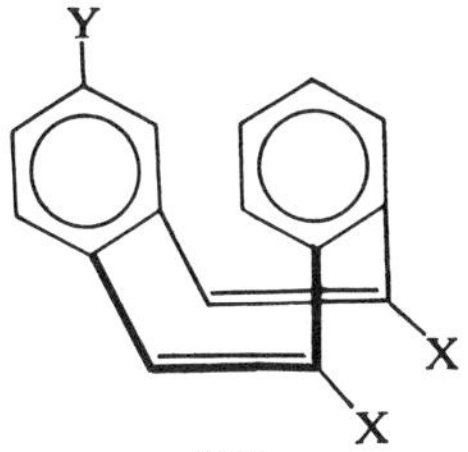

III. METHODS OF CORRELATION OF ABSOLUTE CONFIGURATION

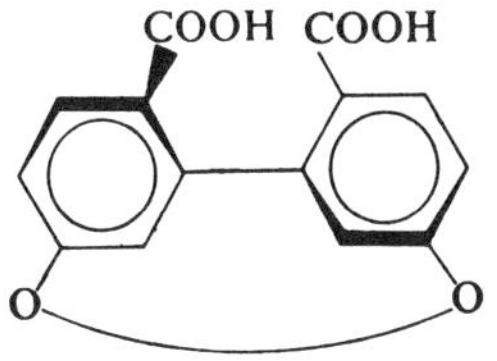

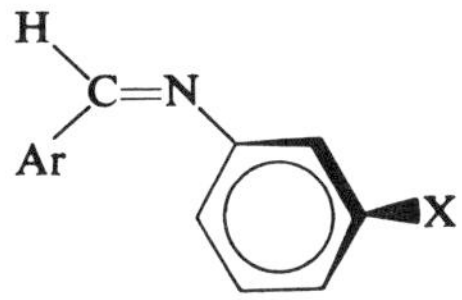

A. Chemical Methods

1. Self-Immolative Asymmetric Syntheses

In a "self-immolative asymmetric synthesis" (5), dissymmetry associated with one chiral grouping is destroyed while a new dissymmetric grouping is created. Where this method has been applied to configurational correlations, the individual steps in the conversion of centro- and axial dissymmetry must occur by known mechanistic pathways. Provided the gross structures of the pathways (transition states) correlating centro-asymmetry and molecular dissymmetry are well enough understood, configurational assignments obtained in this fashion are relatively secure.

a. Biphenyls. Berson and Greenbaum (11) have correlated (−)-thebaine (**16**) of known absolute configuration with (R)-(+)-α-phenyldihydrothebaine and (R)-(+)-δ-phenyldihydrothebaine (**17**).* Through reaction of (−)-thebaine (**16**) with phenylmagnesium bromide, a biphenyl moiety was created in highly stereospecific fashion. On the basis of model considerations, transition state A, in which the pyrrolidine ring is nearly planar, appears to be by far the least strained pathway leading from (−)-**16** to the two (+)-phenyldihydrothebaines (**17**). From inspection of the chirality of the biphenyl precursor in A, the biphenyl moiety of either diastereoisomer (+)-**17** can be assigned the R configuration.†

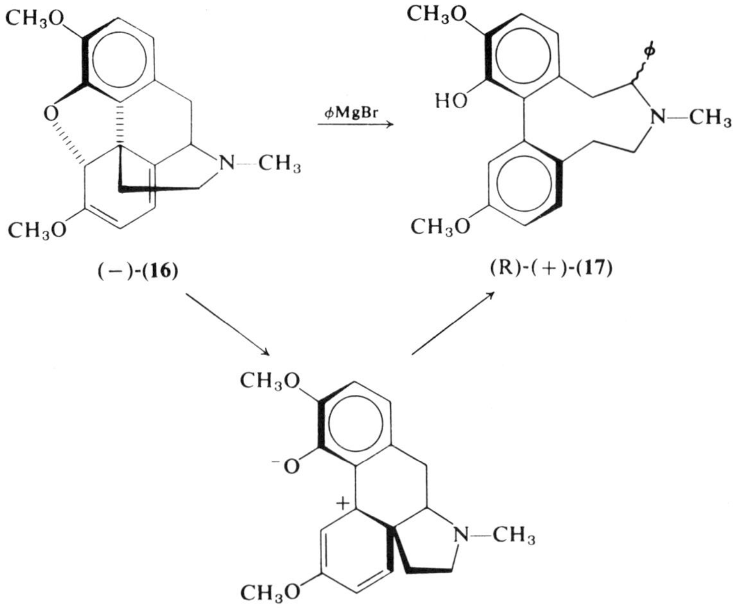

Utilizing a different approach, Mislow and co-workers (12) discovered that incomplete reduction of the racemic bridged biphenyl ketone **18** with (S)-(+)-2-octanol in the presence of aluminum t-butoxide effected a kinetic resolution, giving a mixture of dextrorotatory ketone **18** and levorotatory

---

*The symbol R refers to the configuration of the biphenyl moiety. The benzylic carbon next to nitrogen may be either R or S; both diastereoisomers are formed.

†By considering the fiducial groups in accordance with the revised selection rule for axial chirality (1), (+)-**17** is here described as the R enantiomer. According to an earlier nomenclature rule (2), (+)-**17** had been described as the S enantiomer. The 1966 Cahn-Ingold-Prelog system for naming axially dissymmetric enantiomers, such as (+)-**17**, as well as planar dissymmetric compounds is summarized in the Appendix (p. 59).

alcohol **19**. From considerations of nonbonded interactions in the transition states of the asymmetric hydride transfer reaction, the *S* configuration was assigned to dextrorotatory **18**. The bridged ketone (*S*)-**18** was reduced more slowly than was (*R*)-**18** because of unfavorable steric compression in transition state A, in which the bulkier group (L) of the alcohol opposes the closer of the two methylene groups during the hydride transfer. In the favored transition state B the bulkier group (L) opposes the more distant methylene and so (*R*)-**18** reacted faster.

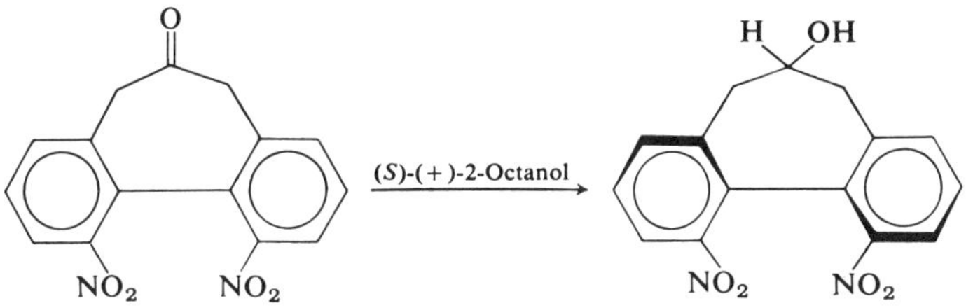

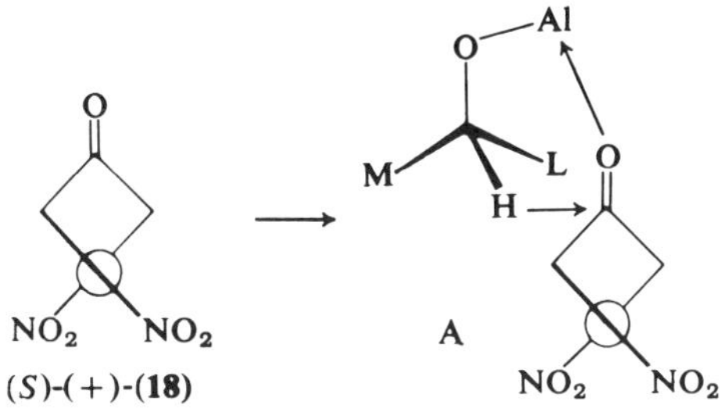

The partially asymmetric Meerwein-Ponndorf reduction method for the determination of absolute configuration has also been successfully applied to bridged binaphthyls (13,14), bridged biphenyls (12,15), and doubly bridged biphenyls (16). The generalization has been made that asymmetric reduction of a bridged biphenyl with (+)-2-octanol leads to an excess of the *S* enantiomer in the residual biphenyl ketone (16).

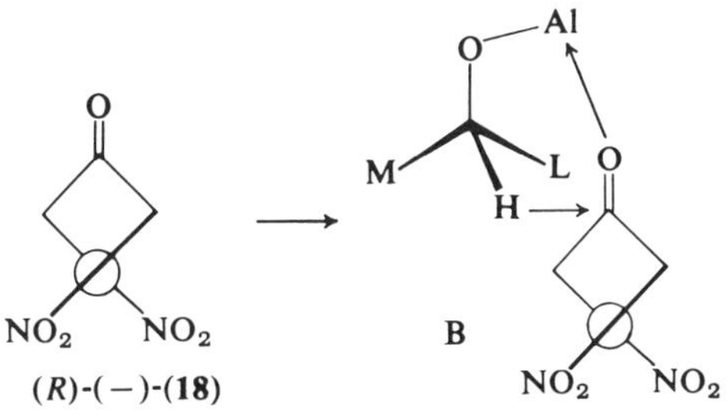

b. Allenes. Several stereospecific rearrangements of optically active acetylenic alcohols of known absolute configuration have been studied. Landor and Taylor-Smith (17) synthesized active 3-methyl-3-*t*-butyl-1-chloroallene (**22**) by a stereospecific rearrangement of optically active methyl-*t*-butyl-ethynylcarbinol (**20**) with thionyl chloride in dioxane. These conditions presumably favor an S_Ni' rearrangement in which the chlorine is delivered on the same side as the leaving oxygen. The determination of the absolute configuration of acetylenic alcohol **20** by Evans and Landor (18) has made possible the correlation of the centro-dissymmetry of alcohol **20** with the axial dissymmetry of allene **22** on the basis of the assumed stereochemistry of the rearrangement.

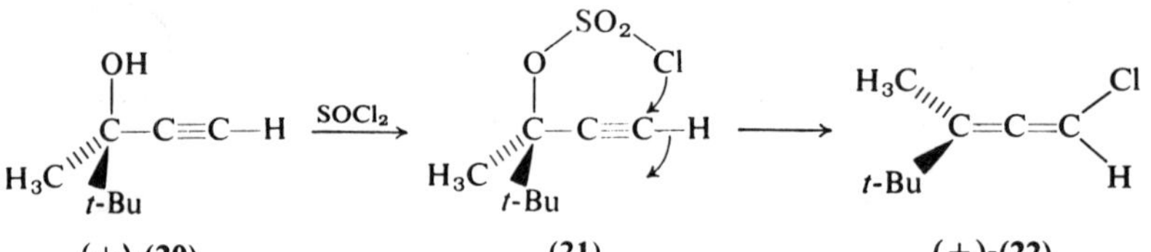

Jones, Loder, and Whiting (20a) and Evans, Landor, and Regan (21) have rearranged vinyl propargyl ethers (**23**) of known absolute configuration to optically active allenes. In the cyclic process of the Claisen rearrangement, the new carbon–carbon bond will be formed on the same side as the old carbon–oxygen bond. The stereospecificity makes the self-immolative asymmetric synthesis suitable for determining absolute configuration. Accordingly, levorotatory allene **24** was shown to be *R*.

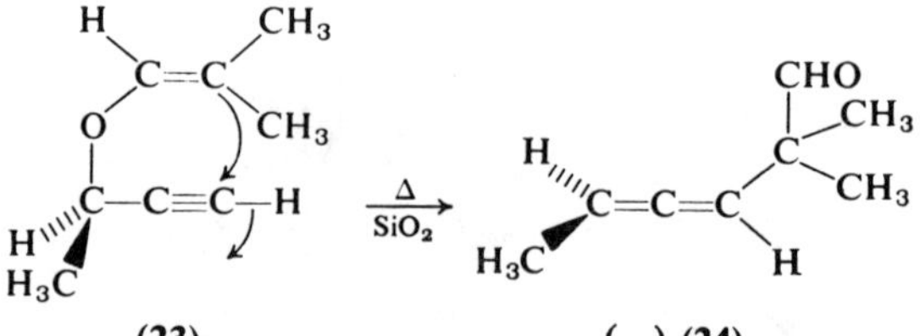

Agosta (22) has related the allene, (−)-glutinic acid (**25**) to norcamphor (**27**). Reaction of cyclopentadiene with the optically active glutinic acid resulted in formation of a diastereomeric mixture which could be separated into its components. The *exo-syn* adduct **26** could be identified because it formed an intramolecular anhydride, but not an iodolactone. Selective hydrogenation of the endocyclic double bond with palladium on carbon catalyst, ozonolysis of the exocyclic double bond, and decarboxylation served to convert (−)-**26** to (−)-**27** of known absolute configuration. The correlation shows (−)-**25** to have the *R* configuration.

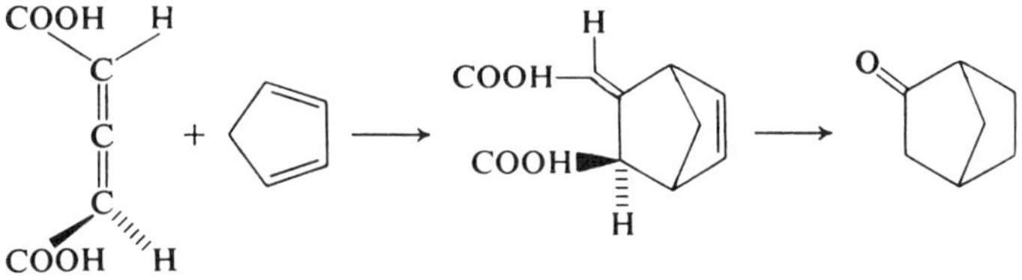

$$\underset{(-)-(25)}{\overset{\text{COOH}\quad\text{H}}{\underset{\text{COOH}\quad\text{H}}{\text{C}=\text{C}=\text{C}}}} + \bigcirc \longrightarrow \underset{(-)-(26)}{\overset{\text{COOH}}{\underset{\text{COOH}}{\bigcirc}}} \longrightarrow \underset{(-)-(27)}{\bigcirc}$$

Gianni (23) attempted to determine the absolute configuration of allene **28** through bromolactone formation. Treatment of the allene with bromine yielded a single stereoisomer. Attempts to ozonize this or its lithium aluminum hydride reduction product to an α-naphthylmandelic acid failed. A possible configurational correlation was proposed by the author on the basis of the fact that the dextrorotatory compounds **28** and **29** had positive optical rotatory dispersion curves, as did (*R*)-(+)-α-naphthylmandelic acid (**30**); however, a correlation of this type is clearly unsafe.

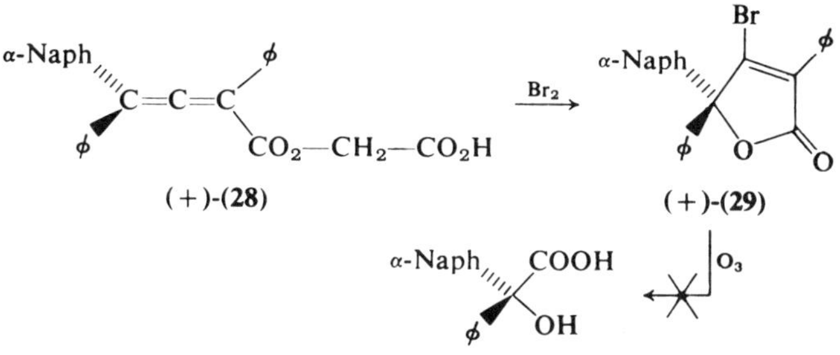

$$\underset{(+)-(28)}{\overset{\alpha\text{-Naph}}{\underset{\phi}{\text{C}=\text{C}=\text{C}}}\overset{\phi}{\underset{\text{CO}_2-\text{CH}_2-\text{CO}_2\text{H}}{}}} \xrightarrow{\text{Br}_2} \underset{(+)-(29)}{\overset{\alpha\text{-Naph}}{\underset{\phi}{\bigcirc}}\overset{\text{Br}}{\underset{\text{O}}{\bigcirc}}\phi}$$

$$\underset{(+)-(30)}{\overset{\alpha\text{-Naph}}{\underset{\phi}{\text{C}}}\overset{\text{COOH}}{\underset{\text{OH}}{}}} \xleftarrow{\text{O}_3}$$

Shingu, Hagishita, and Nakagawa (24) were, however, able to determine the absolute configuration of a series of dextrorotatory phenylallenecarboxylic acids (**31**) by cleavage of their bromolactones (**32**) with potassium permanganate followed by conversion to hydroxyacids (**34**) of known absolute configuration. It should be noted that in **34** when R is *t*-butyl, the long wavelength rotation is negative, but a positive Cotton effect is observed in the 220-nm region.

The semihydrogenation of a suitably substituted optically active allene led to a simple method for determining absolute configuration. Crombie and Jenkins (25), assuming *cis*-addition of hydrogen from a catalyst to the least hindered side of the allene, correlated (*R*)-(−)-**37** with (*R*)-(−)-**35**.

$$\underset{(+)\text{-}(\mathbf{31})}{\overset{R}{\underset{\phi}{\diagdown}}C=C=C\overset{H}{\underset{COOH}{\diagdown}}} \longrightarrow \underset{(+)\text{-}(\mathbf{32})}{\overset{R}{\underset{\phi}{\diagdown}}\overset{Br}{\diagdown}\diagup\diagdown_{O}^{O}}$$

$$\underset{(+)\text{-}(\mathbf{34})}{\overset{R}{\underset{\phi}{\diagdown}}\overset{COOH}{\underset{OH}{\diagdown}}} \longleftarrow \underset{(\mathbf{33})}{\overset{R}{\underset{\phi}{\diagdown}}\overset{COOH}{\underset{OCOCOOH}{\diagdown}}} \overset{KMnO_4}{\longleftarrow}$$

R = Me, Et, H, t-butyl

Jones, Wilson, and Tutweiler (26) found that *trans*-2,3-disubstituted cyclopropanecarboxylic acids may be converted to allenes and that the stereochemistry of this type of ring opening is apparently controlled by steric repulsions of the *trans* groups. Accordingly, rearrangement of (−)-*trans*-2,3-diphenylcyclopropanecarboxylic acid (**38**), ultimately via the carbene **39**, afforded (S)-(+)-1,3-diphenylallene (**40**), whose absolute configuration had been predicted by Mason and Vane (27). Knowing the absolute configuration of the cyclopropanecarboxylic acid **38**, Jones and Wilson (28) argued that because of lesser steric repulsion of the phenyl groups, rearrangement path A is favored over path B.

Extending the model to the rearrangement of (+)-*trans*-2,3-dimethylcyclopropanecarboxylic acid (**41**) of known configuration, the dextrorotatory allene **42** was predicted to have the S configuration (29). This prediction was verified by independent assignments of Caserio and co-workers (30) and Brewster (31) to this allene.

Moore and Bach (32) have determined the absolute configuration of 1,2-cyclononadiene (**45**) from the cyclopropane–allene conversion. Stereospecific addition of dibromocarbene to (S)-(+)-*trans*-cyclooctene (**43**) resulted in formation of (1R:2R)-(−)-dibromocyclopropane (**44**). Treatment of

$$\underset{(-)\text{-}(\mathbf{35})}{\overset{CH_3}{\underset{CH_3-CH_2}{\diagdown}}C=C=C\overset{CO_2R}{\underset{H}{\diagdown}}} \overset{H_2}{\longrightarrow} \underset{(-)\text{-}(\mathbf{36})\ (cis)}{CH_3-CH_2\overset{CH_3}{\underset{H}{\diagdown}}C-C\overset{CO_2R}{\underset{H}{\diagdown}}}$$

$$\underset{(-)\text{-}(\mathbf{37})}{\overset{CH_3}{\underset{CH_3-CH_2}{\diagdown}}\overset{CH_2-CH_2-CO_2R}{\underset{H}{\diagdown}}} \overset{H_2}{\longleftarrow}$$

(−)-(38)

(39)

(*S*)-(+)-(40)

(*R*)-(−)-(40)

(+)-(41) (+)-(42)

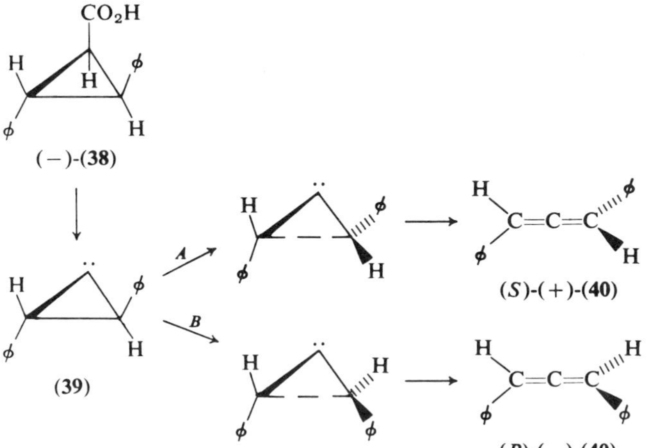

44 with methyllithium resulted in rearrangement leading to levorotatory 1,2-cyclononadiene (**45**). Consideration of steric repulsions of the *trans* methylene groups in the conversion to the allene according to the model of Jones and Wilson (28) and Walbrick et al. (29) resulted in assignment of the *R* configuration to levorotatory 1,2-cyclononadiene.

(+)-(43) (−)-(44)

(−)-(45)

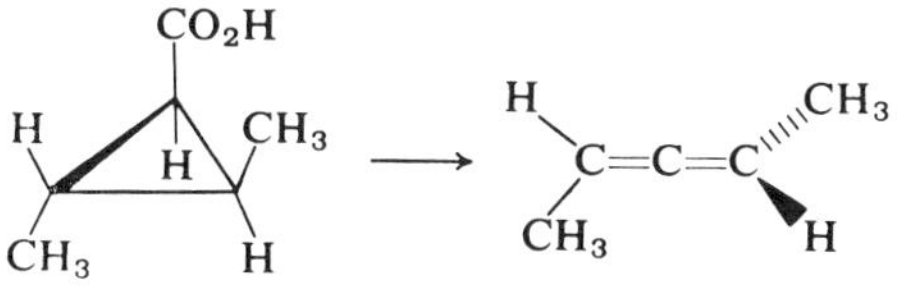

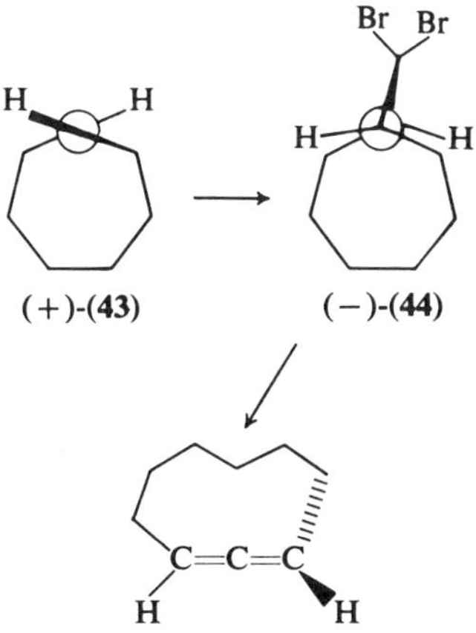

c. Alkylidenecycloalkanes. In 1909 Perkin, Pope, and Wallach (33) succeeded in resolving 4-methylcyclohexylideneacetic acid (**46**). The absolute configuration of this compound was determined in 1966 by Gerlach (34), who correlated the axial dissymmetry of α-deuterio-**46** with the centro-dissymmetry of (−)-isoborneol (**50**).

cis Addition of hydrogen to (*S*)-(+)-**46** gave the two possible reduction products **47** and **48** in the ratio of 1:2. The *cis* and *trans* relationships in **47** and **48** were determined by chemical correlation with the known *cis*- and *trans*-4-methylcyclohexylcarboxylic acids.

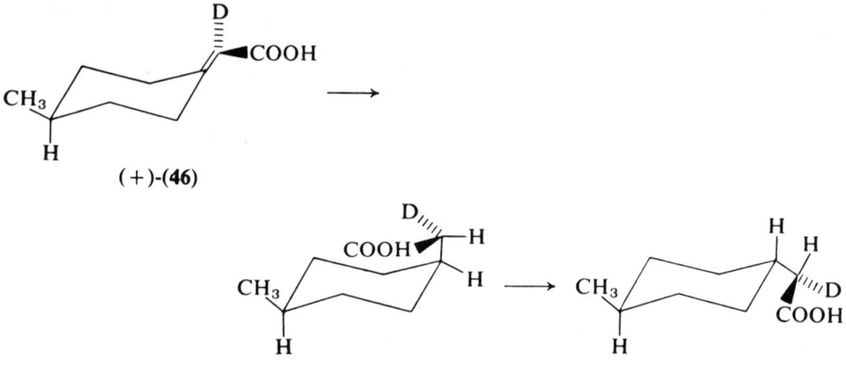

(+)-(**46**)

cis-(+)-(**47**)   *trans*-(+)-(**48**)

In a parallel experiment (*S*)-(+)-*cis*-α-^{2}H-4-methylcyclohexylcarbinol (**51**) was prepared by asymmetric synthesis from *cis*-4-methylcyclohexylmethanal (**49**) and 2-^{2}H-isobornyloxymagnesium bromide (**50**). Conversion

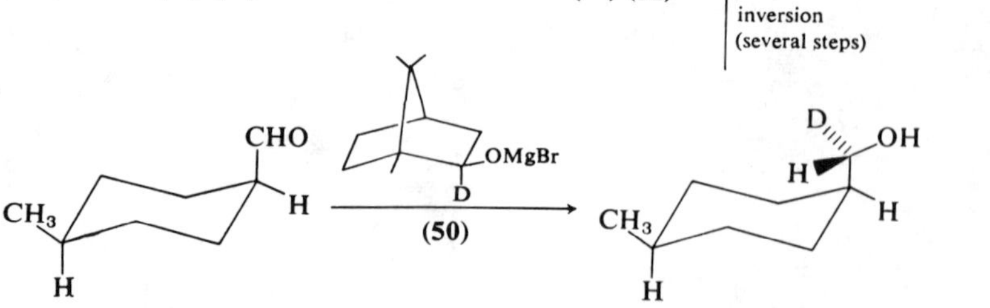

cis-(+)-(**47**)   (−)-(**52**)

inversion (several steps)

cis-(**49**)   (**50**)   (+)-(**51**)

of cis-(R)-(+)-**47** and cis-(S)-(+)-**51** by conventional methods of known stereochemical course to (R)-(−)-**52** led to determination of the absolute configuration of cis-(+)-**47**; and since hydrogenation of **46** occurs in a cis fashion, the configuration of (+)-**46** is deduced to be S as shown.

Brewster and Privett (35a) have determined the absolute configuration of (S)-(+)-1-benzylidene-4-methylcyclohexane (**55**) by a particularly elegant route. Treatment of ketone **53** of known absolute configuration with benzaldehyde resulted in formation of vinyl ketone **54**. The configuration about the olefin function **54** was determined to be transoid by comparison of ultraviolet and NMR data of the cisoid (shorter wavelength, lower extinction coefficient) and transoid compounds. Because the reduction of (−)-**54** with 3:1 aluminum chloride: lithium aluminum hydride had been shown to proceed without change of configuration about the olefin, the absolute configuration of (+)-**55** could be assigned as S.

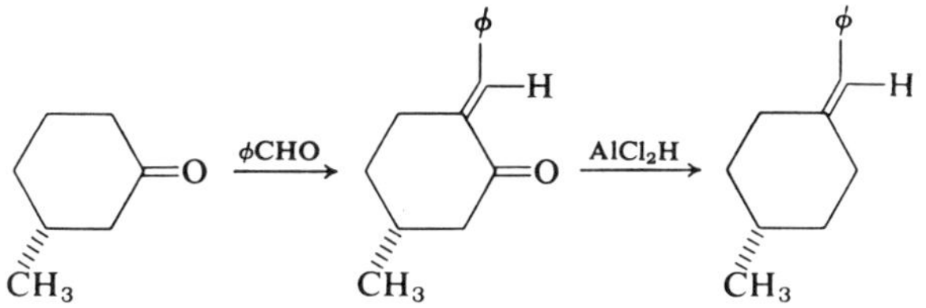

Lyle and Pelosi (36) have determined the absolute configuration of the structurally analogous (+)-1-methyl-2,6-diphenyl-4-piperidone oxime (**56**) to be syn-R by utilizing the known stereospecificity of the Beckman rearrangement. The rearrangement of dextrorotatory oxime **56** afforded lactam **57**, which on acid hydrolysis gave cinnamic acid and the levorotatory diamine **58**.

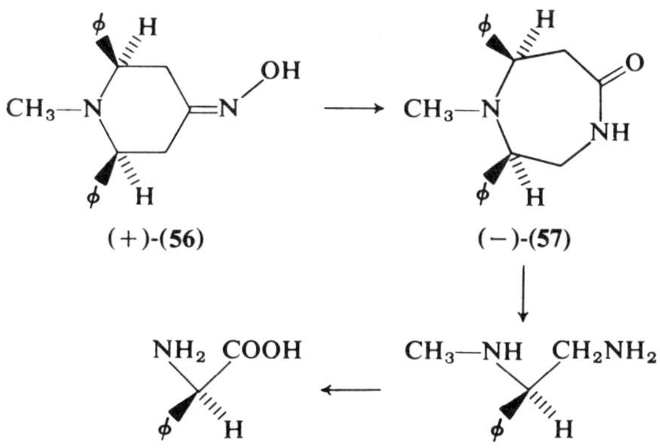

The diamine of identical sign was synthesized from (R)-$(-)$-α-aminophenylacetic acid (**59**). Because the substituent *anti* to the hydroxyl group of the oxime is known to migrate, the absolute configuration of (+)-**56** follows. This oxime may, of course, be considered in terms of its chiral centers, rather than as an axially dissymmetric molecule. (1)

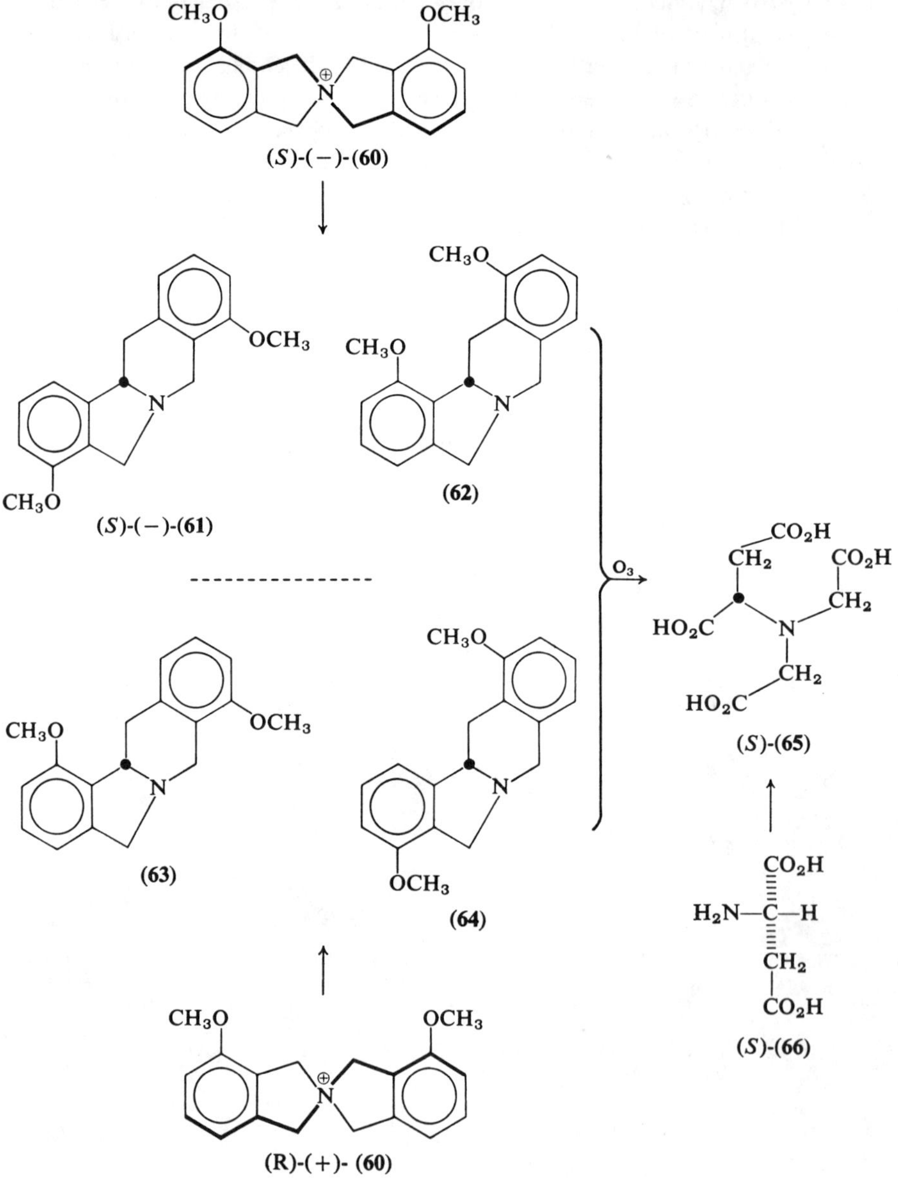

d. Spiranes. Brewster and Jones (37) utilized an asymmetric Stevens' rearrangement of the levorotatory quaternary ammonium salt **60** with sodium hydride in diglyme to convert chiral nitrogen to chiral carbon. By showing the absolute configuration of the chiral center in the rearrangement product to be S by ozonolysis to the derivative **65** of aspartic acid **66**, the product of the Stevens' rearrangement had to be one of four positional isomers **61–64**. NMR evidence indicated a methoxyl distribution compatible only with diastereomer **61**, which could be formed only from the S configuration of **60**. The assignment as (S)-$(-)$-**60*** agreed with Lowe's rule (see p. 55, ref. 74), the helix conductor model, and the Eyring-Jones model of optical activity.

e. trans-Cycloalkenes. By viewing it down the double bond one may consider *trans*-cyclooctene (**43**) an axially dissymmetric compound (38); alternatively, it may also be considered in terms of planar dissymmetry (1). The

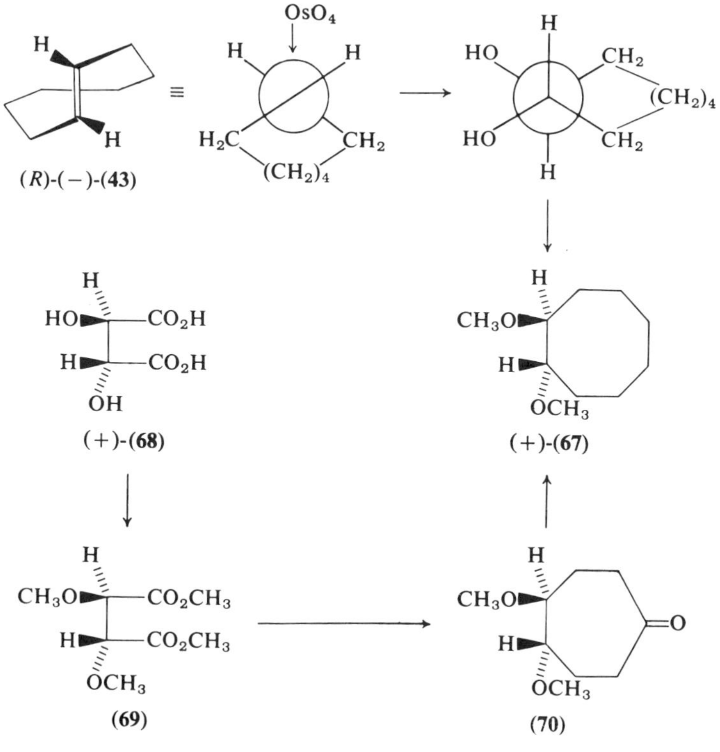

*The present assignment of $(-)$-**60** as S results from consideration of this spirane as a centro-dissymmetric molecule (1). Brewster and Jones (37) treated $(-)$-**60** as an axially chiral molecule and described its configuration as R (2). (See Appendix, p. 62.)

absolute configuration of this olefin has been determined by means of a self-immolative asymmetric synthesis by Cope and Mehta (39). *cis* Addition of osmium tetroxide to (−)-*trans*-cyclooctene (**43**) (from the side not hindered by the methylene groups) followed by methylation resulted in (+)-*trans*-1,2-dimethoxycyclooctane (**67**). In a parallel experiment (+)-tartaric acid (**68**) was esterified with methanolic hydrogen chloride and methylated with methyl iodide and silver oxide to give **69**. The diester was homologated twice by standard procedures, and Dieckmann condensation then yielded the cycloheptanone **70**. Reduction of the cyanohydrin of **70** and subsequent treatment with sodium nitrite in acetic acid converted it to a cyclooctanone, which by Wolff-Kishner reduction gave (1S:2S)-(+)-1,2-dimethoxycyclooctane (**67**) identical with that formed from (−)-*trans*-cyclooctene (**43**). This correlation shows that levorotatory *trans*-cyclooctene has the R configuration.

2. Conservative Asymmetric Syntheses

In a conservative asymmetric synthesis (5) a new dissymmetric grouping is formed as the result of an interaction with another dissymmetric grouping* which retains its stereochemical integrity. A variety of empirical rules are available to predict the outcome of particular asymmetric syntheses; but usually they are applicable only if strong steric or dipolar factors come into play, for, otherwise, small effects may alter the transition state to the point where the pertinent rule will no longer be obeyed.

An alternative approach in utilizing conservative asymmetric syntheses for prediction of absolute configuration involves evaluation of the steric and dipolar factors present in an assumed transition state. Configurational correlations made in this manner cannot be considered valid without corroborative evidence (40). The dramatic effects that change in solvent (41) or temperature (42) can have on the selectivity of an asymmetric synthesis will serve to document this reservation.

a. Biphenyls. Mislow, Prelog, and Scherrer (43) determined the absolute configuration of bridged binaphthyl **71** through an asymmetric atrolactic acid synthesis by extension of Prelog's rule (44). The phenylglyoxylate **72** of levorotatory hydroxybinaphthyl **71** was treated with methyl magnesium iodide to give the atrolactate, which upon hydrolysis resulted in an excess of (R)-(−)-atrolactic acid. An asymmetric synthesis occurred because of the presence of the active binaphthyl moiety. From the known configuration of the atrolactic acid formed in excess and by application of Prelog's rule to interpret the steric course of the asymmetric synthesis, the R configuration was assigned to the levorotatory alcohol **71**.

*A suitable dissymmetric grouping might be a dissymmetric catalyst, solvent, reagent, etc.

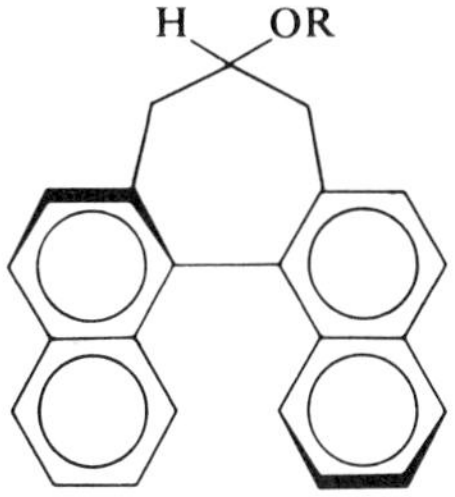

(−)-(71) R = H
(72) R = CO—CO-φ

b. Allenes. Tömösközi and Bestman (45) have found that a partial asymmetric synthesis of allenes could be effected by addition of acid chlorides to dissymmetric ylids. In order to predict the absolute configuration of an allenic acid produced in this reaction, the authors assume the applicability of a Cram-Prelog type rule in the addition to the ylid **73** and invoke steric and electronic requirements to assign preferred conformations to the intermediates **74–75**. When (−)-2-octanol was used in preparing a dissymmetric ylid (**73**), the levorotatory allenic acid (**76**) obtained was assigned the S configuration.

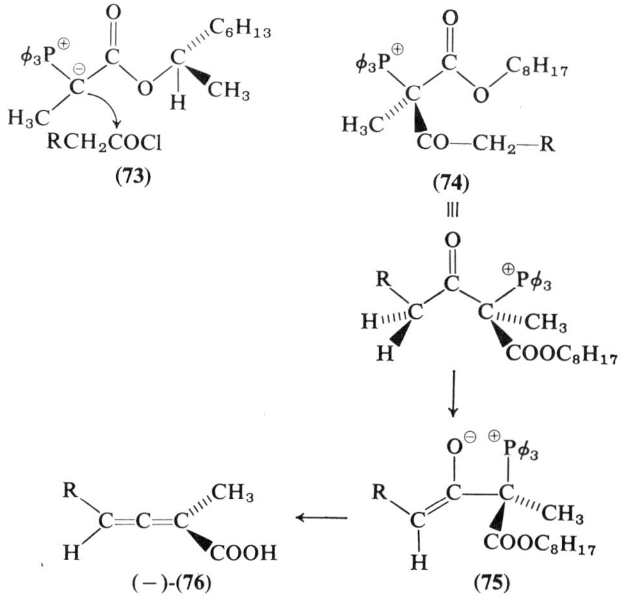

Asymmetric reduction has also been used to determine the absolute configuration of allenes. Evans, Landor, and Regan (21) used optically active lithium dimenthoxyaluminum hydride to reduce alkenynol (**77**) to allene (**79**). After the absolute configuration of **79** was determined by means of a stereospecific Claisen rearrangement, a stereochemical model of an aluminum complex (**78**) could be proposed that fitted the results of the reduction. The model **78** was then used to predict the results of similar hydride reductions.

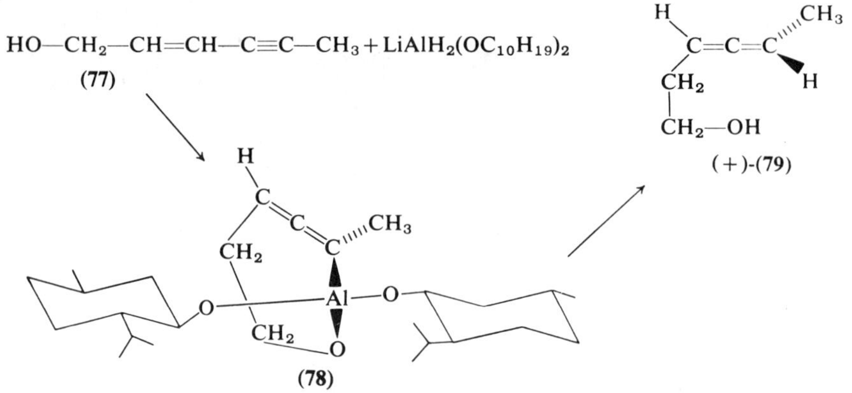

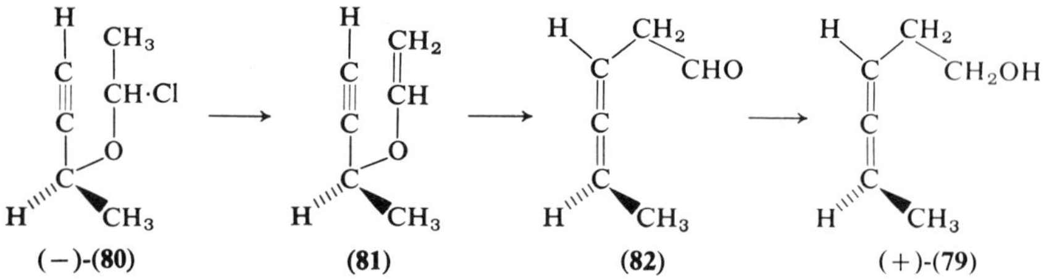

$$HO-CH_2-CH=CH-C\equiv C-CH_3 + LiAlH_2(OC_{10}H_{19})_2$$
(**77**)

$$\begin{array}{c} H \quad\quad CH_3 \\ C=C=C \\ CH_2 \quad\quad H \\ | \\ CH_2-OH \end{array}$$
(+)-(**79**)

(**78**)

(−)-(**80**)   (**81**)   (**82**)   (+)-(**79**)

Using this method Landor and co-workers (46) have determined the absolute configuration of the naturally occurring allene (−)-marasine (**84**) and (−)-9-methylmarasine (**85**). The asymmetric reduction of non-2-en-4,6,8-triyn-1-ol (**83**) with a lithium aluminum hydride sugar complex afforded (−)-marasine (**84**). The absolute stereochemistry of this allene was predicted to be R by assuming similar stereochemistry for the transition states in reduction of **77** and **83**.

Waters and Caserio (30) have found that partial asymmetric hydroboration of racemic 1,3-dimethylallene (**42**) with (+)-tetra-3-pinanyldiborane (diisopinocampheylborane) resulted in an excess of unreacted levorotatory **42**. A similar reaction with 1,3-diphenylallene (**40**) led to (R)-(−)-**40**, whose absolute configuration was determined by Mason and Vane (27). Assuming the transition states for reduction of **42** and **40** to be similar, levorotatory **42**

$$HO-CH_2-CH=CH-C\equiv C-C\equiv C-C\equiv C-R$$
(83)

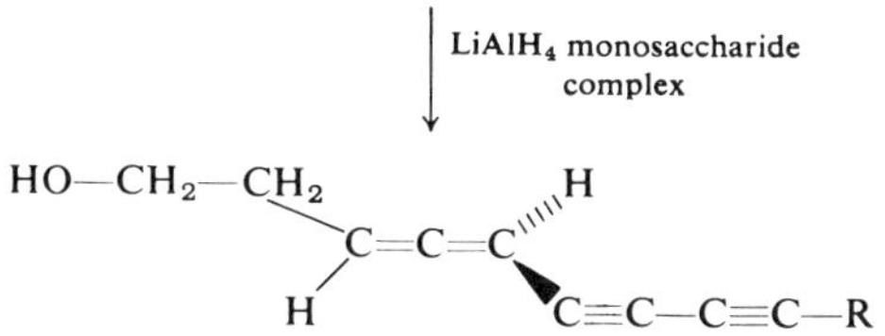

↓ LiAlH$_4$ monosaccharide complex

HO—CH$_2$—CH$_2$\\C=C=C\\H ... C≡C—C≡C—R

(−)-**(84)** R = H, marasine
(−)-**(85)** R = CH$_3$, methylmarasine

also has the *R* configuration. This assignment is in agreement with that of Jones and Walbrick (29) and with Lowe's rule (see ref. 74, p. 55), relating configuration with relative polarizability of substituents.

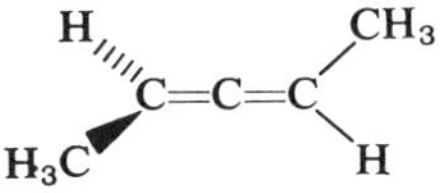

(−)-**(42)** 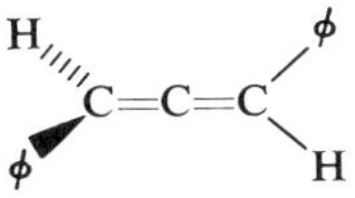   (−)-**(40)**

c. Spiranes. Gerlach (47) has synthesized spiro[4.4]nonane-1,6-dione **(88)** in optically active form and determined the chirality of the enantiomers by chemical correlation. The key step in Gerlach's correlation was the determination of absolute configuration of a secondary alcohol **(86)** using a conservative asymmetric synthesis according to the method of Horeau (48).

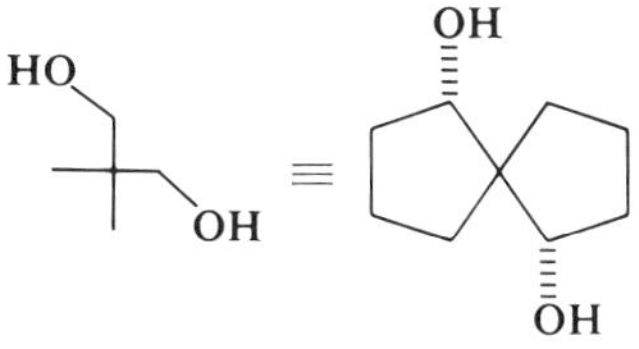

(−)-**(86)** 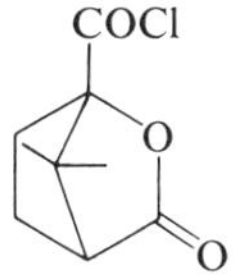   **(87)**

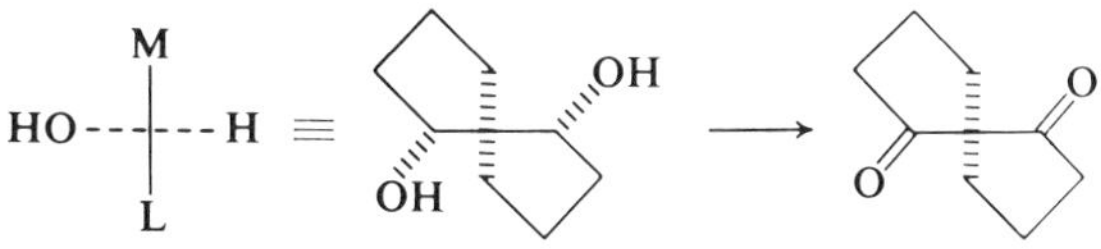

(89)   (1*R* : 6*R*)-(−)-**(86)**   (−)-**(88)**

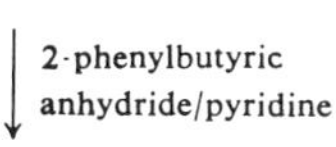

↓ 2-phenylbutyric anhydride/pyridine

(+)-2-phenylbutyric acid in excess

Acetylation of racemic *trans,trans*-spiro[4.4]nonane-1,6-diol (**86**) with (−)-camphanic acid chloride (**87**) in pyridine afforded two diastereomeric esters which were separated by chromatography. From the higher melting ester optically pure levorotatory *trans,trans*-diol **86** was obtained. Gerlach's recognition that the hydroxyl groups of **86** were equivalent as a result of symmetry, and the use of Horeau's method for determining the absolute configuration of secondary alcohols showed the configuration of the levorotatory diol **86** to be as in formula **89** (48).

The levorotatory diol **86** had, accordingly, the $1R:6R$ chirality. Although a chiral center exists at C-5 in addition to the chiral centers at C-1 and C-6, it is dependent on the configuration of the two others and need not be specified. When the two asymmetric centers at C-1 and C-6 were eliminated, the center at C-5 remained and a molecule of the spiro type was obtained. Accordingly, oxidation of (−)-**86** resulted in formation of (−)-**88**, whose absolute configuration is thus known to be S. This agrees with the prediction based on Lowe's rule (74).

d. Carbodiimides. Schlögl and Mechtler (49) have effected a kinetic resolution of N,N'-diferrocenylcarbodiimide (**90**) by partial reaction of racemic **90** with (R)-(+)- or (S)-(−)-6,6′dinitrodiphenic acid in anhydrous benzene to form the acylurea. Unreacted carbodiimide recovered was optically active. From stereochemical considerations of the assumed preferred transition state for addition of the carboxyl group to the —N=C— structure, it was predicted that (−)-**90** has the R configuration.

$$\text{Fc} \diagdown \text{N}=\text{C}=\text{N} \diagup \text{Fc}$$

$$\text{Fc} = C_5H_5FeC_5H_4\text{—}$$

$$(-)\text{-}(\mathbf{90})$$

e. Paracyclophanes. Falk and Schlögl (50) assigned the S configuration to the (+)-[2.2]paracyclophanecarboxylic acid (**92**) from comparisons based on the similar topologies of **92** and methylferrocene-α-carboxylic acid (**91**). Kinetic resolution of the anhydride of racemic **91** with (−)-α-phenethylamine affords the dextrorotatory acid **91**, whose absolute configuration* has been established as $1S$ by two independent methods (51). Similarly, reaction of the

---

*The varied approaches to determination of absolute configuration of centrally chiral metallocenes (1) have been reviewed by Professor Schlögl in *Topics in Stereochemistry*, Vol. 1 (52). The methods used to assign metallocene chirality parallel, in their ingenuity, those summarized in the present review.

anhydride of racemic [2.2]paracyclophanecarboxylic acid (**92**) with (−)-phenethylamine yielded the dextrorotatory acid **92**. The *S* configuration has also been assigned to **92** on the assumption that the topology of the reaction sites in **91** and **92** determine the course of the kinetic resolution.*

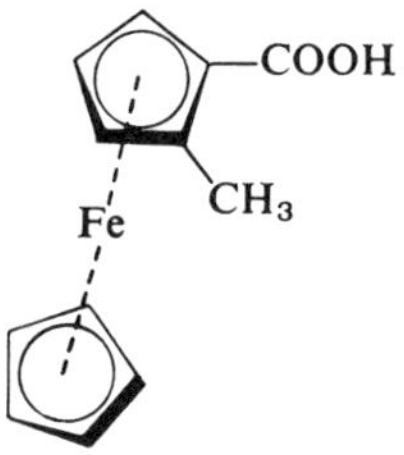

(1*S*)-(+)-(**91**) 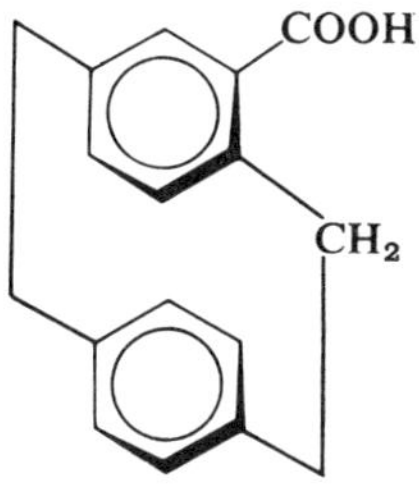   (*S*)-(+)-(**92**)

3. Direct Chemical Correlation

a. Biaryls. With the availability of configurational standards, it became possible to correlate the absolute configurations of biphenyls by modifying the ring substituents (53). For example, the biphenyls derived from **93** all have the same chirality (54).

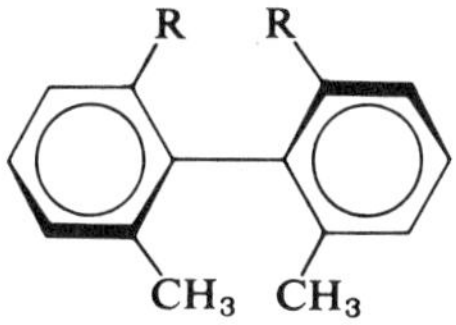

(**93**)

R	Sign	Chirality
COOCH$_3$	+	S
CH$_2$OH	−	S
CH$_2$Br	+	S

Mislow and co-workers (13–14) have correlated some binaphthyl derivatives with bridged binaphthyls by direct chemical correlation; for example, bridged binaphthyl (*S*)-(−)-**95** was readily obtained from binaphthyl (*S*)-(−)-**94**. Analogous correlations have been made with bianthracyl derivatives by Badger and co-workers (56–57).

---

*Professor Schlögl has informed us that more recent results employing Horeau's method for determination of the absolute configurations of cyclic carbinols and utilizing circular dichroism studies have confirmed the configuration established for **92**.

Yamada and co-workers (58–59) were able to intercorrelate chemically a series of biphenyl, binaphthyl and bianthracyl derivatives with a configurational standard whose absolute configuration was determined by X-ray analysis.

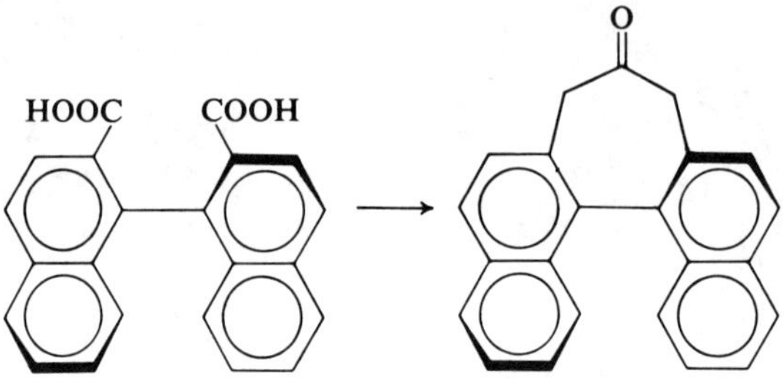

(S)-(−)-(94)   (S)-(−)-(95)

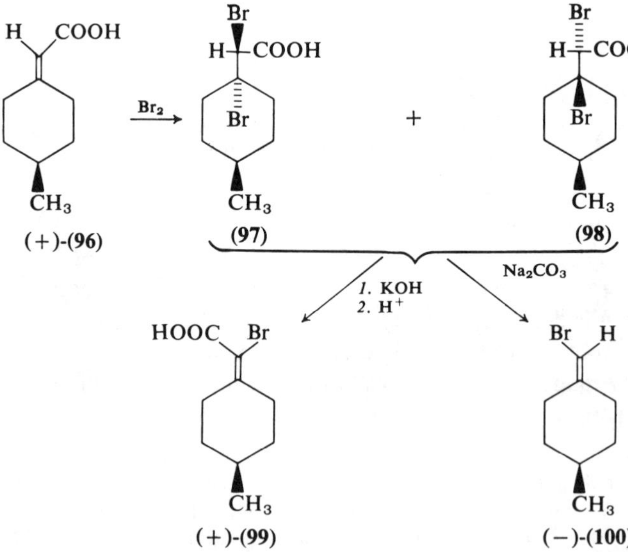

b. Alkylidenecycloalkanes. The absolute configuration of two alkylidenecyclohexanes has been determined by chemical correlation. In 1911 Perkin and Pope (60) prepared two optically active dibromides by *trans* addition of bromine to (+)-4-methylcyclohexylideneacetic acid (**96**). Treatment of the bromide mixture **97–98** with potassium hydroxide resulted in *trans* elimination of hydrogen bromide to form (+)-**99**; similarly, use of sodium carbonate resulted mainly in *trans* elimination of hydrogen bromide and carbon dioxide to form (−)-**100**. Based on the stereochemistry of these reaction sequences and the absolute configuration of **96**, Gerlach (34) assigned absolute configurations to **99** and **100**. Brewster and Privett (35a) also assigned the absolute configurations of **96**, **99**, and **100** on the basis of the relative polarizabilities of substituents and signs of rotation.*

c. Spiranes. Spiranes, because of an asymmetric tetracoordinate atom, can be correlated by standard methods with other molecules containing asymmetric atoms. Krow and Hill (61) have determined the absolute configuration of a 2,7-diazaspiro[4.4]nonane derivative (**101**). Their approach was to synthesize **101** from a centro-dissymmetric intermediate which could be correlated with 2-methyl-2-ethylsuccinic acid (**102**). Therefore, (R)-(+)-**102** was converted to its succinimide and thence to its amine and methanesulfonamide (**103**). At the same time, levorotatory **104** was converted by successive reactions with diazomethane, lithium aluminum hydride, and methanesulfonyl chloride into an NOO-trimethylsulfonate, which on treatment with sodium sulfide afforded the dextrorotatory thioether **105**. Raney nickel desulfurization of **105** afforded dextrorotatory **103**. These reactions establish the R configuration for levorotatory **104**. In order to convert **104** to spirane **101**, the ammonium salt of **104** was pyrolyzed to the imide **106**. This on reduction with lithium aluminum hydride followed by treatment with p-toluenesulfonyl chloride gave (R)-(−)-**101**. The assignment* of the R configuration to **101** thus rests on its direct correlation with (R)-(+)-2-methyl-2-ethylsuccinic acid (**102**).

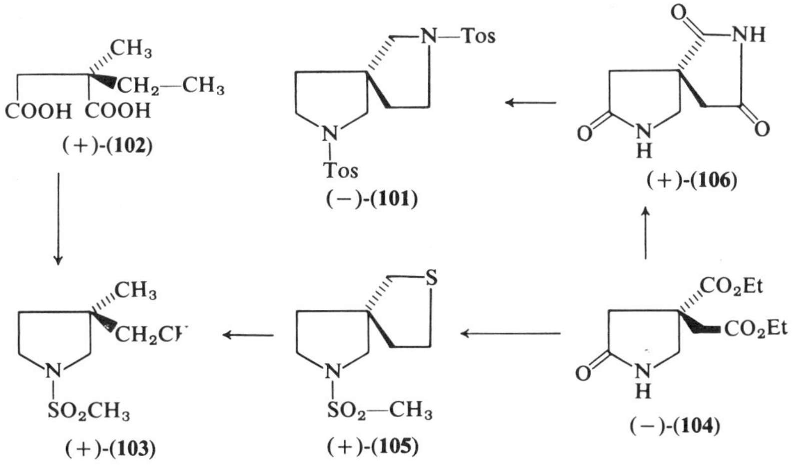

*See Lowe's rule, ref. 74, p. 55.
*The present assignment of (−)-**101** as R results from consideration of this spirane in terms of its chiral center (1). Krow and Hill (61) treated (−)-**101** as an axially chiral molecule and described its configuration as S (2). (See Appendix, p. 62.)

B. Physical Methods

1. Thermal Analysis

Siegel and Mislow (62) were able to correlate several optically active diphenic acids (**107**) by the method of quasi-racemates. The configurationally related pair formed solid solutions, while the nonisomorphic pair formed a 1:1 quasi-racemic compound. Using this method, (S)-(−)-6,6′-dinitro-2,2′-diphenic acid of established configuration (12) has been intercorrelated with (S)-(−)-6,6′-dichloro-, and (S)-(+)-6,6′-dimethyl-2,2′-diphenic acid. An attempt to correlate 1,1′-binaphthyls with hindered biphenyls by this method was unsuccessful (64).

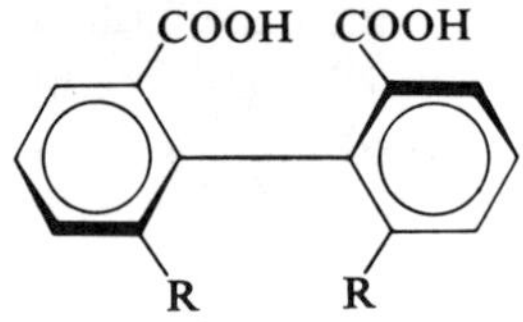

(**107**)

R	Sign	Chirality
Cl	−	S
NO_2	−	S
CH_3	+	S

2. Optical methods

An optical displacement rule proposed in the biaryl series (64) states that a symmetrically substituted hindered biaryl has the S configuration if change of the open to a bridged biaryl system produces a marked shift of the optical activity in the positive direction. Accordingly, the fact that dextrorotatory dinaphthylamine **108** gives the strongly levorotatory bridged derivative **109** suggests that (+)-**108** has the R configuration.

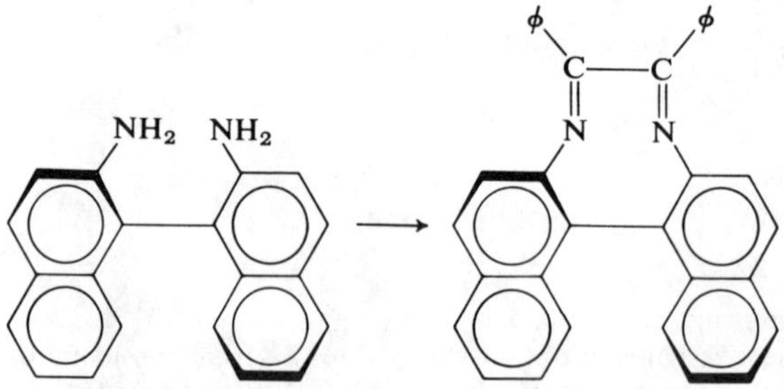

(+)-(**108**)   (−)-(**109**)

Analysis of optical rotatory dispersion curves of atropisomers has shown that the signs of the Cotton effect curves are uniquely determined by the absolute configuration (65). It was found that for 2,2'-bridged biaryls of the R configuration, the sign of the long wavelength Cotton effect was negative for 6,6'-dinitro derivatives, positive for 6,6'-dimethyl and 6,6'-dichloro derivatives, and positive for 1,1'-binaphthyls (66).

It was subsequently shown (67–69) that the signs of the ultraviolet circular dichroism (CD) curves correspond to the signs of the ORD curves, and so CD curves also reflect the chirality of the biphenyl chromophore. The two complementary methods are powerful tools and in the assignment of absolute configuration to atropisomers the use of both techniques is desirable. In those cases where background rotations hide weak Cotton effects, CD will be more informative than ORD. On the other hand, the tail absorption of high intensity short wavelength Cotton effects which cannot be observed with available instrumentation can frequently be detected in the ORD curve.

Mislow and Djerassi (70) have ingeniously correlated the ORD curves of santonide (**110**) and parasantonide (**111**) of known absolute configuration with the optical rotatory dispersion curve of the bridged biphenyl ketone **112**. Because the chirality of the β,γ-unsaturated ketone moiety was known in the natural products, the biphenyl with an ORD curve superimposable with that of the natural product could be predicted to have the same chirality or absolute configuration. The resulting configurational assignment to (+)-**112** as R was consistent with that made based on other chemical evidence (15). Natural products of the biaryl type, such as the aporphines, have also been investigated by optical means (6).

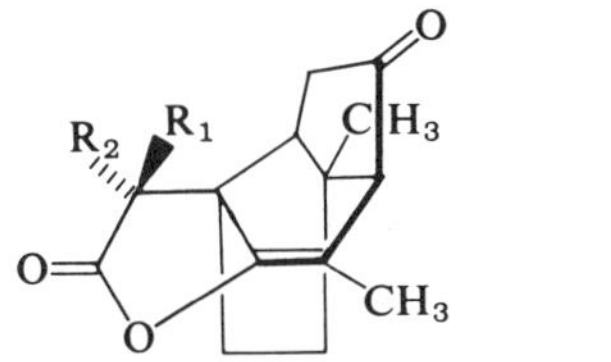

(+)-(**110**) R_1 = H R_2 = CH_3
(+)-(**111**) R_1 = CH_3 R_2 = H

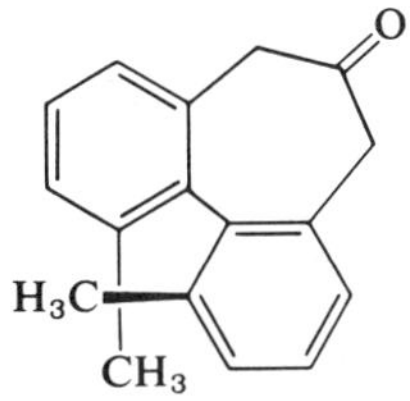

(R)-(+)-(**112**)

C. Theoretical Methods

1. Biaryls

Kuhn and co-workers (71), on the basis of a theoretical model, assigned the S configuration to (+)-6,6'-dinitro-2,2'-diphenic acid (**113**), (+)-6,6'-dimethyl-2,2'-diphenyldiamine (**114**), and (+)-2,2'-dichloro-6,6'-dimethyl-4,4'-diaminobiphenyl (**115**). However, these configurational assignments were later shown to be incorrect (12,55,72), the dextrorotatory isomers having the R configuration.

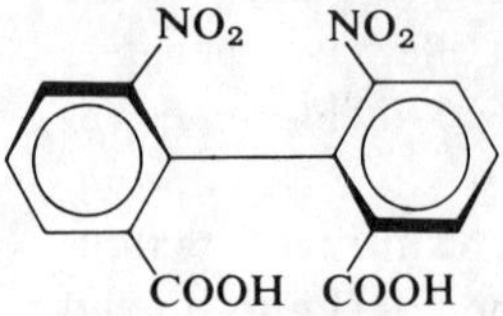

(+)-(113) 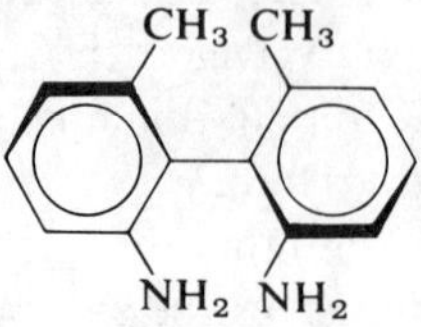 (+)-(114) 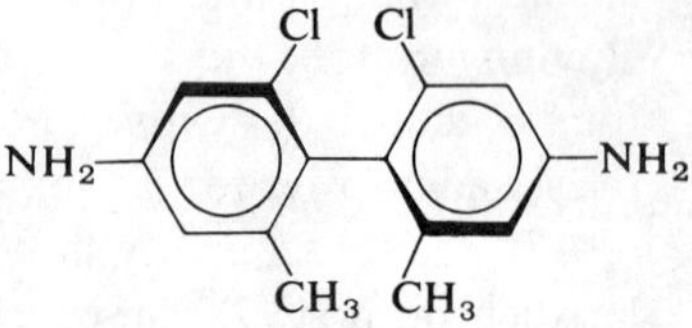 (+)-(115)

Grinter and Mason (73) have analyzed the electronic absorption and CD spectra of bianthryls **116** and **117**. Using a coupled oscillator model and molecular orbital theory, they predicted that the levorotatory enantiomers have the R configuration. This assignment agreed with assignments by Badger and co-workers (56), who used the optical displacement rules discussed earlier (64), and by Yamada and Akimoto (59), who correlated diacid **118** with binaphthyl **126**, whose configuration was proven by anomalous X-ray dispersion analysis.

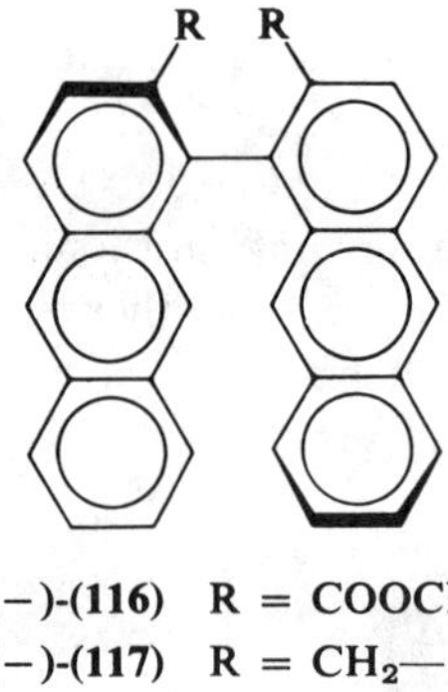

(−)-(116)   R = COOCH$_3$
(−)-(117)   R = CH$_2$—
(−)-(118)   R = COOH

Mislow and co-workers (64), using the method of Kirkwood, correctly predicted the absolute configuration of (+)-**119** to be S. The approach involved a qualitative treatment of the rotation of **119** based on polarizability theory.

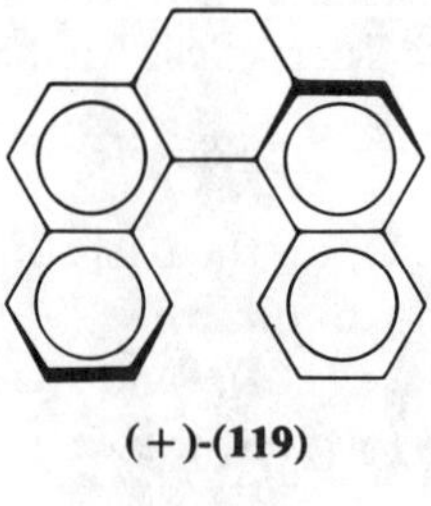

(+)-(119)

2. Allenes (31)

A comparison of calculated and observed circular dichroism curves enabled Mason and Vane (27) to predict the absolute configuration of 1,3-diphenyl allene (**40**). By considering **40** to be two orthogonal styrene chromophores, the signs of the Cotton effects were predicted from the two coupling modes of styrene excitation moments. The levorotatory enantiomer could be assigned the R configuration.

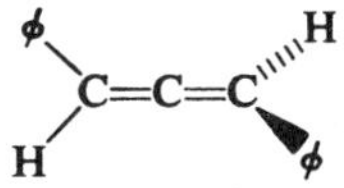

$$\phi\diagdown\diagup H$$
$$C=C=C$$
$$H\diagup\diagdown\phi$$

(−)-(**40**)

Lowe (74) and Brewster (31) have analyzed the rotation of allenes according to a screw pattern of polarizability of substituents. By this treatment the more polarizable substituents are placed at the top of a pair of perpendicular axes. The screw pattern of polarizability is clockwise if the more

TABLE I

Application of Lowe's Rule to Determining the Absolute Configuration of Allenes

Allene	No.	Sign	Screw pattern	Configuration	Refs.
ϕ H--┼--ϕ H	40	+	Clockwise	S	27
COOH H--┼--COOH H	25	+	Clockwise	S	22
C(CH$_3$)$_2$—CHO H$_3$C--┼--H H	24	−	Anticlockwise	R	20
Cl H$_3$C--┼--t-Bu H	22	−	Anticlockwise	S	17
CH$_3$ H--┼--CH$_3$ H	42	+	Clockwise	S	30,31
t-Bu H--┼--t-Bu H	127	+	Clockwise	S	75

polarizable substituent on the horizontal axis is on the right of the vertical axis, anticlockwise if on the left. According to the polarizability theory (31), an allene (**120**) will be dextrorotatory when A is more polarizable than B and X is more polarizable than Y. In Table I Lowe's rule has been applied to those allenes of known absolute configuration whose substituents lack asymmetric

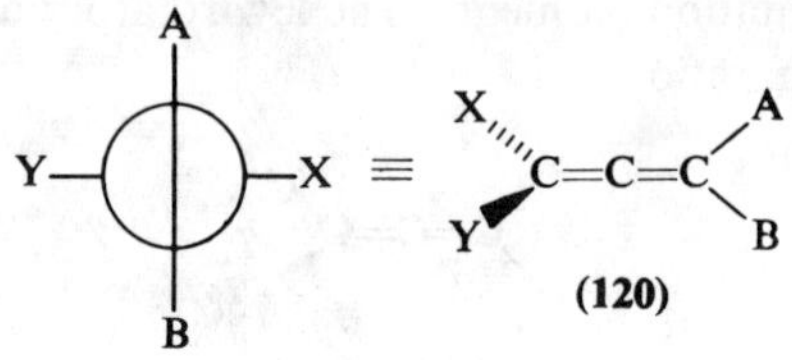

$$Y-\bigodot-X \equiv \underset{Y}{\overset{X}{>}}C=C=C\underset{B}{\overset{A}{<}}$$
(**120**)

conformations. An order of polarizability of substituents Cl > ϕ > COOH > $(CH_3)_2$C—CHO > CH_3 > *t*-butyl > H is used. The absolute configuration of **127** has been inferred by Borden and Corey (75) from application of Lowe's rule.

Predictions of absolute configuration of some naturally occurring diacetylenic allenes from fungi have been tabulated elsewhere by Brewster (31).

3. Alkylidenecycloalkanes

Brewster (35) has applied the polarizability theory to alkylidenecycloalkanes of type **121**. When A is more polarizable than B, **121** will be dextrorotatory, provided some near ultraviolet Cotton effect does not control rotation in the visible region.

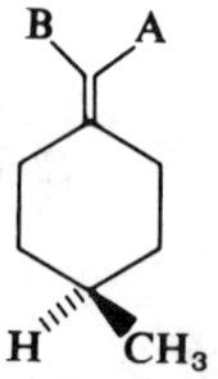

(**121**)

4. Spiranes

Assignment of the S configuration to levorotatory spirane **60** was made in accordance with the prediction of the helix conductor model and the Eyring-Jones model of optical activity (37). The S configuration for (−)-**60*** and the S configuration for (−)-**88** could also be predicted using Lowe's rule (74). Although spirane **101** failed to obey Lowe's rule (61), it has been pointed out by Brewster and Jones (37) that this failure might be due to swamping by

* See footnotes pp. 43, 51, and Appendix, p. 62.

long range effects associated with the *p*-toluenesulfonamide groupings. Spirane (*R*)-(+)-**105**, with which (*R*)-(−)-**101*** was correlated (61), did fit Lowe's rule by showing a small positive rotation above 330 nm.

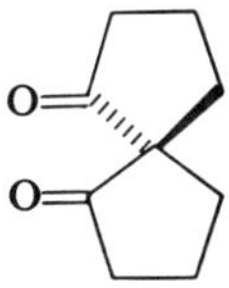

(−)-(**88**) 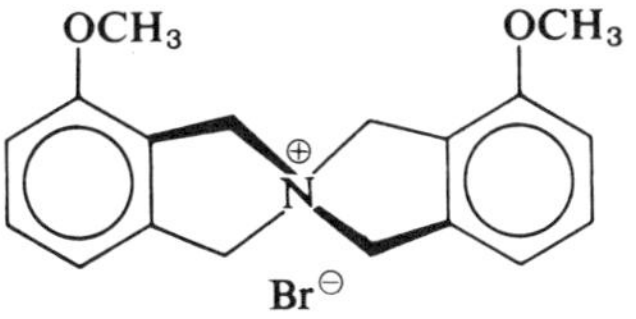   (−)-(**60**) 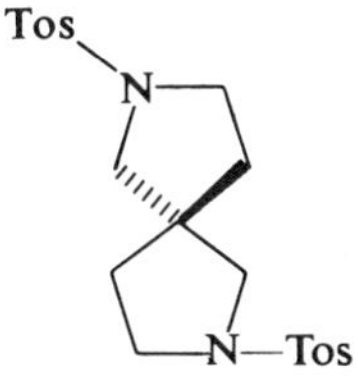   (−)-(**101**) 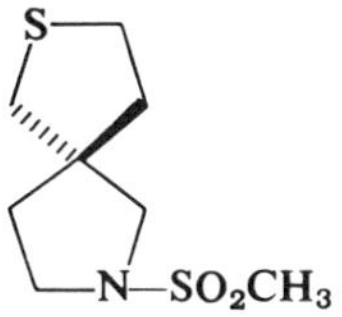   (+)-(**105**)

Attempts were made by Kuhn and Bein and by Lowry (76), who used a molecular model of M. Born, to predict the absolute configuration of spiranes, e.g., **122** and **123**. Conflicting predictions correlating chiralities and signs of rotation were advanced, however.

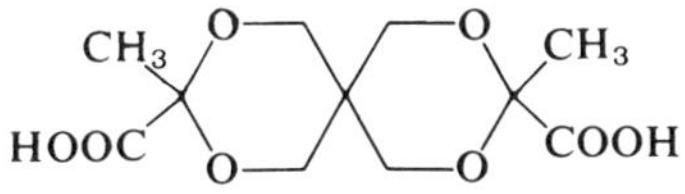

(**122**) 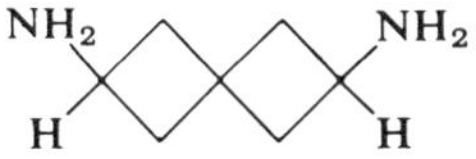   (**123**)

5. *trans*-Cycloalkenes

An optical model used by Moscowitz and Mislow (38) resulted in an incorrect assignment of the *S* configuration to levorotatory *trans*-cyclooctene (−)-(**43**). The correct absolute configuration was shown to be *R* by Cope and Mehta (39).

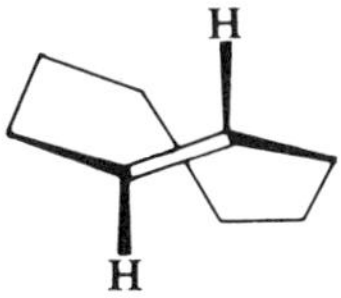

(−)-(**43**)

6. Phenanthrenes

The chiralities of certain 3,4-benzophenanthrene derivatives have been determined from the signs of the Cotton effects associated with certain CD bands. Mason and co-workers (77,78) assigned the *M* configuration to levorotatory phenanthrene **124** and the *P* configuration to dextrorotatory **125** by

*See footnotes pp. 43, 51, and Appendix, p. 62.

analysis of circular dichroism bands in conjunction with molecular orbital theory.

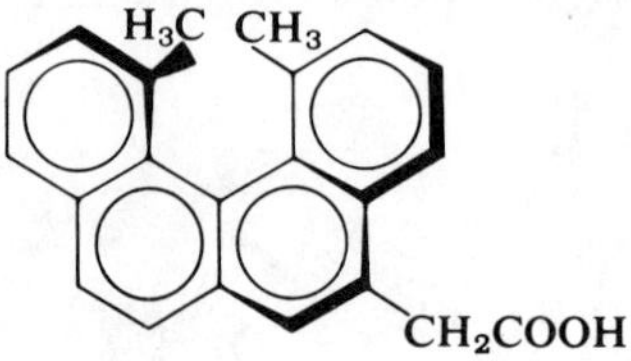

(−)-(124) 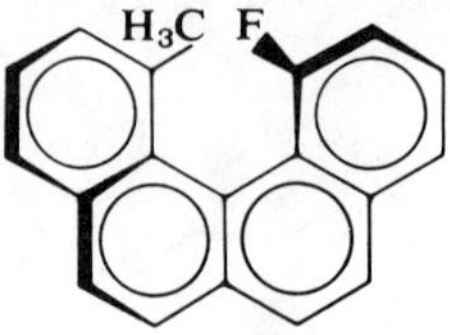   (+)-(125)

7. Hexahelicene

Tinoco and Woody (79) derived the expected absorption and rotation of polarized light for a free electron constrained to move on a helix. They predicted positive rotational strength for the longest wavelength absorption band of a right-handed helix. The assignment of absolute configuration to **9** as (P)-(+)-hexahelicene (79,80) according to this method was in agreement with that of Fitts and Kirkwood (81), who used polarizability theory. Moscowitz (82), using detailed Hückel molecular orbital calculations, reached the opposite conclusion as to the helical sense of the dextrorotatory enantiomer.

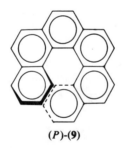

(P)-(9)

D. X-Ray Analysis

Although ordinary X-ray crystallography does not provide the information necessary to establish the chirality of one of a pair of enantiomers, the technique of anomalous X-ray scattering has been used to overcome this difficulty (83). Yamada and co-workers (58) have used this technique to determine the absolute configuration of binaphthyl **126**. Crystallization of **126** in bromobenzene resulted in incorporation of one equivalent of bromobenzene into the crystal lattice. The anomalous dispersion effect of bromine was then used in order to determine the absolute configuration of (+)-**126** as R.

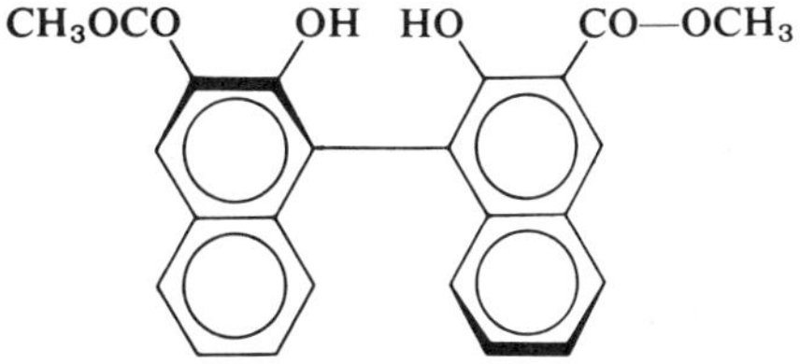

(R)-$(+)$-(**126**)

Horrocks and co-workers (84), in a direct X-ray determination of the absolute configuration of a coordination compound, cobalt(salicylaldehyde)$_2$-$(+)$-2,2′-diamino-6,6′-dimethylbiphenyl, showed the dextrorotatory biphenyl moiety **114** to exist in the R configuration. This assignment confirmed the determination of Mislow and co-workers (55).

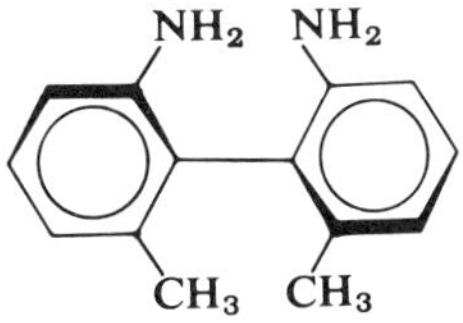

$(+)$-(**114**)

Acknowledgments

The author wishes to thank Professor Richard K. Hill for his help and encouragement. Gratefully acknowledged is the kindness of Professors J. H. Brewster, K. Schlögl, M. C. Caserio, W. R. Moore, R. D. Bach, and Dr. H. Gerlach for supplying results of their work prior to publication, of Dr. H. Falk for bringing his earlier review (85) to our attention, and of Dr. H. Westen in connection with matters of nomenclature.

IV. APPENDIX: SPECIFICATION OF MOLECULAR CHIRALITY

The chirality of axial or planar molecules is specified according to a sequence-rule method developed by R. S. Cahn, C. Ingold, and V. Prelog (1). The approach of this nomenclature system is, firstly, to arrange ligands associated with an element of chirality into a sequence. This sequence is then used to trace a chiral path. Finally, the chiral sense of the path, designated in a prescribed way, classifies the element of chirality.

The sequence-rule specification of molecular chirality is embodied in a series of well-defined rules. In so far as these rules bear upon assignments of chirality in the preceding review, they are here defined and discussed.

A. Definition of Rules

The general procedure for assigning chirality to configurational isomers is outlined below.

1. Ligancy Complementation

All atoms other than hydrogen are complemented to quadriligancy by providing, respectively, one or two duplicate representations of any ligands which are doubly or triply bonded and then a necessary number of phantom atoms of atomic number zero.

2. Factorization Rule

Overall chirality is factorized into elements, which are treated in the order: centers, axes, and finally planes, as far as necessary.

3. Sequence Rule

The ligands associated with an element of chirality are ordered by comparing them at each step in bond-by-bond explorations, from the element, along the successive bonds of each ligand, and, where the ligands branch, first along branch-paths providing highest precedence to their respective ligands, the explorations being continued to total ordering by use of the following standard subrules, each to exhaustion in turn:

1. Near end of axis or side of plane precedes further.
2. Higher atomic number precedes lower.

4. Chirality Rule

Among ligands of highest precedence, the path of their sequence is followed from the preferred side of the model, that is, the side remote from the group of lowest precedence, and, according to whether the path turns to right or left, the element is assigned the chiral label R or S.

5. Helicity Rule

According to whether the identified helix is left- or right-handed, it is designated "minus" and denoted by M, or designated "plus" and denoted by P.

Helicity is a special case of chirality. A helix is characterized by a helical sense, a helical axis, and a pitch (ratio of axially linear to angular properties). When a molecular structure has these three properties, the helical model is an easy model to use. Right- or left-handed helicity associates, respectively, a right- or left-handed turn with axial translation away from the observer. To reverse the helix with respect to the observer makes no difference.

PLANAR AND AXIALLY DISSYMMETRIC MOLECULES 61

6. *Conformational Chirality*

Molecular conformations can be described, in terms of partial conformations about single bonds. In order to do this, first, the spatial relations of two sets of groups attached to a single bond are specified. Next, a fiducial group is chosen from each set by means of the *conformational selection rule*. In the case of **114** the pairs of groups nearest together (circled) that lie one pair on each of the planes intersecting along the axis are considered. The atom of each pair of highest priority is the carbon bearing the amino group. These carbons are designated as the fiducial groups. The smallest torsional angle between the fiducial groups specifies the pitch and sense of the conformational helix.

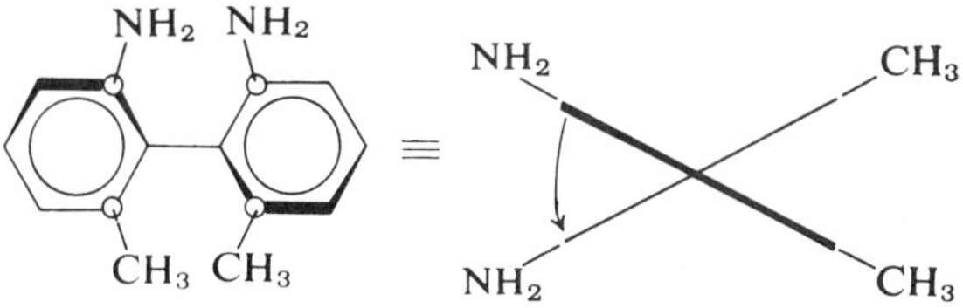

M-(**114**)   Minus torsion angle

a. Conformational Selection Rules

a. If all the groups of a set are different, the group most preferred by the standard sequence subrules is fiducial.

b. If an identity among the groups of a set leaves one group unique, the unique one is fiducial.

b. Conformational Helix.
The single bond about which conformation is to be specified is made the axis, and the smallest torsion angle between the fiducial groups is made to define the pitch and sense of the helix.

Right- and left-handed helices, thus defined, correspond to (+) and (−) torsion angles and are designated by the *helicity rule* as *P* ("plus") and *M* ("minus"), respectively. Biphenyl **114**, whose fiducial groups describe a (−) torsion angle, is designated as the *M* conformation. It should be noted that in the areas of conformations and secondary structures (helicity), chirality can be specified in some cases using *either* the chirality rule *or* the helicity rule.

B. Application of the Sequence-rule Method

1. *Central Chirality—Procedural Consequences of Symmetry*

Centrally chiral molecules of no symmetry (C_1) are represented by the asymmetric carbon atom of type Cabcd. Another symmetry class to which chiral centers can belong contains one twofold axis of symmetry (C_2). It is now recognized that to class C_2 belong the centrally chiral spiranes **60, 88,** and **101**, which previously (2) were treated as axially chiral molecules.

a. Spiranes. In order to specify the chirality of spirane **88**, the first member of the sequence is taken as *either* carbonyl group adjacent to the chiral center. The other neighbor carbonyl then is the second member of the sequence. Of the other atoms adjacent to the chiral center, that which occupies the same ring as the preferred carbonyl is the third member of the sequence. The sequence is thus completed, and it shows **88** to be represented in the *S* configuration.

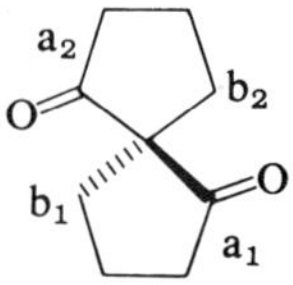

(*S*)-(**88**)

b. Alkylidenecycloalkanes. The piperidone oxime **56** also falls within the scope of central chirality, rather than axial chirality as originally supposed (2). When the chiral centers of the ring have been assigned, the remaining matter is one of geometrical isomerism (*cis* or *trans*), outside the province of the sequence rule. Structure **56** is shown in the *syn-R* form.

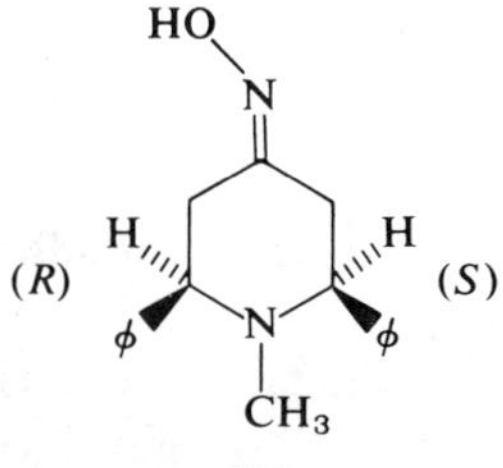

(**56**)

2. Selection Rule for Axial Chirality

The fiducial groups shall be chosen from the pairs nearest together of groups, directly bonded to atoms on the axis, that lie one pair in each of the planes of atoms that intersect along the axis.

a. Biaryl (C_1). The five main areas of application of the concept of axial chirality are allenes, alkylidene-cycloalkanes, spiranes, biaryls, and adamantoids.

As the selection rule is applied to biaryl **128**, the fiducial groups are the *ortho-* carbons in the order shown and the configuration is (*R*)-**128**.

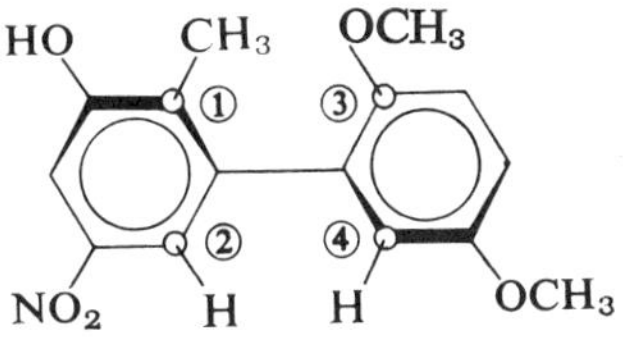

PLANAR AND AXIALLY DISSYMMETRIC MOLECULES

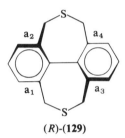

(R)-(128)

b. Biaryl (D_2). To allot a symbol of chirality to biaryl **129** (16), any of the a-atoms is chosen as first in sequence. The other a-atom of the near end is second, and the a-atom of the far end, from which exploration round the ring would lead to the first atom of the sequence, is third. Structure **129** has the R configuration.

(R)-(129)

3. Selection Rule for Planar Chirality

Of atoms directly bound to atoms in the plane, that most preferred by the standard subrules, the pilot atom, marks the side of the plane from which, under the chirality rule, an in-plane sequence is observed; and the sequence starts with the in-plane atom directly bound to the pilot atom and continues to and through other atoms, by way of a succession of bonds along that in-plane path, which at each branch leads to the atom more preferred by the standard subrules.

a. Paracyclophanes. In the specification of planar chirality to **92**, the first step is selection of a plane of chirality. The natural plane of the aromatic system and atoms directly bound to it is chosen. The second step is to identify the preferred side of the plane, a "nearer" side, containing a locus P from which observation of the model is made in order to apply the chirality rule. In determining P, each atom among atoms of the set directly bound to the atoms of the plane is considered in accordance with the sequence rules. The preferred atom, P, is designated as the pilot atom. The third step is to pass from P to the in-plane atom to which it is directly bound. This atom becomes the highest atom of the in-plane sequence. The second atom in the sequence is the in-plane atom b directly bound to it, which is most preferred by the standard subrules. This sequence is continued until the chirality rule can be applied. In **92** as shown, the chiral path defines a right-handed turn and so the chiral label is *R*.

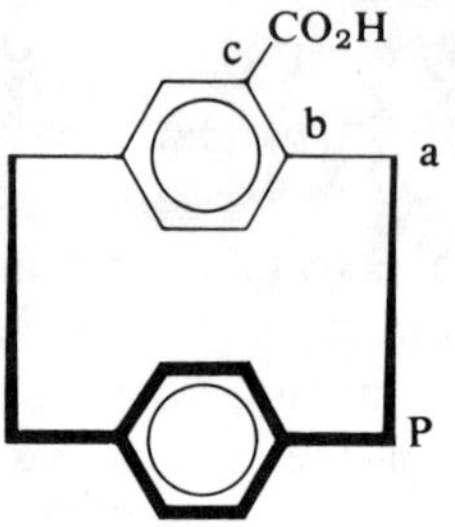

(*R*)-(**92**)

b. trans-Cycloalkenes. *trans*-Cyclooctene (**43**) offers another example of planar chirality. There are two equivalent in-plane sequences, one of them marked as shown. The chiral label of (*R*)-**43** follows from the right-handed chiral path of the in-plane sequence.

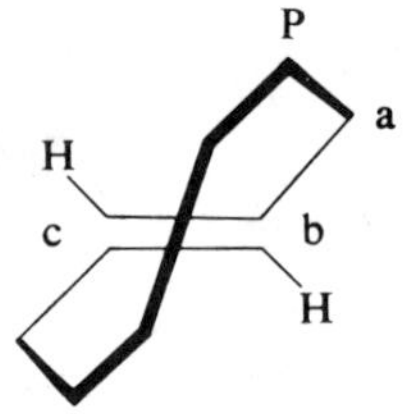

(*R*)-(**43**)

4. Secondary Structures—Helicity

The helicenes, rather than being considered as axially chiral molecules, can be treated as secondary structures. For hexahelicene (**9**) as shown, the benzene rings form a left-handed helix, and so the symbol *M* applies.

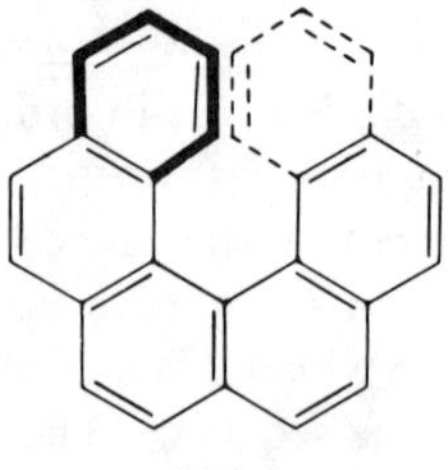

(*M*)-**9**

5. Conformation—Planar and Axial Chirality

An alternative approach to specification of axial or planar chirality which can be used where applicable is the procedure discussed earlier for conformations. In the biaryl series, R and S by the former correspond, respectively, to M and P by the latter. One result of the choice of symbols for designating helicity is the mnemonic that, in alphabetical order, R and S correspond to M and P, respectively.

C. Chirality—Special Nomenclature Problems

1. Chiral Symbol

It is recommended that the symbols of chirality be placed in front of the otherwise complete name, in parentheses, accompanied by any necessary indications of location, as, for instance in (R)-[6-nitro-(S)-6'-sec-butyl]diphenic acid.

2. Rotational Symbol

The sign of rotation is inserted directly after the symbol of chirality, as: (S)-(+)-6,6'-dinitrodiphenic acid.

3. Chiral Space-model

Where clarity is desired when the symbols R and S are applied to axes or planes of chirality, the italic letters a (for axial) and p (for planar) can be inserted immediately before the corresponding R's or S's, as: (aS)-6,6'-dinitrodiphenic acid.

4. Secondary Chiral Structures

The assignment to secondary structures of chiralities R and S or M and P can be signalized by the prefix *sec*, to be placed immediately before the chiral symbol. Hexahelicene would thus be designated as *secM* or *secP*.

REFERENCES

1. R. S. Cahn, C. Ingold, and V. Prelog, *Angew. Chem.*, **78**, 413 (1966), *Angew. Chem. Intern. Ed. Engl.*, **5**, 385 (1966).
2. R. S. Cahn, C. Ingold, and V. Prelog, *Experientia*, **12**, 81 (1956).
3. E. Eliel, *Stereochemistry of Carbon Compounds*, McGraw-Hill, New York, 1962.
4. E. Barnett, *Stereochemistry*, Pitman, Toronto, 1950, Chap. 4 and refs. therein.
5. K. Mislow, *Introduction to Stereochemistry*, Benjamin, New York, 1965.

6. P. Crabbé, "Recent Applications of Optical Rotatory Dispersion and Optical Circular Dichroism in Organic Chemistry," in *Topics in Stereochemistry*, Vol. 1, N. L. Allinger and E. L. Eliel, Eds. Interscience, New York, 1968, pp. 93-198 and especially refs. 87-100 therein.
7. H. G. Leeman and S. Fabbri, *Helv. Chim. Acta*, **42**, 2696 (1959).
8. P. Crabbé, *ORD and CD in Organic Chemistry*, Holden-Day, San Francisco, 1965, p. 253.
9. V. I. Minkin, Yu. A. Zhdanov, E. A. Medyantseva, and V. I. Naddaka, *J. Org. Chem. USSR*, **2**, 368 (1966).
10. K. Mislow and H. D. Perlmutter, *J. Amer. Chem. Soc.*, **84**, 3591 (1962).
11. (a) J. Berson and M. A. Greenbaum, *J. Amer. Chem. Soc.*, **80**, 445, 653 (1958); (b) See S. R. Johns, J. A. Lamberton, A. A. Sioumis, and H. Suares, *Chem. Commun.*, **1969**, 646, for the derivation of optically active biphenyls from the alkaloid schelhammeridine.
12. P. Newman, P. Rutkin, and K. Mislow, *J. Amer. Chem. Soc.*, **80**, 465 (1958). See also K. Mislow, *Trans. N.Y. Acad. Sci.*, **19**, 298 (1957); K. Mislow and P. Newman, *J. Amer. Chem. Soc.*, **79**, 1769 (1957); K. Mislow, P. Rutkin, and A. K. Lazarus, *ibid.*, **79**, 2974 (1957).
13. K. Mislow and F. A. McGinn, *J. Amer. Chem. Soc.*, **80**, 6036 (1958).
14. K. Mislow and P. A. Grasemann, *J. Org. Chem.*, **23**, 2027 (1958).
15. K. Mislow, *Angew. Chem.*, **70**, 683 (1958).
16. K. Mislow, M. A. W. Glass, H. B. Hopps, W. Simon, and G. H. Wahl, Jr., *J. Amer. Chem. Soc.*, **86**, 1710 (1964).
17. S. R. Landor and R. Taylor-Smith, *Proc. Chem. Soc.*, **1959**, 154.
18. R. J. D. Evans and S. Landor, *Proc. Chem. Soc.*, **1962**, 182; *J. Chem. Soc.*, **1965**, 2553; see also ref. 19.
19. E. L. Eliel, *Tetrahedron Letters*, No. 8, 16 (1960).
20. (a) E. R. H. Jones, J. D. Loder, and M. C. Whiting, *Proc. Chem. Soc.*, **1960**, 180; (b) see D. K. Black and S. R. Landor, *J. Chem. Soc.*, **1965**, 6784 for a discussion of Claisen-Cope rearrangements of propargyl-vinyl systems.
21. R. J. Evans, S. R. Landor, and J. P. Regan, *Chem. Commun.*, **1965**, 397.
22. W. Agosta, *J. Amer. Chem. Soc.*, **84**, 110 (1962); *ibid.*, **86**, 2638 (1964).
23. M. H. Gianni, *Dissertation Abstr.*, **21**, 2474 (1961).
24. K. Shingu, S. Hagishita, and M. Nakagawa, *Tetrahedron Letters*, **1967**, 4371.
25. L. Crombie and P. A. Jenkins, *Chem. Commun.*, **1967**, 870.
26. W. M. Jones, J. W. Wilson, Jr., and F. B. Tutweiler, *J. Amer. Chem. Soc.*, **85**, 3309 (1963).
27. S. F. Mason and G. W. Vane, *Tetrahedron Letters*, **1965**, 1593.
28. W. M. Jones, and J. W. Wilson, Jr., *Tetrahedron Letters*, **1965**, 1587.
29. W. M. Jones and J. M. Walbrick, *Tetrahedron Letters*, **1968**, 5229.
30. W. L. Waters and M. Caserio, *Tetrahedron Letters*, **1968**, 5233; W. L. Waters, W. S. Linn, and M. C. Caserio, *J. Amer. Chem. Soc.*, **90**, 6741 (1968).
31. J. H. Brewster, "Helix Models of Optical Activity," in *Topics in Stereochemistry*, Vol. 2, N. L. Allinger and E. L. Eliel, Eds., Interscience, New York, 1968, pp. 33-39 In this review of helical models of optical activity, Brewster has discussed theoretical predictions of absolute configuration of allenes.
32. W. R. Moore and R. D. Bach, personal communication.
33. W. H. Perkin, W. J. Pope, and O. Wallach, *Ann. Chem.*, **371**, 180 (1909); *J. Chem. Soc.*, **1909**, 1789.
34. H. Gerlach, *Helv. Chim. Acta*, **49**, 1291 (1966).

35. (a) J. H. Brewster and J. E. Privett, *J. Amer. Chem. Soc.*, **88**, 1419 (1966); (b) see also J. F. Kistner, *Dissertation Abstr.*, **28**, 4499-B (1968) for a similar determination of absolute configuration of alkylidenecycloalkanes.
36. G. G. Lyle and E. T. Pelosi, *J. Amer. Chem. Soc.*, **88**, 5276 (1966).
37. J. Brewster and R. S. Jones, *J. Org. Chem.*, **34**, 354 (1969).
38. A. Moscowitz and K. Mislow, *J. Amer. Chem. Soc.*, **84**, 4605 (1962).
39. A. C. Cope and A. S. Mehta, *J. Amer. Chem. Soc.*, **86**, 1268 (1964).
40. D. R. Boyd and M. A. McKervey, *Quart. Rev. (London)*, **22**, 95 (1968).
41. Y. Inouye, S. Inamasu, M. Ohno, T. Sugita, and H. M. Walborsky, *J. Amer. Chem. Soc.*, **83**, 2962 (1961).
42. H. Pracejus, *Ann. Chem.*, **634**, 9, 23 (1960).
43. K. Mislow, V. Prelog, and H. Scherrer, *Helv. Chim. Acta*, **41**, 1410 (1958).
44. V. Prelog, *Bull. Soc. Chim. France*, **1956**, 987.
45. I. Tömösközi and H. Bestman, *Tetrahedron Letters*, **1964**, 1293.
46. (a) S. R. Landor, B. J. Miller, J. P. Regan, and A. R. Tatchell *Chem. Commun.*, **1966**, 585; (b) S. R. Landor, B. J. Miller, and A. R. Tatchell, *J. Chem. Soc.*, C, **1967**, 196; (c) see S. R. Landor and N. Punja, *Tetrahedron Letters*, **1966**, 4905 for the absolute configuration of laballenic acid.
47. H. Gerlach, *Helv. Chim. Acta*, **51**, 1587 (1968).
48. A. Horeau, *Tetrahedron Letters*, **1961**, 506.
49. K. Schlögl and H. Mechtler, *Angew. Chem.*, **78**, 606 (1966), *Angew. Chem. Intern. Ed. Engl.*, **5**, 596 (1966).
50. H. Falk and K. Schlögl, *Angew. Chem.*, **80**, 405 (1968), *Angew. Chem. Intern. Ed. Engl.*, **7**, 383 (1968).
51. K. Schlögl, *Fortschr. Chem. Forsch.*, **6**, 479 (1966), G. Haller and K. Schlögl, *Monatsh. Chem.*, **98**, 2044 (1967), O. L. Carter, A. T. McPhail, and G. A. Sim , *J. Chem. Soc.*, A, **1967**, 365; cf. G. Haller and K. Schlögl, *Monatsh. Chem.*, **98**, 603 (1967).
52. K. Schlögl, "Stereochemistry of Metallocenes," in *Topics in Stereochemistry*, Vol. 1, N. L. Allinger and E. L. Eliel, Eds., Interscience, New York, 1967, pp. 39–91.
53. K. Mislow, *Ann. N.Y. Acad. Sci.*, **93**, 457 (1962), and references cited therein. See refs. 12, 13, 15, 54–55, and 58–59, for chemical correlations of biphenyls. See refs. 13 and 14 for chemical correlations of binaphthyl derivatives.
54. G. Wittig and H. Zimmerman, *Chem. Ber.*, **86**, 629 (1953).
55. F. A. McGinn, A. K. Lazarus, M. Siegel, J. E. Ricci, and K. Mislow, *J. Amer. Chem. Soc.*, **80**, 476 (1958).
56. G. M. Badger, R. J. Drewer, and G. E. Lewis, *J. Chem. Soc.*, **1962**, 4268; see also refs. 50 and 65.
57. G. M. Badger, P. R. Jefferies, and R. W. L. Kimber, *J. Chem. Soc.*, **1957**, 1837.
58. H. Akimoto, T. Shiori, Y. Iitaka, and S. Yamada, *Tetrahedron Letters*, **1968**, 97.
59. S. Yamada and H. Akimoto, *Tetrahedron Letters*, **1968**, 3967.
60. W. H. Perkin and W. J. Pope, *J. Chem. Soc.*, **1911**, 1510.
61. G. Krow and R. K. Hill, *Chem. Commun.*, **1968**, 430.
62. M. Siegel and K. Mislow, *J. Amer. Chem. Soc.*, **80**, 473 (1958); see also ref. 63.
63. J. T. Melillo and K. Mislow, *J. Org. Chem.*, **30**, 2149 (1965).
64. D. D. Fitts, M. Siegel, and K. Mislow, *J. Amer. Chem. Soc.*, **80**, 480 (1958).
65. K. Mislow, M. A. W. Glass, R. E. O'Brien, P. Rutkin, D. Steinberg, and C. Djerassi, *J. Amer. Chem. Soc.*, **82**, 4740 (1960).
66. K. Mislow, M. A. W. Glass, R. E. O'Brien, P. Rutkin, D. Steinberg, J. Weiss, and C. Djerassi, *J. Amer. Chem. Soc.*, **84**, 1455 (1962).

67. K. Mislow, E. Bunnenberg, R. Records, K. Wellman, and C. Djerassi, *J. Amer. Chem. Soc.*, **85**, 1342 (1963).
68. E. Bunnenberg, C. Djerassi, K. Mislow, and A. Moscowitz, *J. Amer. Chem. Soc.*, **84**, 2823, 5003 (1962).
69. N. Musso, W. Steckelberg, *Ann. Chem.*, **693**, 187 (1966).
70. K. Mislow and C. Djerassi, *J. Amer. Chem. Soc.*, **82**, 5247 (1960).
71. W. Kuhn and K. Bein, *Z. Phys. Chem. (Leipzig)*, **24(B)**, 335 (1934); W. Kuhn and R. Rometsch, *Helv. Chim. Acta*, **27**, 1080, 1346 (1944); W. Kuhn, *Z. Elektrochem.*, **62**, 28 (1958); see also W. Kuhn, *ibid.*, **56**, 506 (1952).
72. D. W. Slocum and K. Mislow, *J. Org. Chem.*, **30**, 2152 (1965).
73. R. Grinter and S. F. Mason, *Trans. Faraday Soc.*, **60**, 274 (1964).
74. G. Lowe, *Chem. Commun.*, **1965**, 411.
75. W. T. Borden and E. J. Corey, *Tetrahedron Letters*, **1969**, 313.
76. T. Martin Lowry, *Optical Rotatory Power*, Dover, New York, 1964, pp. 371, 391.
77. C. M. Kemp and S. F. Mason, *Chem. Commun.*, **1965**, 559.
78. C. M. Kemp and S. F. Mason, *Tetrahedron*, **1966**, 629.
79. I. Tinoco and R. Woody, *J. Chem. Phys.*, **40**, 160 (1964).
80. O. E. Weigang, J. A. Turner, and P. A. Trouard, *J. Chem. Phys.*, **45**, 1126 (1966).
81. D. Fitts and J. Kirkwood, *J. Amer. Chem. Soc.*, **77**, 4940 (1955).
82. (a) A. Moscowitz, *Advan. Chem. Phys.*, **4**, 67 (1962); (b) A. Moscowitz, PhD thesis, Harvard University (1957).
83. J. M. Bijvoet, A. F. Peerdeman, and A. J. van Bommel, *Nature*, **168**, 271 (1951).
84. L. H. Pignolet, R. P. Taylor, and W. DeW. Horrocks, Jr., *Chem. Commun.*, **1968**, 1443.
85. H. Falk, *Österr. Chem.-Z.*, **66**, 242 (1965).

Polypeptide Stereochemistry

MURRAY GOODMAN, ANTONIO S. VERDINI,*
NAM S. CHOI and YUKIO MASUDA†

Department of Chemistry, Polytechnic Institute of Brooklyn, Brooklyn, New York

I.	Introduction	70
II.	Helical Polypeptides	70
	A. Ultraviolet Absorption Studies	73
	B. Optical Rotatory Dispersion and Circular Dichroism Studies	82
	C. Nuclear Magnetic Resonance Studies	93
	D. Infrared Studies	97
III.	Some Special Aspects of Helical Structures	105
	A. Non-Hydrogen Bonding Polypeptides	105
	B. Aromatic Polypeptides	110
IV.	"Random Coil" Polypeptides	113
V.	β-Structures for Polypeptides	116
VI.	Conformational Transformations	124
	A. Helix–Coil and Helix–Helix Transitions	124
	1. The Zimm-Bragg Model	125
	2. Lifson-Roig Model	127
	B. Kinetics of Transitions	131
	C. Thermodynamics of Transitions	134
	D. Helix–Helix Transition in Poly-L-proline	137
VII.	Calculation of Most Probable Conformations	138
	A. Construction of Steric Maps	141
	B. Potential Functions	144
	1. Torsional Potentials	144
	2. Van der Waals Interactions	145
	3. Electrostatic Interactions	146
	4. Hydrogen Bonding	147
	5. Other Interactions	147
	C. Conformations of Homopolypeptides	148
	1. Polyglycine	148
	2. Poly-L-alanine	152
	3. Poly-L-valine	153
	4. Poly-β-methyl L-aspartate and Poly-γ-methyl L-glutamate	154
	5. Other Polypeptides	154
VIII.	Conclusions	156
	References	157

* Postdoctoral Fellow on leave from University of Padua.

† Postdoctoral Fellow on leave from Japan Women's University, Tokyo.

I. INTRODUCTION

The ever increasing success which protein chemists have had in unraveling the three-dimensional structure of proteins has pointed to the coexistence of helical, β, and other nonhelical sequences in these natural systems. Whale myoglobin contains more than 70% α-helical regions, most of the other globular proteins studied contain much smaller percentages of helicity. Just recently (1969) there has been an announcement of the complicated molecular geometry found in insulin in the solid state. The molecule appears to be made up of three folded dimer chains of the primary insulin structure connected by two zinc atoms.

It is evident to us that a substantial understanding of the forces which contribute to the stereochemistry of proteins can be gained from the study of model systems. Synthetic polypeptides offer an attractive means of determining conformations and interactions which relate to protein structure and function.

The development of sophisticated instrumentation has now evolved to the stage where it is possible to examine absorption spectral properties of biopolymers extending into the far ultraviolet region. Proteins and their synthetic polypeptide analogs have also been investigated by 220 MHz nmr spectroscopy. These advances allow quite subtle stereochemical questions to be answered. For example, we are able to consider the origin of hypochromism for helical polypeptides in their $\pi \rightarrow \pi^*$ transitions for the peptide linkage. We are beginning to understand the enhancement of the amide $n \rightarrow \pi^*$ transition when the polypeptide is stereoregular. The stereochemist can also observe very detailed environmental differences for protons by nmr. Helix–coil transitions can be carefully studied and conformational properties have been elucidated by the magnetic resonance technique.

This review of polypeptide stereochemistry deals mainly with structure from the static and dynamic standpoints, with methods of measuring structure and with calculations aimed at predicting most probable conformations.

II. HELICAL POLYPEPTIDES

Helical structures are common features of proteins, nucleic acids, and polysaccharides, as well as of many synthetic polypeptides. The study of the helical structures of model polypeptides in the solid state and in solution is of great significance in that it leads to a clear interpretation of results obtained in conformational studies of fibrous and globular proteins.

Polypeptides are polyamides composed of α-amino acids of the general structure

$$\left[HN-CH(R)-CO \right]_n$$

where R can be any of the various side chains found in α-amino acids.

Pauling and Corey (1,2) proposed the α-helix as one of the more stable secondary structures for folded polypeptide chains on the basic assumptions that:

a. the peptide group is in a *trans*-planar conformation,

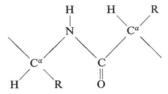

b. the residues in the chain are all stereochemically equivalent,

c. each peptide C=O and N—H group is hydrogen bonded with a nearly linear disposition of the N—H···O atoms, and

d. the length of N—H···O bonds is about 2.75 Å and the linear arrangement of carbonyls and NH groups makes an angle of 12° with the helical axis.

Thus, the α-helix has 3.7 residues per turn with intrachain hydrogen bonds between N—H and C=O groups four residues apart forming 13-membered rings. The residue translation (height per residue) is 1.5 Å along the axis and the pitch is 5.4 Å.

The α-helix, designated also as 3.7_{13} helix, can be either right or left handed. When it is built from optically active residues, these two forms are different structures which gave rise to different X-ray diffraction patterns (Fig. 1).

The existence of the α-helix was experimentally confirmed more than 15 years ago, by X-ray diffraction studies on the synthetic polypeptides poly-γ-benzyl and poly-γ-methyl L-glutamate (3–5) and poly-L-alanine (6–10).

The X-ray photographs of these polymers showed a strong meridional reflection with a spacing of 1.5 Å, which corresponds to the residue translation in the α-helix. A residue translation represents the vertical rise in proceeding from any residue to a subsequent residue. Subsequent X-ray studies on poly-L-alanine and several other polypeptides have confirmed the α-helical structure (11–15).

In general, it may be stated that in cases where a layer line corresponding

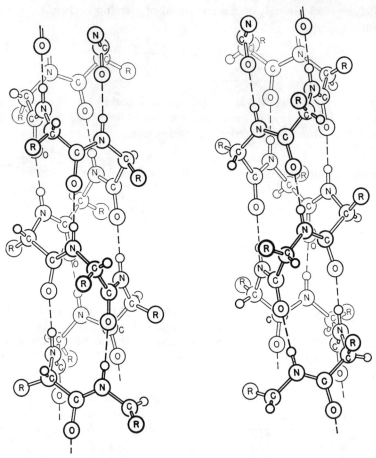

Fig. 1. The α-helix built from L-amino acids: (left) left-handed and (right) right-handed (1,2).

to 5.4 Å and a meridional streak of 1.47–1.50 Å are observed, it is reasonable to assume that α-helices are present.

The α-helix, the most common secondary structure of polypeptides, is not the only helical conformation experimentally found. A different type of helix, the so-called ω-helix, with four residues per turn, an axial translation of 1.325 Å, and a pitch of 5.3 Å, was observed in thermally treated fibers of poly-β-benzyl L-aspartate (16), in oriented films of poly-S-benzylthio-L-cysteine (17), and in liquid crystalline solutions of poly-β-benzyl L-aspartate in m-cresol (18).

Because of steric effects, the peptide units deviate from planarity by about 5°. This is consistent with recent findings that the ω-form with planar peptide units (Fig. 2) is not allowed (19,20). The ω-helix is, in essence, a distorted α-helix where a stable structure can be achieved only by relaxing the requirements for linearity of hydrogen bonds and the planarity of peptide groups. This structure is less stable than an α-helix.

Essentially rigid structures, such as helices, and their conformational changes in solution are commonly studied by means of ultraviolet absorption (uv), optical rotatory dispersion (ORD), and circular dichroism (CD).

A. Ultraviolet Absorption Studies

Polypeptides are composed of amide chromophores in a dissymmetric environment. It is well known that the amide group exhibits several electronic transitions in the ultraviolet spectral region (21–30).

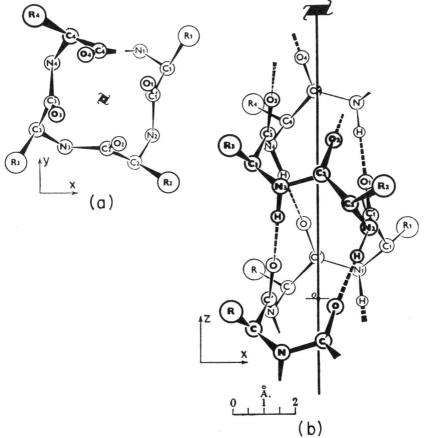

Fig. 2. Representations of ω-helix observed in poly-β-benzyl L-aspartate and poly-S-benzylthio-L-cysteine: (a) top view; (b) side view (6).

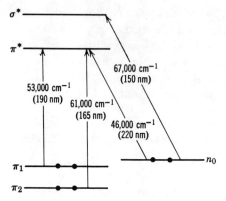

Fig. 3. The nature of the electronic transitions of amide chromophore based on observed transitions in simple amides (31).

According to the more recent interpretations of the amide spectrum, four bands are present in the 150–220 nm wavelength range. The electronic absorption spectrum and the calculated energy levels for singlet orbitals of the amide chromophore in this region are depicted in Figures 3 and 4 (31–33). Since the spectroscopic investigations of polypeptides have not been extended below 185 nm, the amide transitions of interest are:

a. the weak $n_0 \to \pi^*$ electronic transition at about 220 nm, which involves promotion from the $2p_y$ lone-pair orbitals of the amide oxygen and

b. the $\pi_1 \to \pi^*$ transition which is associated with an intense band at about 190 nm.

Figure 5 illustrates the molecular orbitals of the amide group, showing the

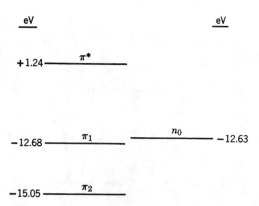

Fig. 4. Singlet orbitals of the amide chromophore based on the self-consistent field calculations (32,33).

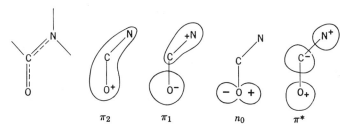

Fig. 5. Representations of amide group molecular orbitals. Only the upper lobes are shown for the π orbitals (84).

symmetry relationship among some of the simplest LCAO wavefunctions of amides.

The band at 220 nm is assigned as $n_0 \to \pi^*$ on the basis of its low intensity, its frequency shift in going from nonpolar aliphatic to polar hydroxylic solvents, and its optical rotatory dispersion and circular dichroism spectra (30,34–40).

The high-energy band at 190 nm is assigned as $\pi_1 \to \pi^*$ on the basis of theoretical treatments which have placed the $\pi_1 \to \pi^*$ electronic transition in the 170–190 nm region, with a polarization direction quite near that measured on myristamide (Fig. 6). The exact position of the $\pi_1 \to \pi^*$ band depends upon the number of alkyl substituents on the amide nitrogen and shifts toward higher wavelength going from amides to peptides to N,N-disubstituted amides (28,29).

A moderately strong band centered between the 220 nm and 170–195 nm

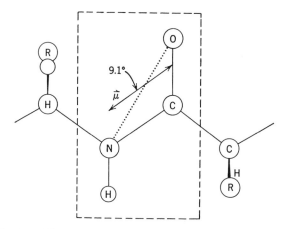

Fig. 6. Polarization of the $\pi_1 \to \pi^*$ absorption band of a peptide link. The angle of 9.1° is based on values for myristamide found by Peterson and Simpson (26).

bands has been recently observed in the absorption spectra of simple N-substituted amides (28,29,41).

The occurrence of this band in amides is rather general. Coming just below the $\pi \to \pi^*$ band, it is very much reminiscent of the olefin "mystery band" which recently was assigned as $\pi \to CH\sigma^*$. On the basis of calculations on formamide, Basch and colleagues concluded that the new band near 200 nm in amides involves an $n \to \sigma^*$ transition† (28). Since the wavefunctions and excited states of polypeptides are derived from those of the component amide chromophores, caution must be exercised in the interpretation of spectra in the far ultraviolet wavelength range. For a randomly coiled polypeptide, in which the various segments of the chain are in constant motion when the molecules are in solution, the uv spectrum is expected to show the same features as that of simple N-substituted amides (see below). In a long polypeptide chain with an α-helix conformation, the uv spectra show features strikingly different from those of simple N-substituted amides and those of polypeptides in random or β-conformations (see Sect. V). Figure 7 shows the ultraviolet absorption spectra of poly-L-lysine hydrochloride in helical, random, and β-conformations in aqueous solution (42). The existence of the three structures depends upon the pH and temperature of the aqueous solutions.

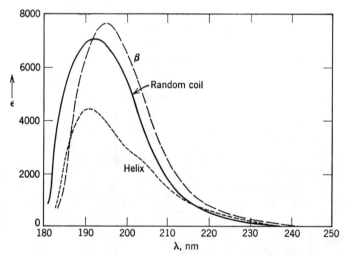

Fig. 7. Absorption spectra in $\pi_1 \to \pi^*$ region for poly-L-lysine in aqueous solution: random coil, pH 6.0, 25°; helix, pH 10.8, 25°; β-form, pH 10.8, 52° (42).

†The σ^* orbital of the $n \to \sigma^*$ excitation is almost totally composed of Rydberg orbitals (molecular orbitals formed from atomic orbitals of high quantum number) on C, N, and O.

The distinguishing features of the spectrum for helical polypeptides are:

a. a strong absorption maximum near 190 nm followed by a definite shoulder near 206 nm, and

b. a marked hypochromism of the band near 190 nm in comparison with the corresponding bands of the random and β-structures.

The maximum at 190 nm and the shoulder at 206 nm are associated with the exciton splitting of the $\pi \to \pi^*$ electronic transition of the peptide chromophores in a helical array (31). Because of its importance in this and in the following discussions, the phenomenon of exciton resonance coupling between the moments of the $\pi \to \pi^*$ electronic transitions of the peptide chromophores in helical structures will be briefly discussed in the sequel.

Let us consider the electronic states of a "bimer,"† i.e., a simple molecule consisting of two identical chromophoric groups, A and B, in close proximity. For each isolated chromophore (monomer) there are two states, a ground state ϕ^0 and an excited state ϕ^*. For the bimer the ground state is $\psi^0 = \phi_A^0 \phi_B^0$ but there are now two excited states: $\psi_+^* = \phi_A^0 \phi_B^* + \phi_A^* \phi_B^0$ and $\psi_-^* = \phi_A^0 \phi_B^* - \phi_A^* \phi_B^0$. The two excited bimer states do not have the same energy.

In the monomer there is one absorption band corresponding to the $\phi^0 \to \phi^*$ transition, while in the bimer two bands appear, corresponding to $\psi^0 \to \psi_+^*$ and $\psi^0 \to \psi_-^*$ transitions.

The origin of the band splitting may be seen pictorially in the energy-level diagram in Figure 8. The energy of the state ψ_+^* is higher than that of the

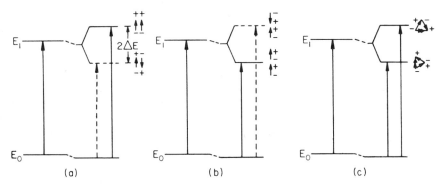

Fig. 8. A schematic representation for energy levels of chromophoric dimers (bimers) indicating exciton splitting when the transition dipole moments are aligned (*a*) side by side, (*b*) head-to-tail and tail-to-tail, and (*c*) in a herringbone pattern. The allowed transitions are shown by full lines, while the forbidden transitions are shown by dotted lines (31).

† We use the term "bimer" instead of dimer because we wish to call attention to the interactions of two chromophoric groups.

monomer state because of the predominantly repulsive energy of the alignment of the dipoles for this excited state. In the state ψ_-^*, on the other hand, the dipoles are in an attractive array and, consequently, this excited state of the bimer has a lower energy than that of the excited monomer.

The band splitting, labeled $2\Delta E$ in the diagram, is dependent on the strength of the electrostatic interaction between the excited monomeric units. The splitting thus depends on the relative orientation position, and magnitudes of the dipoles. For a "trimer" there will be, in general, three excited energy levels and for the "N-mer," N energy levels, so that when N is large, a continuous band of levels is formed.

In an infinitely long, rigid helix only transitions to levels for which the vector sum of the transition dipole moments does not vanish are allowed. Figure 9 shows the orientation of dipoles in helical structures.

In case 1 the sum of the many individual dipoles oriented essentially head-to-tail is a single dipole along the helix axis. The transition from the ground state to this state is thus polarized along the helix axis and lies at longer wavelengths than the corresponding transition in the random coil.

In case 2 the sum of the individual dipoles (which are oriented head-to-head) is a dipole perpendicular to the helix axis. The transition is polarized perpendicular to the axis of the molecules and lies at shorter wavelengths than the transition in the random coil.

These theoretical predictions are largely confirmed by experiments (24,43–46). Figure 10 shows a polarized ultraviolet absorption spectrum of an oriented film of poly-γ-methyl L-glutamate in α-helical form. The 190 nm ($\pi \rightarrow \pi^*$) band splits into two bands of opposite polarization relative to the helix axis. The band at lower energy (206 nm) is polarized parallel to the

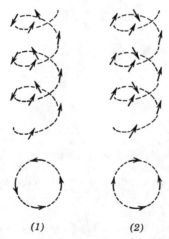

Fig. 9. Schematic representation for the orientation of dipoles in helical arrays.

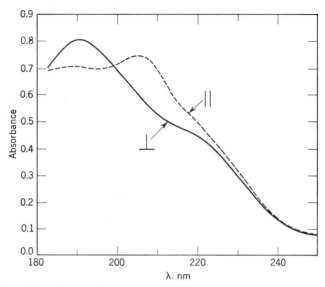

Fig. 10. Parallel and perpendicular polarized absorption spectra of oriented films of poly-γ-methyl L-glutamate in a helical form (46).

helix axis, while the band at higher energy (191 nm) is polarized perpendicular to the helix axis.

Polarized absorption spectra of oriented helical polypeptides in the far ultraviolet not only show bands which are of the predicted polarization and intensity ratio, but also give evidence for a weak perpendicularly polarized band at about 222 nm, which is assigned to the amide $n \rightarrow \pi^*$ transition. The hypochromic effect in the $\pi \rightarrow \pi^*$ band of the uv absorption spectra of helical polypeptides was also theoretically predicted (24). Table I summarizes the molar residue absorptivities for some polypeptides in the α-helical and random coil forms. Hypochromism has been defined as "a situation in which the integrated intensity of an absorption band from an assemblage of chromophores is less than the sum of its parts" (31). It is related to a parallel alignment of peptide chromophores in the helical conformation. Hypochromism (Fig. 11) arises from intereaction between the transition moment of the lowest-energy absorption band in one residue with transition moments of higher-energy transitions of adjoining residues. Hypochromicity was also observed in uv studies of poly-γ-methyl D,L-glutamate, in which the helical content has been estimated to be about 90%. This phenomenon is not peculiar to high molecular weight polypeptides, but is also present in small peptides (oligomers). The finding that hypochromicity of oligomers is close in magnitude to that of high molecular weight polypeptides suggested the use of the absorption measurements at 190 nm as a method for evaluating the

TABLE I

Molar Residue Absorptivities[a,b] for Water-Soluble Polypeptides in the α-Helical and Random Forms

	Molar absorptivity				Partial oscillator strength[c]
	Max.	190 nm	197 nm	205 nm	
Random coil					
Poly-L-glutamic acid					
No added salt	7100	7000	6550	3500	0.115
Ionic strength 0.0004		(7200)	(6600)	(4850)	—
Ionic strength 2.0		(6050)	(5350)	(3700)	—
Poly-L-lysine (water)	7100(6600)	6900(6500)		3300	0.106
Average		6950	—	3400	0.1104
Poly-L-aspartic acid	(6350)	(6350)			—
α-Helix					
Poly-L-glutamic acid	4200	4200	3300	2150	0.0778
Poly-L-lysine	4400	4400	3500	2300	0.0720
Average		4300	3400	2200	0.0749

[a] All figures from work of Rosenheck and Doty (42) and Rosenheck (unpublished), except values in parentheses, which are taken or interpolated from McDiarmid (44) or McDiarmid and Doty (45).
[b] All values corrected for side-chain contributions.
[c] From integrated absorption intensity to absorption maximum:

$$\int_{\text{partial}} = 4.33 \times 10^{-9} \int_0^{\nu_{\max}} \epsilon\, d\bar{\nu}$$

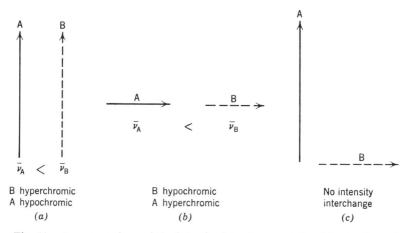

Fig. 11. Representations of the intensity interchange relationships in absorption spectra of ordered structures. The interaction between two adjacent chromophores is shown where the transition dipole moments are arrayed (a) side by side, (b) head-to-tail, and (c) perpendicular to each other; $\bar{\nu}_A$ and $\bar{\nu}_B$ denote the frequencies of the two excitations (31).

helix content of proteins, in which generally only short segments of polypeptide chains attain the helical structure.

Ultraviolet spectral studies on peptide oligomers provide important information on the critical number of amino acid residues necessary to start the folding of a helical conformation (47–49). Figure 12 shows the results obtained with γ-methyl L-glutamate and β-methyl L-aspartate oligomers.

In the case of γ-methyl L-glutamate peptides, hypochromicity of the absorption band centered at 189–190 nm commences at the nonamer and increases in magnitude for the higher members of the series. In the case of β-methyl L-aspartate oligomers, hypochromicity was first noted at about the dodecamer. Since the left-handed poly-β-methyl L-aspartate helix has less stability than the poly-γ-methyl L-glutamate helix toward both heat and helix-breaking solvents, a longer chain length is required in the aspartate oligomers for helix formation to occur.

In a polypeptide chain, the proximity of the peptide chromophores to the chiral centers of the adjacent amino acid residues and their interaction gives rise to a dissymmetry dependent on conformation and distinct from that determining the rotatory properties of the free amino acids. The peptide chromophore belongs to the class of inherently symmetric chromophores which acquire optical activity from asymmetric perturbations of their environment. The relationship between molecular structure and optical activity is very delicate because a slight structural variation often results in a very significant change in the asymmetric environment of the chromophore.

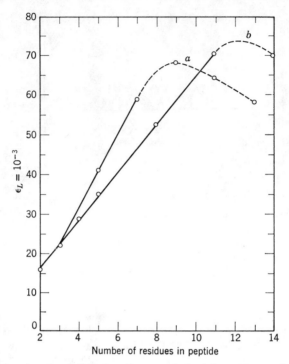

Fig. 12. Total molar absorptivity for oligomers derived from (a) γ-methyl L-glutamate and (b) β-methyl L-aspartate in trifluoroethanol as a function of chain length (47).

B. Optical Rotatory Dispersion and Circular Dichroism Studies

Before the advent of proper instrumentation to measure optical rotatory dispersion (ORD) and circular dichroism (CD) in the far ultraviolet, it was necessary to employ derived constants for rotatory dispersion. Many workers had devised equations relating optical activity to wavelength (50–58). Among the most successful of these are the Drude and Moffitt-Yang treatments.

The rotatory dispersion curves of random coil polypeptides (see Sect. IV) in spectral regions far from optically active electronic transitions (300–600 nm) can be described by the one-term Drude equation:

$$[\alpha]_\lambda = A/(\lambda^2 - \lambda_c^2)$$

where λ is the wavelength of incident radiation, λ_c is the wavelength characteristic of the optically active electronic transitions, and A is a constant related to the rotatory strength of the transitions. From a linear plot of $[\alpha]_\lambda \cdot \lambda^2$ versus $[\alpha]_\lambda$, these constants are evaluated. The magnitude of $[\alpha]_\lambda$ for polypeptides in the random-coil conformation increases monotonically with decreasing wavelength and obeys the Drude equation extremely well. For helical polypeptides the equation fails to fit the observed complex dispersion because of the contributions to the optical rotation of optically active electronic transitions produced by the interaction of identical chromophores in a rigid, helical array. As postulated by Moffitt, this interaction causes any transition of an individual chromophore to be split into two transitions that are characteristic of the helix as a whole (59–61).

The complex dispersion data of helical polypeptides fit the two-term phenomenological equation proposed by Moffitt and Yang in 1956 (60).

$$[m'] = \frac{a_0 \lambda_0^2}{\lambda^2 - \lambda_0^2} + \frac{b_0 \lambda_0^4}{(\lambda^2 - \lambda_0^2)^2}$$

where $[m']$ is the reduced mean rotation per residue, λ_0 is a constant wavelength generally taken as 212 nm, and α_0 is a parameter governed by conformational and environmental effects, whereas b_0 depends on the conformation alone. From a plot of $[m'](\lambda^2 - \lambda_0^2)/\lambda_0^2$ against $\lambda_0^2/(\lambda^2 - \lambda_0^2)$, b_0 is obtained directly from the slope and a_0 from the intercept of the resulting straight line. The b_0 parameter was at one time widely used for estimating the helical content of polypeptides and proteins. For a right-handed helix, b_0 has a value of -630 to -700; for the random-coil conformation this value is near zero. The b_0 criterion for helicity is not suitable for polypeptides containing chromophoric side chains (poly-L-tyrosine, poly-L-phenylalanine, poly-L-tryptophan, and poly-L-histidine). In these cases, the Moffitt-Yang parameter assumes "anomalous" values because of the rotational contributions from the side chains. The b_0 value varies also as a function of the refractive index of the solvent, as demonstrated in a detailed ORD investigation on poly-γ-benzyl L-glutamate in different solvent systems (62).

The sign of the Moffitt parameter can be related to the sense of the helix, and the positive values around +630 exhibited by poly-β-benzyl L-aspartate in some solvents have been associated with a left-handed helical structure (63,64).

Again, the sign of b_0 alone is not a secure indication of helical sense (65). To ascertain the sense of the helix, it is necessary to prepare "random" copolymers containing residues of one amino acid for which the helical sense of the corresponding homopolypeptide is well known and residues of the amino acid for which the helical sense of the corresponding homopolypeptide is unknown. A linear change in b_0 with copolymer composition provides evidence that both polypeptides are of the same helical sense; a nonlinear change may be associated with opposite helical senses for the two homopolypeptides under consideration (66–70).

As noted above, now that instruments for the direct observation of ultraviolet Cotton effects of polypeptides and proteins are available, the use of an inferential treatment, such as the Moffitt-Yang equation, is no longer necessary. The direct estimation of helical content from observation of the

Cotton effect, rather than from Drude and Moffit-Yang constants, has many advantages and allows more accurate structural assignments. The bands of the ORD spectra recorded in the far ultraviolet region are associated with the electronic transitions and can be separated into contributions from the helical, extended, and disordered conformations. Furthermore, the very large rotations found in the Cotton effect region increase the sensitivity of the measurements.

However, some applications remain for the derived-constant treatment of Moffitt and Yang. This equation is useful in situations in which the far ultraviolet is obscured by solvent absorption. In fully characterized systems, such as the α-helical proteins from muscle, it is a simple and rapid method of monitoring samples (71–73).

Rigid helical structures can be conveniently identified and studied by ORD and CD techniques. Excellent reviews on the basic theory of optical rotation and circular dichroism are available (74–77). We indicate here only the rationale underlying the two methods and illustrate ORD and CD spectra of polypeptides in helical conformations. Both ORD and CD spectra have their origins in the absorption bands of optically active compounds. Thus, the two phenomena are intimately related.

Figure 13 shows the influence of an optically active sample on a beam of plane polarized light. The incident radiation, represented by the electric vector $E_0 \cdot \sin \omega t$ may be resolved into two equal components of left and right circularly polarized light, E_L and E_R. The two components travel through the sample with different velocities, giving rise to a phase shift between them which is observed as optical rotation (designated by α in the figure).

According to the Fresnel model, optical rotation is produced by differences in refractive index for right (n_R) and left (n_L) circularly polarized light:

$$[\Phi]_\lambda = (18 \, MW/\rho\lambda)(n_L - n_R)$$

where $[\Phi]$ is molecular rotation (deg.), MW is molecular weight, ρ is density or concentration (g/ml), and λ is wavelength *in vacuo* (cm).

Optical rotatory dispersion is the change of optical rotation with wavelength (52). The two circularly polarized components also undergo different attenuation, since the sample exhibits different transmittances, T_L and T_R, for the two components. Thus, unequal attenuation gives rise to the elliptical polarization of the emergent beam (Fig. 13). Circular dichroism may be expressed as the difference in absorbance, ΔA, for the circularly polarized components or as the ellipticity, θ, of the emergent beam (θ is defined as the arc tangent of the semiminor axis b divided by the semimajor axis a). The two methods of expressing circular dichroism are related by a constant.

$$[\theta_\lambda] = 3300(\epsilon_L - \epsilon_R)$$

where ϵ_L and ϵ_R are the molecular extinction coefficients for the left and

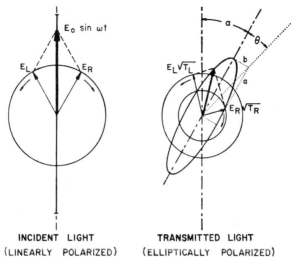

Fig. 13. The conversion of incident linearly polarized light into elliptically polarized light by an optically active medium. The incident radiation is denoted by the electric vector $E_0 \sin \omega_t$ resolved into two equal right and left circularly polarized light components E_L and E_R, respectively. T_L and T_R indicate the transmittances of these two components. $E_L\sqrt{T_L}$ and $E_R\sqrt{T_R}$ represent the transmitted electric vectors from the respective circular components.

right circularly polarized light, respectively. In contrast to ordinary absorption, circular dichroism can be positive or negative (78–81).

The circular dichroism spectra of complex molecules in solution generally consists of broad bands which occur only at those wavelengths at which ordinary absorption occurs (82) (Fig. 14). The optical rotation varies with the wavelength in the vicinity of an isolated absorption band in a manner shown in Figure 15.

The ORD curves show characteristic peaks and troughs at each absorption band (and consequently at each CD band). The phenomena of circular dichroism and its associated dispersion curve are termed a *Cotton effect*. Figure 16 shows typical positive and negative Cotton effects and the relation between the circular dichroism and rotatory dispersion of each polarity. The curves refer to a single isolated optically active absorption band centered at a wavelength of λ_K. A positive or negative Cotton effect is defined by whether the peak or trough, respectively, is on the long wavelength side.

In the case of ORD, the rotation persists at wavelengths far from the peak and trough of the Cotton effect. When an optically active substance presents several positive and negative Cotton effects, each superimposed upon the background provided by the tails of the others, it is very difficult to identify precisely the shape, position, and amplitudes of the bands in the ORD spectra. Consequently, the structural information can be hard to unravel.

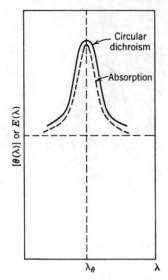
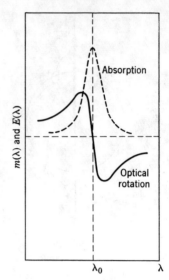

Fig. 14. The relationship between an idealized absorption band and a positive circular dichroism band (79).

Fig. 15. The relationship between an idealized absorption band and a negative optical rotatory dispersion band (79).

In the CD spectra, on the other hand, ellipticity bands are not generally set upon background rotations from more distant absorption bands of the same chromophore or of other atoms in the same molecule. Even when a number of bands fall very close to one another, computer analysis frequently

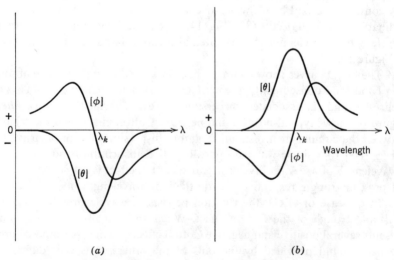

Fig. 16. Another effect for an isolated absorption band (idealized).

allows resolution of the bands and unequivocal assignment of their positions, signs, and amplitudes.

The close correspondence between ORD and CD phenomena is demonstrated by the existence of mathematical transforms (Kronig-Kramer dispersion relations) which allow calculation of the circular dichroism spectrum from the rotatory dispersion curve and vice versa, without assumptions about the band shapes.

We now consider the Cotton effects of the polypeptide chain in a helical conformation (83). The ORD results in the far ultraviolet region show that α-helical polypeptides characteristically exhibit a rotatory trough at 233 nm, a crossover point (zero rotation) at about 225 nm, a shoulder near 215 nm, and a peak at 198 nm (52,71). The value at the depth of the trough ($[m']_{233}$ = $-15,000$ deg-cm^2/decimole), which is peculiar to the right-handed α-helical conformation, is often used as a measure of helical content in polypeptides and proteins (Fig. 17) (84–86).

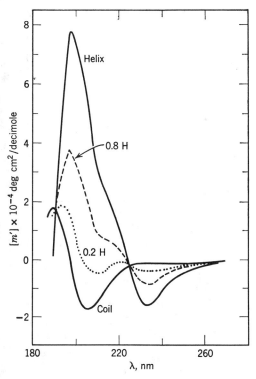

Fig. 17. Optical rotatory dispersion spectra for poly-L-glutamic acid at pH 7.3. Helix, dioxane-water 50:50 (v/v); "random coil," water. The dotted lines denote calculated curves for helix fractions 0.2 and 0.8 (52).

CD spectra of right-handed helical polypeptides show a negative $n \to \pi^*$ band at 222 nm, a negative band polarized parallel to the helix axis at 206–207 nm, and a positive band polarized perpendicularly to the helix axis at 190 nm (78–83). Figures 18 and 19 show the characteristic spectrum of poly-L-glutamic acid and poly-L-lysine in α-helical conformation. The negative and positive Cotton effects at about 206 nm and 190 nm originate from the two components of the split $\pi \to \pi^*$ transition of the peptide chromophore. The $n \to \pi^*$ transition (which is electronically forbidden) acquires large rotational strength because a perturbation from a dissymmetric distribution of charge, i.e., the static field of the groups in the α-helix, gives rise to a component of the electronic transition moment in the direction of the magnetic transition moment. Thus, their vector product becomes finite. The

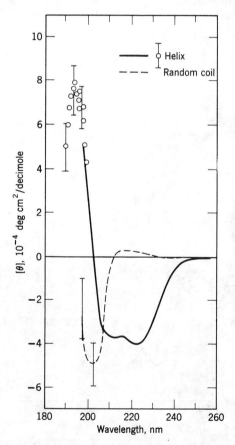

Fig. 18. The CD spectra for poly-α-L-glutamic acid in 0.1 M NaF at pH 4.3 (helix) and pH 7.6 (nonhelix) (78).

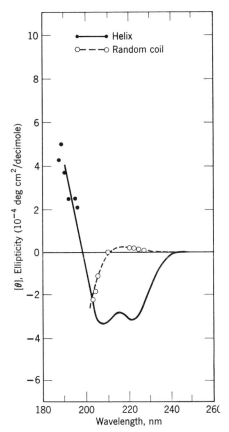

Fig. 19. The CD spectra for poly-L-lysine in 0.1 M NaF at pH 10.6–10.8 (helix) and pH 6.7 (nonhelix) (78).

$n \to \pi^*$ transition can consequently give rise to a very similar Cotton effect in the absence of α-helices, even in small molecules such as the cyclic dipeptide L-alanylgycyl anhydride, the lactam L-3-aminopyrrolidine-2-one, and similar cyclic molecules containing one or two amide groups (30,34–40,87,88).

As a consequence, it is necessary to emphasize that this Cotton effect alone cannot be regarded as diagnostic of an α-helix.

Among helical polypeptides for which ORD, CD, and uv spectra have been reported, poly-L-alanine is particularly important because this polypeptide does not contain chromophores in the side chain with absorptions that overlap those of the peptide chromophore (89,90). The bands of the resolved CD spectrum which are, without ambiguity, those of the peptide chromophore have been assigned to specific electronic transitions and characterized in terms of dipole strengths, rotational strengths, and anisotropies.

These results are shown in Figures 20 and 21. Optical rotatory dispersion and circular dichroism studies have led to the conclusion that many L-polypeptides have a primary helix with a right-handed screw sense. Poly-β-benzyl L-aspartate and poly-β-methyl L-aspartate represent an exception to the general rule and exist as a left-handed primary helix in helix-supporting solvents such as chloroform and methylene dichloride (63,64,68–70). The peculiar behavior of these polymers has attracted considerable attention. The problem of the stability and screw sense of the α-helix of poly-L-aspartic acid (91) and its esters has been studied by several workers, notably Bradbury and co-workers, Hashimoto and co-workers, and Goodman and co-workers (91–108). The CD curve for poly-β-methyl L-aspartate shows Cotton effects

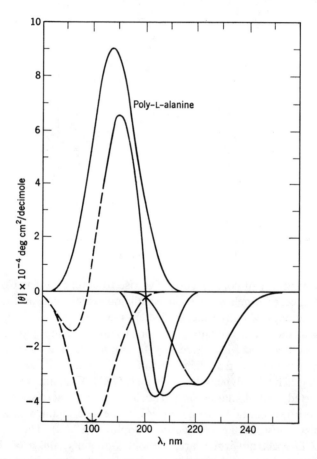

Fig. 20. Resolution of experimental CD curve (boldface) for poly-L-alanine in TFE–TFA (98.5:1.5, v/v) (89).

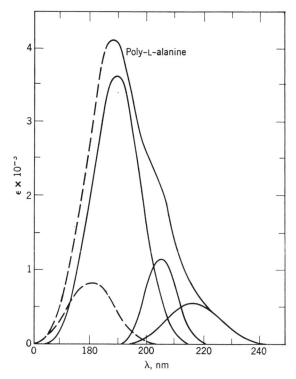

Fig. 21. Resolution of experimental uv spectrum (boldface) for poly-L-alanine in TFE–TFA (98.5:1.5, v/v) (89).

which are reversed in sign and differ considerably from those of the right-handed helix (109). The Cotton effect at 206 nm largely disappears. As pointed out before, a left-handed helix of L-amino acids is not the mirror image of a right-handed helix of L-amino acids. Accordingly, an exact reversal of sign of ellipticity bands with retention of magnitude is not possible. The differences lie in the side-chain ester groups whose electronic transitions couple with the peptide transitions in the wavelength range of interest.

The results of conformation studies on poly-L-aspartate esters are summarized in Table II.

The agreement between the above results and minimum potential energy analysis of helical structures of these esters carried out by Scheraga and co-workers is good, except in the case of poly-β-ethyl L-aspartate (110,111). This polymer appears to exist in a right-handed helical conformation in chloroform, while the α-helical conformation of lowest energy provided by theory is the left-handed one. Conformational studies of copolymers of benzyl and methyl L-aspartate with nitrobenzyl aspartate show that factors such as

TABLE II

Experimentally Observed Helix Sense of Poly-L-Aspartic Acid and of its Esters under Helix-Supporting Conditions

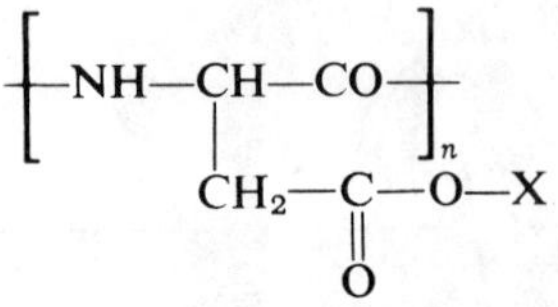

$$\mathrm{[-NH-CH-CO-]}_n$$
$$\mathrm{\ \ \ \ \ \ \ \ |}$$
$$\mathrm{CH_2-C(=O)-O-X}$$

X	Helix sense	Solvent	Refs.
—H	Right	Water, pH 4.3	91
—CH$_3$	Left	Chloroform	103
		Chloroform	99
—CH$_2$—CH$_3$	Right	Chloroform	103
—CH$_2$—CH$_2$—CH$_3$	Right	Chloroform	103
—CH(CH$_3$)$_2$	Right	Chloroform	103
—CH$_2$—CH(CH$_3$)(CH$_2$—CH$_3$)	Right	Dioxane	101
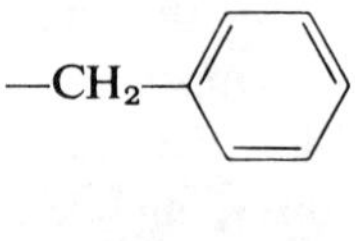 —CH$_2$—C$_6$H$_5$ (benzyl)	Left	Chloroform	105,106
		Chloroform	63,64
		Chloroform	68,69
		Chloroform	92,93,103
		Chloroform	95
		Methylene dichloride	97
		Methylene dichloride	92
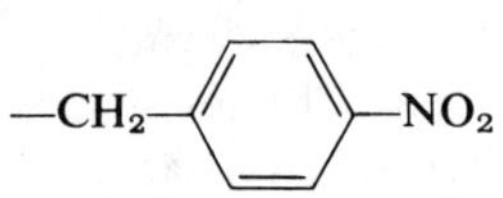 —CH$_2$—C$_6$H$_4$—NO$_2$ (p)	Right	Chloroform	68–70,96,98
		Chloroform	105
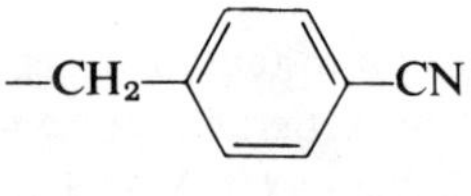 —CH$_2$—C$_6$H$_4$—CN (p)	Right	Chloroform	105–108
		Dimethylformamide	105
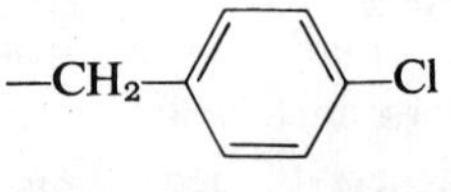 —CH$_2$—C$_6$H$_4$—Cl (p)	Right	Chloroform	105–108
		Dimethylformamide	105

(*Continued*)

TABLE II—(*Continued*)

X	Helix sense	Solvent	Refs.
—CH$_2$—C$_6$H$_4$—CH$_3$	Right	Chloroform Dimethylformamide	105–108 105
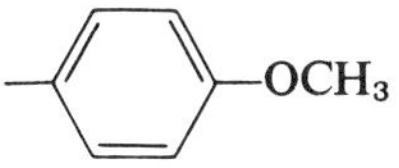 —C$_6$H$_4$—OCH$_3$	Right	Hexafluoroacetone trihydrate	102
—CH$_2$—CH$_2$—C$_6$H$_5$	Right	Chloroform Chloroform	108 103

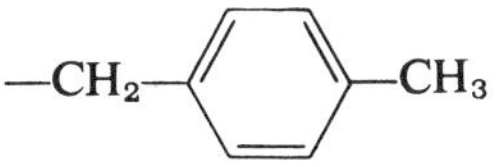

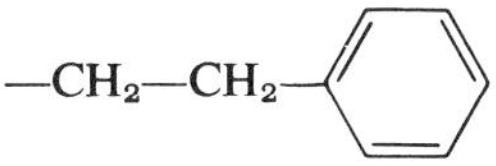

temperature and solvent play an important role in determining the conformation of these polypeptide chains in solution (100).

C. Nuclear Magnetic Resonance Studies

Helical structures, as well as helix–coil conformational transitions, have been identified and studied by nmr spectroscopy (112–117). The nmr spectrum of poly-γ-benzyl L-glutamate in helix-breaking solvents (trifluoroacetic acid) is normal, showing separate resonance absorption peaks for all the different types of protons (amide NH, phenyl group, α-CH, glutamic β- and γ-CH$_2$, benzylic CH$_2$) at different values of the applied magnetic field, measured in parts per million from a standard reference signal. On the other hand, the nmr spectrum of this polypeptide dissolved in helix-supporting solvents, such as CDCl$_3$ or trifluoroethanol (TFE), is featureless with an extreme broadening of the peaks, although on addition of appropriate quantities of trifluoroacetic acid broad bands appear (Fig. 22). A helix-supporting solvent is defined as a solvent in which a helical conformation is known to exist on the basis of spectroscopic and/or hydrodynamic studies.

The shape of the resonance peaks is determined by two relaxation times, T_1 and T_2; T_1 (spin–lattice relaxation time) measures the rate at which energy is transferred from the excited state to the surrounding environment, while T_2 (spin–spin relaxation time) measures the rate of redistribution of protons within the spin system. The relaxation times are in turn dependent on the correlation times, which are a measure of the rates of intra- and intermolecular motions in the system.

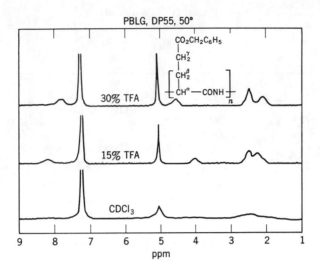

Fig. 22. Poly-γ-benzyl L-glutamate ($\overline{DP}$ 55), 100-MHz nmr spectra in $CDCl_3$, $CDCl_3$–TFA (85:15), and in $CDCl_3$–TFA (70:30). The concentration of solute in each case was 10% (112).

High viscosity, low temperature, and increased size and rigidity of the molecules reduce the motions in the system; hence the correlation times increase, the relaxation times become shorter, and line-broadening occurs.

The broadening of the nmr spectra of the polypeptides in the α-helical form in helix-supporting solvents such as $CDCl_3$, CCl_4, and TFE is a general phenomenon and is ascribed to the circumstance that the protons of the polypeptide, which form part of a rigid framework of atoms extending throughout the molecule, resemble those in solids in giving extremely broadened bands. It was suggested that the broadening of the nmr bands is probably due to a very efficient energy exchange between the protons which results in very short spin–spin relaxation times. As these rigid frameworks break down by addition of the helix-breaking solvent, this type of energy exchange becomes less efficient, the spin–spin relaxation time increases, and the bands sharpen.

As a consequence, the intensity of the signals in the nmr spectra develops gradually for the protons of the side chains and then later for those of the backbone as the helix-breaking solvent concentration is increased. A helix-breaking solvent, such as dichloroacetic acid (DCA) or trifluoroacetic acid (TFA), is defined as a solvent that destroys secondary structures by solvating the polypeptide chain. Evidence for the effect of such solvents has been obtained by spectroscopic and/or hydrodynamic studies. Thus, for poly-γ-benzyl L-glutamate (116) sharp bands for the various types of protons appear progressively from the periphery of the molecule inwards in the following order: phenyl, benzyl-CH_2, β and γ-CH_2, and, finally, α-CH and N—H.

By nmr studies it is possible to obtain useful information not only on the mechanism of the helix–coil transitions (118), but also on the flexibility of the side chains, side-chain–side-chain interactions and side-chain–main-chain interactions.

Helix–coil transitions have been observed with a large number of polypeptides in α-helical conformation, such as poly-L-leucine, poly-L-phenylalanine, poly-γ-alkyl L-glutamates (alkyl equals methyl, t-butyl or benzyl), poly-β-alkyl L-aspartates (alkyl equals methyl or benzyl), poly-L-methionine, and a copolymer of γ-methyl L-glutamate and L-methionine, as well as with other ordered structures (118–122). Poly-β-benzyl L-aspartate and sequential polypeptides composed of L- and D-γ-t-butyl-glutamyl, L- and D-γ-benzyl-glutamyl, and γ-benzyl-D-glutamyl and L-leucyl residues have also been shown to exhibit conformational transitions (122).

Nuclear magnetic resonance at 100 and 220 MHz has been shown to provide a sensitive measure of helix content in various polypeptides, since the backbone α-CH and peptide NH proton resonances both show separate peaks for atoms in helical and random coil environments (123,124).

The resonances of the α-CH and peptide NH protons are of primary importance since these protons are most sensitive to changes in the conformation of the polypeptide backbone. In Figure 23 are reported the 100-MHz nmr spectra of the α-CH and peptide NH protons of poly-L-methionine at various concentrations of TFA in $CDCl_3$ in the range of the helix–coil transition.

The spectra show that the ratio of the areas of α-CH and peptide NH separate peaks varies as a function of the concentration of TFA. As the TFA concentration is increased, the lower field peak grows at the expense of the upper field peak in the case of the α-CH proton; vice versa for the peptide NH proton.

Since TFA favors the "random coil" conformation, we conclude that the lower field α-CH peak and the upper field peptide NH peak correspond to the random coil forms of the polypeptides, while the upper field α-CH and the lower field peptide NH peaks are those of the helical forms.

The appearance of separate helix and random coil peaks for the α-CH and peptide NH backbone proton resonances is a general phenomenon which seems not to depend on such factors as the size and the nature of the side chains, the molecular weight of the polypeptide, or the nature of the solvent system employed. Furthermore, the detection of separate helix and random coil resonances makes possible the determination of the helix and random coil content of polypeptides by measurement of peak areas.

The observation of separate helix and random coil peaks also permits the

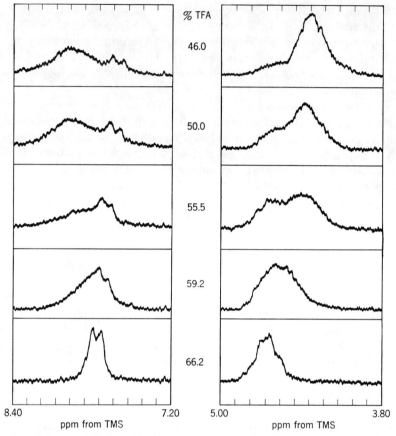

Fig. 23. Poly-L-methionine, 100-MHz nmr spectra showing α-CH and peptide NH regions in various TFA–CDCl$_3$ mixed solvent systems (123).

determination of minimum limits for the lifetimes, τ, of the protons in the two environments, since

$$\tau \geq \frac{1}{\omega_\alpha - \omega_\beta},$$

where $\omega_\alpha - \omega_\beta$ is the chemical shift difference in Hertz between helix and random coil protons. The lifetimes of the helix and random coil portions of poly-L-alanine have lower limits of about 10^{-2} sec (124). The above nmr results imply $\tau_{1/2}$ for the helix-to-coil transition to be longer than 10^{-2} sec. On the other hand, studies of the transitions based on fast reaction kinetics show $\tau_{1/2}$ for the transition to be faster than 10^{-5} sec.

In the case of poly-L-alanine, a plot of the per cent helicity (obtained from the peptide NH proton resonances) as a function of TFA concentration shows that this polypeptide is more than 50% helical in 100% TFA (123). The discrepancy with the helix content estimated by the use of the Moffitt-Yang parameter b_0 is rather perplexing. At present, no clearcut explanation of the helical content by the two methods is possible. It is essential to ascertain if similar differences can be observed with other polypeptides. It is conceivable that poly-L-alanine is a special case.

The chemical shift behavior of the helical peaks has been analyzed in order to obtain some insight into the nature of the forces involved in the helix–coil transition induced by TFA (124). The fact that neither the helix peptide NH peak chemical shift nor its linewidth changes appreciably as a function of TFA concentration, whereas the coil NH peak does show a marked solvent dependency, is presumed to be strong evidence that protonation is not responsible for the helix–coil transition, as concluded by Klotz and co-workers (125–129) and by J. H. Bradbury and Fenn (130,131), and that TFA does not strongly interact with the helical peptide $\rangle$N—H$\cdots$O=C$\langle$ hydrogen bonds, as suggested by Stewart and co-workers (132,133). Ferretti and Paolillo suggest "solvent–peptide and peptide–peptide hydrogen bond competition, coupled with a destabilization of the helix by solvation with TFA, is responsible for the helix–random coil transformation" (124). These results are in agreement with nmr, CD, and ORD studies on a number of polypeptides and small model compounds, which do not show charges for the peptide bond of polypeptides in strong acid solvents (112,118,134,135). Of particular interest in this connection is the work of Steigman and co-workers on the behavior of model amides and polypeptides in sulfuric acid and methanesulfonic acid–water mixtures and in dichloroacetic acid–organic solvents mixtures (136,137). These workers concluded that polypeptides, such as poly-γ-ethyl L-glutamate, exist in the coil conformation in dichloroacetic acid because of strong solvation rather than protonation. Furthermore, the order of decreasing efficiency of DCA cosolvents in effecting the coil–helix transition has been interpreted as the ability of the various donors to form hydrogen bonds with DCA to destroy the solvent network: triethylamine > acetic acid $\simeq$ methanol > water $\simeq$ nitrobenzene > nitroethane > formic acid > monochloroacetic acid $\simeq$ cyanoacetic acid $\simeq$ 1,2-dichloroethane > carbon tetrachloride.

D. Infrared Studies

Infrared spectra of polypeptides and proteins exhibit several relatively strong absorption bands which are associated with the CONH grouping, the structural unit common to all molecules of this type.

The position and intensity of these bands change from one conformation of the polypeptide chain to another and these variations have been used for conformational diagnoses since Ambrose and Elliott pointed out certain frequency–conformation correlations (138). Before illustrating the ir results obtained in conformational studies on polypeptides, we will briefly present the characteristic absorption bands of small model compounds, such as secondary amides, because of their similarity to the structural repeat units of polypeptide chains

$$R-\overset{\overset{O}{\|}}{C}-N\overset{R'}{\underset{H}{}}$$

Figure 24 shows the ir spectrum of N-methylacetamide with the indication of the positions of the characteristic amide bands. The frequencies and approximate descriptions of the bands associated with the peptide linkage are summarized in Table III. The two bands designated as amide A and amide B appear in the ir spectra of polypeptides in the 3000-cm^{-1} region, at ca. 3300 cm^{-1} and ca. 3100 cm^{-1}, respectively. These bands disappear on N-deuteration and are associated with stretching motions of hydrogen bonded NH groups. The amide A and B bands are not too sensitive to the chain conformations. However, the dichroism of these bands is useful in distinguishing between the α-helical form and the extended form. The amide A band which, in the case of α-helical poly-L-alanine, appears at 3293 cm^{-1} shows parallel dichroism, whereas the amide A band of the β-form exhibits perpendicular dichroism. The amide I band, which is associated with the stretching vibration of the C=O bond, is present in the spectra of all secondary amides, polypeptides, and proteins. This band and the amide II band, which involves primarily C—N stretching and N—H bending, have been largely used for conformational studies of a number of polypeptides and proteins.

The correlations between the amide I and II bands and the chain conformations (α and β) have been analyzed by Miyazawa and Blout on the basis of the treatment of localized vibrations (139).

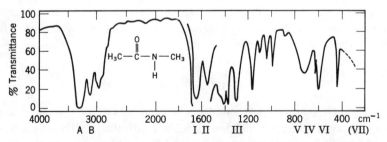

Fig. 24. Assignments of various characteristic amide bands from a capillary film of N-methylacetamide. The amide VII band is not shown (149).

TABLE III

Summary of the Characteristic Bands Associated with the Peptide Linkage

Symmetry	Designation	Approx.[a] frequency	Description[b,c,d]
In plane	A	3300	NH(s), 2 × amide II in Fermi resonance
	B	3100	
	I	1650	C=O(s) (80%), C—N(s) (10%), N—H(b) (10%)
	II	1560	C—N(s) (40%), N—H(b) (60%)
	III	1300	C=O(s) (10%), C—N(s) (30%), N—H(b) (30%), O=C—N(b) 10%), other (20%)
	IV	625	O=C—N(b) (40%), other (60%)
Out of plane	V	725	N—H(b)
	VI	600	C=O(b)
	VII	200	C—N(torsion)

[a] Rounded frequencies based on spectra of N-methylacetamide.
[b] The percentages refer to the potential-energy distribution, based on a simple force field.
[c] Description of the out-of-plane modes is very approximate.
[d] (s), stretching; (b), bending.

For poly-γ-benzyl L-glutamate in the α-helical conformation, both the amide I band parallel to the helix axis and the amide I band perpendicular to the helix axis have been observed at 1650 and 1652 cm^{-1}, respectively (see Fig. 25).

For the α-helix the parallel amide I and the perpendicular amide I frequencies are not much different from each other, so that the apparent dichroism observed for the amide I band of the α-helix is not too strong (140).

A weak parallel amide II band and a strong perpendicular amide II band of poly-γ-benzyl L-glutamate have been observed at 1516 and 1546 cm^{-1}, respectively (see Fig. 25). The perpendicular amide II band of the α-helix at ca. 1550 cm^{-1} is due to the transition moment exactly perpendicular to the helix axis and this band is not overlapped by parallel bands. Accordingly, the apparent dichroism of this band may be used as a useful measure of the orientation of the helix axes in the oriented samples.

Infrared bands from amide III vibrations in the 1200–1300 cm^{-1} region arise from C=O and C—N stretching and NH bending. For the α-helical conformation of poly-L-alanine, a weak parallel band and a strong perpendicular band are present at ca. 1270 and at 1274 cm^{-1}, respectively.

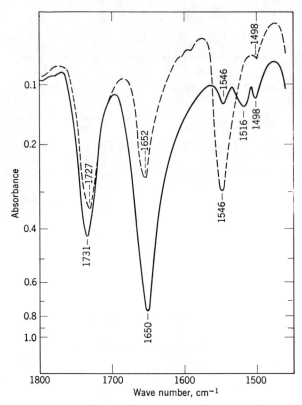

Fig. 25. The polarized infrared spectra for films of poly-γ-benzyl L-glutamate oriented from chloroform solution. The solid line denotes the spectrum recorded with the polarization parallel to the orientation; the dotted line refers to the spectrum with polarization perpendicular to the orientation (142).

These peaks disappear on N-deuteration. Infrared spectra of poly-γ-benzyl L-glutamate, poly-β-benzyl L-aspartate, and their N-deuterated analogs in the region of amide III vibrations (1150–1350 cm^{-1}) are shown in Figure 26. The helical form, parallel-chain β-form, antiparallel-chain β-form, and disordered form of polypeptides may be distinguished by amide I and II bands observed for oriented samples. The ir results for amide I and II bands of polypeptides in these conformations are summarized in Table IV. However, if various conformations coexist in unoriented samples, the fraction of each conformation may not necessarily be estimated by intensity measurements of the amide I and II bands only, because these amide bands of the α-helical form are hardly separated from those of the disordered form.

Characteristic amide IV, V, and VI bands which are present in the low-frequency region might be expected to be responsive to conformational

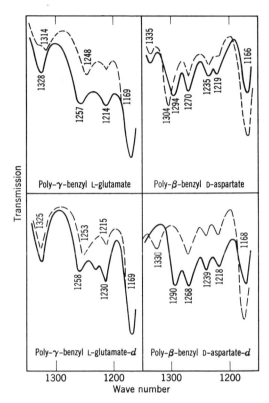

Fig. 26. The polarized infrared spectra for oriented films of poly-γ-benzyl L-glutamate, poly-β-benzyl D-aspartate, and N-deuterated polymers derived from these two materials. The solid line denotes that the polarization is parallel to the direction of orientation, while the dotted line refers to the case where it is perpendicular to the orientation direction (146).

changes; however, amide bands IV and VI are, in fact, found to be insensitive to such changes. On the other hand, the amide V bands in the region 400–800 cm^{-1} have been found to be extremely useful for conformational diagnoses of polypeptide chains (141).

The polarized infrared spectrum of high molecular weight poly-γ-methyl L-glutamate is shown in Figure 27 for the regions 800–400 and 1800–1350 cm^{-1}. On the low-frequency region, a strong perpendicular band is observed at 615 cm^{-1}, which is shifted, on N-deuteration, to 462 cm^{-1} (Fig. 28). These bands arise largely from the N—H and N—D out-of-plane bending modes.

Amide I, II, and V frequencies of α-helical polypeptides such as poly-γ-methyl L-glutamate, poly-γ-ethyl L-glutamate, poly-γ-isoamyl L-glutamate,

TABLE IV

The Frequencies (cm^{-1}) and Relative Intensities of the Amide I and II Bands of Polypeptides in Various Conformations

Conformation	Designation	Amide I[a]	Amide II
Random coil (polyserine)	—	1655 (s)	1535 (s)
α-Helix	$\nu(0)$	1650 (s)	1516 (w)
(poly-γ-benzyl L-glutamate)	$\nu(\theta)$	1652 (m)	1546 (s)
Parallel-chain β	$\nu(0, 0)$	1645 (w)	1530 (s)
(α-keratin)	$\nu(\pi, 0)$	1630 (s)	1550 (m)
Antiparallel-chain β	$\nu(0, \pi)$	1685 (w)	1530 (s)
(polyglycine I)	$\nu(\pi, 0)$	1632 (s)	1540[b]
	$\nu(\pi, \pi)$	1668[b]	1550 (w)

[a] Relative intensities are given in parentheses: (s) = strong; (m) = medium; (w) = weak.
[b] Calculated value.

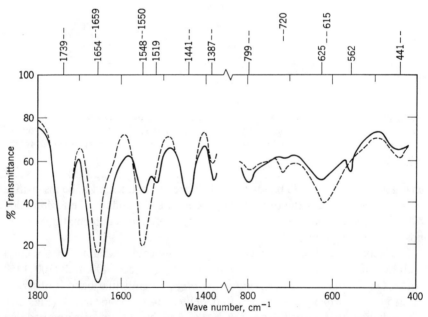

Fig. 27. The polarized infrared spectra for oriented films of poly-γ-methyl L-glutamate ($A/I = 200$). The solid lines refer to the case where the electric vector of the polarized light is parallel to the orientation direction, while the dotted lines describe the situation where the vector is perpendicular to the orientation direction of the film (142).

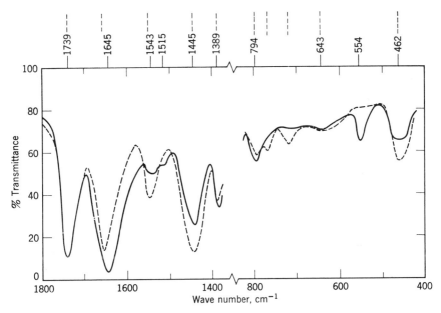

Fig. 28. The polarized infrared spectra for oriented films of poly-γ-methyl L-glutamate-d ($A/I = 200$). The solid line denotes that the polarization is parallel to the orientation; the dotted line denotes that it is perpendicular to the orientation (147).

poly-γ-benzyl L-glutamate, and poly-L-alanine are listed for comparison in Table V together with amide V' frequencies of the N-deuterated derivatives. The frequencies of perpendicular amide V bands of the α-helical form are thus observed in a narrow frequency region of 610–620 cm^{-1} and perpendicular amide V' bands in the region of 455–465 cm^{-1}. Amide bands have

TABLE V

Characteristic Amide Frequencies[a] for the α-Helical Form of Polypeptides

	I(∥)	II(⊥)	V(⊥)	V'(⊥)
Poly-γ-methyl L-glutamate	1658	1550	620	462
Poly-γ-ethyl L-glutamate	1658	1550	620	465
Poly-γ-isoamyl L-glutamate	1656	1550	615	461
Poly-γ-benzyl L-glutamate	1653	1550	615	458
Poly-L-alanine	1658	1548	610	456

[a] Frequencies (cm^{-1}) are listed for the main component of amide I, II, and V bands of undeuterated species and amide V' bands of N-deuterated species.

also been observed at 615 cm^{-1} for the helical form of poly-L-α-amino-*n*-butyric acid, poly-L-norleucine, and poly-L-leucine (142–150).

The infrared spectra of the two crystalline forms of polyglycine have been measured in the region 400–190 cm^{-1}. The bands at 217 cm^{-1} for polyglycine I and at 365 cm^{-1} for polyglycine II are tentatively assigned to the amide VII vibrations (151).

A few other characteristic amide bands have been found in the low-frequency region for polypeptides in α-helical conformations. A strong 375 cm^{-1} band is present in the far ir spectrum of α-helical poly-L-alanine. Poly-L-glutamate esters also show a band at 410 cm^{-1} These bands, which are tentatively assigned to the E vibration, are not present when these polypeptides are in the β-form.

Amide V bands have been used for studying the chain conformations of racemic (a mixture of separately polymerized chains made from the L and

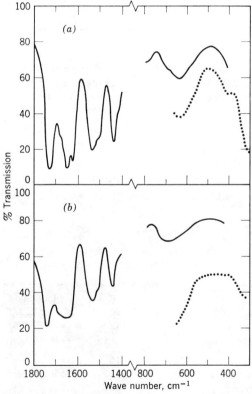

Fig. 29. The infrared spectra of poly-γ-methyl DL-glutamate (*a*) $A/I = 100$ and (*b*) $A/I = 4$. The solid line is obtained using an Hitachi 225 infrared spectrophotometer, while the broken line is from an EPI-L instrument (153).

from the D amino acid) and *meso* polypeptides (in which residues of L and of D configurations are randomly sited along the polymer chain) and polymerization schemes of D,L-copolypeptides (152–154). The racemic form is indicated here with the notation -DL-; the *meso* form with the notation D,L. Figure 29 shows the infrared spectrum of poly-γ-methyl DL-glutamate. The strong amide I band at $1655\ cm^{-1}$, amide II band at $1540\ cm^{-1}$, and the strong amide V band at $635\ cm^{-1}$ indicate that the predominant conformation of this polymer is the α-helix. These results are similar to the case of copoly-γ-methyl D,L-glutamate, which exhibits a strong amide V band at $630\ cm^{-1}$.

III. SOME SPECIAL ASPECTS OF HELICAL STRUCTURES

In addition to specific questions of main chain conformation discussed in the previous section, we have considered perturbations by side chains which alter main chain helical sense, i.e., esters of poly-L-glutamic acid as compared with poly-L-aspartic acids. In this part of our review, we present two other features of structure that are common in proteins. The first deals with the effects of having the imino acid L-proline as a substantial component of the chain; in the extreme we will discuss effects in the homopolymer poly-L-proline. Naturally, the most dramatic differences from other typical poly-α-amino acids result from the absence of hydrogen bonding. The second special feature we consider is based upon aromatic side chains. There is much overlap in the electronic transitions stemming from the aromatic and amide groups. This complicates the assignment of conformation based upon spectral measurements described above.

A. Non-Hydrogen Bonding Polypeptides

Two chemically indistinguishable forms of poly-L-proline have been observed (155). It is found that each form has a characteristic configurational pattern which is reflected in its hydrodynamic and optical rotatory properties. If we look at the poly-L-proline chain we see that hydrogen bonding is impossible. Therefore, any stable configuration must be due to rotational restrictions at bonds along the polyproline chain. Rotation about the $N-C_2$ bond is impossible due to the pyrrolidine ring. Rotation about the $N-C_1$ bond is restricted by the partial double bond character of the imide bond. The two possible configurations of this bond are *cis* and *trans*. The C_1-C_2 bond may assume two positions for each of the two configurations of the imide bond. These conformations are known as *cis'* and *trans'*. We see that there are four possible combinations of the configurations of the imide and C_1—C_2 bonds. Of these four possible isomers only two have been observed:

Poly-L-*proline I*, a right-handed helix containing $3\frac{1}{8}$ residues per turn with a unit translational distance of 1.85 Å. In poly-L-proline I the imide bond is *cis* and the C_1—C_2 bond *trans'* (Traub helix) (156,157).

Poly-L-*proline II*, a left-handed helix containing 3 residues per turn with a unit translational distance of 3.12 Å (158,159). The imide bond is *trans* and the C_1—C_2 bond *trans'* (Cowan-McGavin helix).

The latter structure is very similar to that of poly-L-hydroxyproline which exists in only one form (160). In the solid state each chain is linked to the other six adjacent chains by O—H···O=C hydrogen bonds. The study of the optical properties of poly-L-proline II is a logical first step in the interpretation of the optical properties of collagen since both have similar conformational features (161).

It is expected that the left- and right-handed helices will contribute to the inherent optical rotation of L-proline with opposite signs. This is what is observed. Poly-L-proline I has $[\alpha]_D^{25} = +50°$, while poly-L-proline II has $[\alpha]_D^{25} = -540°$. The hydrodynamic properties are also different because poly-L-proline II is much more extended than the I form (162–165).

The two forms of poly-L-proline are interconvertible under certain conditions. Poly-L-proline I mutarotates to poly-L-proline II in solvents such as water, acetic acid, benzyl alcohol, and chloroethanol. These changes may be reversed in 90% 1-propanol–10% acetic acid. Forms I and II may be interconverted back and forth indefinitely with no change in properties (165). The kinetics of the reactions I → II and II → I have been thoroughly studied and found to be first order in polymer concentration (162,165,166). The transition is much slower than the helix–coil transition in poly-α-amino acids.

ORD and CD studies demonstrate clearly that poly-L-proline in forms I and II has a conformation in solution distinctly different from the α-helix (167–175). The far ultraviolet ORD curve of poly-L-proline II in water displays a single large negative Cotton effect centered at 205 nm (Fig. 30). The study of the optical properties of form I of poly-L-proline has been hampered since propionic acid, which stabilizes form I, is opaque in the far ultraviolet.

However, Bovey and Hood have reported ORD measurements on high molecular weight polyproline in trifluoroethanol, in which they observed that the conformation of the polypeptide immediately after dissolution corresponds to form I of polyproline. The ORD spectrum shows a large positive extremum at about 223 nm with a residue rotation $[R]_{223} \sim +47{,}500°$ cm^2 decimole^{-1} and a negative extremum at about 208 nm with a residue rotation $[R]_{208}$ $-90{,}000°$ cm^2 decimole^{-1} (174).

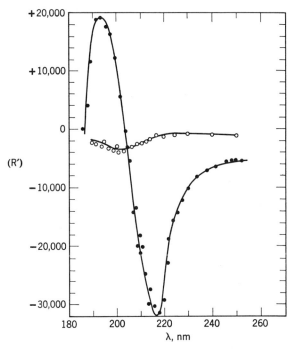

Fig. 30. The optical rotatory dispersion of poly-L-proline II (filled-in circles) and L-proline (open circles) in water (167).

Poly-L-proline form I immediately after solution in trifluoroethanol gives a CD spectrum showing a weak negative band at 232 nm (assigned to an $n \to \pi^*$ electronic transition), a large positive band at 215 nm, and a weaker negative band at 199 nm, which, in accord with theoretical calculations, arise from an exciton splitting of the $\pi \to \pi^*$ transition of form I poly-L-proline in the Traub right-handed helix.

As the isomerization proceeds (form I $\to$ form II), the positive band becomes weaker and the initial negative band is displaced by a stronger negative band at about 206 nm. The spectrum of poly-L-proline, after equilibration, shows a weak positive band at 226 nm in addition to the negative 206 nm band; these bands have been interpreted as representing exciton split components of the form II peptide $\pi \to \pi^*$ transition (left-handed helix of Cowan and McGavin) (Fig. 31). Table VI summarizes the CD results of detailed studies of the conformational isomerizations of poly-L-proline and a closely related polypeptide, poly-L-acetoxyproline, in trifluoroethanol (176).

Infrared spectra of polypeptides containing pyrrolidine rings are, in general, not as well understood as the spectra of the previously discussed α, β, and random coil conformations. The nitrogen atoms of pyrrolidine rings

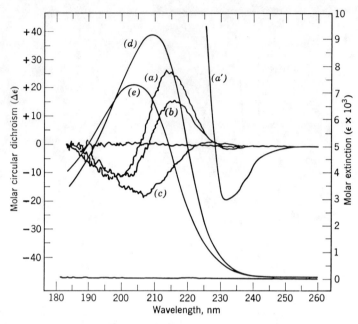

Fig. 31. CD and uv spectral measurements of poly-L-proline in trifluoroethanol. The time for each curve is recorded from dissolution until the peak or trough is observed: (a) 13 min, 18 min; (b) 56 min, 60 min; (c) 19 hr; (a') 10 min (10-fold path length for cell); (d) 15 min; (e) 19 hr (175).

have no hydrogen atoms and, therefore, there are no amide I and II bands in the usual sense. The spectra of poly-L-proline I and II have been investigated in detail by Isemura and colleagues (177). Differentiation between the two forms is possible, even though the above correlations and calculations do not apply. As mentioned above, poly-L-hydroxyproline does not exhibit the mutarotation properties of poly-L-proline, except when it is dissolved in 6 M LiBr. This difference has been attributed to the fact that poly-L-hydroxyproline occurs as an intertwined triple chain in solution. Such a conclusion is justified by X-ray diffraction analysis, which demonstrates that the ordered structure of this polypeptide (in the so-called form A) is composed of three intertwined helical chains held together by O—H···O interchain hydrogen bonds.

Since collagen contains a high percentage of proline, hydroxyproline, and glycine, extensive conformational studies on synthethic polypeptides containing these amino acid residues were performed in order to obtain useful information on the molecular structure of the protein (178–183).

"Random" copolymerization of mixtures of N-carboxyanhydrides of proline, glycine, and hydroxyproline usually yield complex materials whose

TABLE VI

Polymer[a]	Absorption		CD band 1		CD band 2		CD band 3	
	λ_{max}	$\Delta\epsilon$	λ_{ext}	$\Delta\epsilon$	λ_{ext}	$\Delta\epsilon$	λ_{ext}	$\Delta\epsilon$
Poly-L-proline I	210	9000	232	−1.8	215	+30.0	190	−10.5
Poly-L-proline II	202	7200			226	ca. +2	206	−18.0
Poly-L-acetoxyproline I	208	9300	229	−5.0	212	+25.0	ca. 190	[b]
Poly-L-acetoxyproline II	200	7600			226	+1.35	206	−9.2

[a] Data for form I are extrapolated to zero time; for form II, data are taken at times (>24 hr) at which no further changes are observable. All data are for trifluoroethanol solutions. Wavelengths are in nanometers.
[b] Difficult to measure, but much weaker than band 2 or the corresponding band of poly-L-proline II.

properties cannot be usefully interpreted. In recent years, interest has turned particularly to sequential polytripeptides of which every third residue is either proline, hydroxyproline, or glycine, in accordance with features of the composition of collagen which are believed to play an important part in determining its structure (184–187).

There is evidence that several polytripeptides do indeed have collagen-like conformations, and, recently, a very detailed X-ray analysis of the polymer (Gly-Pro-Pro)$_n$ has been completed. This polytripeptide and other polyhexapeptides are models of collagen with the sequences (Gly-Pro-Ala-Gly-Pro-Pro)$_n$, (Gly-Ala-Pro-Gly-Pro-Pro)$_n$, (Gly-Ala-Ala-Gly-Pro-Pro)$_n$, and (Gly-Ala-Pro-Gly-Pro-Ala)$_n$. They have the same triple helical conformation with one NH···O interchain hydrogen bond per tripeptide. On the basis of these results, it was suggested that collagen itself probably has this structure (178).

B. Aromatic Polypeptides

As we have seen before, in polypeptides (and proteins) the peptide chromophores present two electronic transitions, $n \to \pi^*$ and $\pi \to \pi^*$, in the accessible isotropic absorption region and produce characteristic Cotton

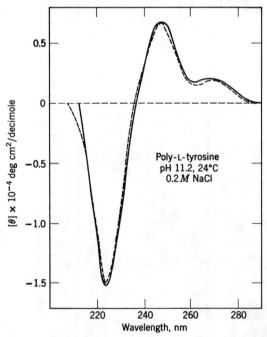

Fig. 32. CD spectrum of helical poly-L-tyrosine. The solid line represents the experimentally measured curve; the dashed line refers to calculated gaussian bands (188).

effects which can be related to secondary structure. In polypeptides with side chains absorbing in the wavelength range of the peptide chromophore, the identification of helical structures is a difficult task. Side-chain aromatic chromophores in poly-L-phenylalanine (**1**), poly-L-tryptophan (**2**), and poly-L-tyrosine (**3**) also produce Cotton effects which are set upon the background

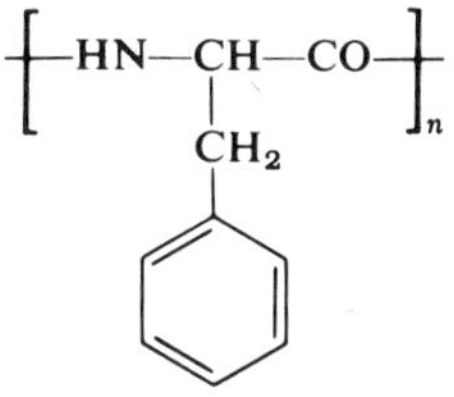

$$\left[-HN-CH-CO-\right]_n$$
$$|$$
$$CH_2$$
$$|$$
$$C_6H_5$$

(**1**)

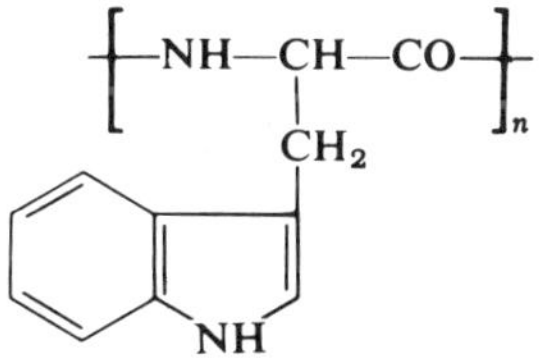

$$\left[-NH-CH-CO-\right]_n$$
$$|$$
$$CH_2$$
$$|$$
$$indolyl$$

(**2**)

$$\left[-NH-CH-CO-\right]_n$$
$$|$$
$$CH_2$$
$$|$$
$$C_6H_4-OH$$

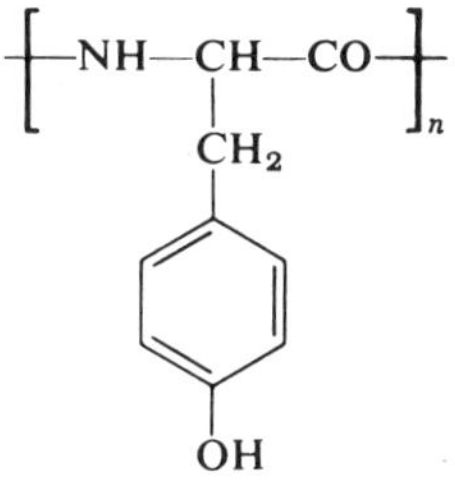

(**3**)

rotation provided by tails of all other Cotton effects. The background rotation may be sufficiently large to obscure small Cotton effects. The conformation of polypeptides with aromatic side chains is in part determined by electronic and steric interactions between side-chain chromophores and the optically active centers in the polypeptide main chain.

Circular dichroism is the technique of preference (see above) for conformational studies of these polypeptides, since asymmetric absorptions alone contribute to the effect. The CD spectrum of poly-L-tyrosine in 0.1 N NaClO$_4$ (pH = 10.8) (Fig. 32) reveals two positive dichroic bands at 275 and 247 nm and a negative band at 229 nm which correspond to absorptions in the tyrosyl side chain (188). Furthermore, a positive shoulder at 215 nm and a strong positive peak at 200 nm are present.

This CD pattern is clearly distorted from that of the right-handed α-helix, because the electronic transitions in the tyrosyl side chains overlap strongly with peptide transitions.

The screw sense of the helix of poly-L-tyrosine is still an open question. Energy calculations demonstrate a lower energy in the right-handed than in the left-handed α-helical form (189). This result is in agreement with uv, ORD, and CD studies (190), but in disagreement with a recent study of dipole moments and rotational relaxation times of poly-L-tyrosine in quinoline, which shows that the polypeptide exists as a left-handed α-helix (191). It was suggested that poly-L-tyrosine under suitable conditions can be left-handed or right-handed and that solvent–polymer interactions may convert the polypeptide from one conformation to another (190).

CD and ORD studies on block and random copolymers of L-phenylalanine indicate an α-helical conformation (192–195). Weak Cotton effects in the wavelength region where the phenyl side chain chromophore exhibits a number of absorption maxima (300–225 nm) have been detected and attributed to aromatic side-chain chromophores existing in a dissymmetric environment.

CD studies of poly-L-tryptophan in ethylene glycol monomethyl ether show a rather complex spectrum in the region 260–300 nm and two positive bands and a single negative band below 250 nm. The positions of the bands between 260 and 300 nm are approximately the same as those of the absorption peaks of the indole chromophore, thus suggesting their origin (Fig. 33). Although an X-ray diffraction study of this polymer demonstrates that the polypeptide is α-helical in the solid state, it is not possible to interpret the spectrum in the region below 250 nm in terms of α-helical conformation. However, CD measurements on various random copolymers of L-tryptophan and γ-ethyl L-glutamate show clearly that, on decreasing the L-tryptophan content, the typical spectrum of the right-handed α-helix becomes more and more evident (196,197).

Poly-L-p-aminophenylalanine, an amino analog of poly-L-tyrosine, also shows strong conformational aromatic side chain effects (198). This polypeptide exists as a random coil between pH 1.08 and 2.56 and as an α-helix at pH > 2.78.

We have observed Cotton effects in azopolypeptides derived from L-p-(phenylazo)phenylalanine-γ-benzyl L-glutamate residues and L-p-(p'-hydroxyphenylazo)phenylalanine and N-(3-hydroxypropyl)-L-glutamine (199–201). These Cotton effects contain dichroic bands attributable to the azoaromatic chromophore. Poly-L-p-(phenylazo)phenylalanine and its copolymers exist as right-handed α-helices in dioxane, as random coils in TFA-dioxane mixed solvent, and as extended ordered polyelectrolytes in nearly pure TFA. Under the last named conditions, we believe the polypeptide is composed of highly

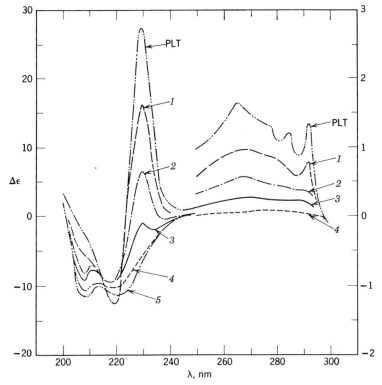

Fig. 33. CD spectra for copolymers of L-tryptophan and γ-ethyl L-glutamate. An expanded scale was used to show the CD curves in the 250–330 nm region more clearly. PLT denotes poly-L-tryptophan; Curves 1–5 refer to copolymers where molefractions L-tryptophan are: 0.84, 0.68, 0.50, 0.32 and 0.16 respectively (196).

protonated azoaromatic groups. These ionic groups repel each other to form the fully extended chain.

Poly-L-p-(p'-hydroxyphenylazo)phenylalanine and its copolymers exist as right-handed α-helices in trimethyl phosphate; the homopolymer also exists as a right-handed α-helix in aqueous solutions between pH 10 and 11.9.

IV. "RANDOM COIL" POLYPEPTIDES

The concept of a "random" or "disordered" polypeptide chain is important in a discussion of the conformational transitions of poly-α-amino acids and the related phenomenon of the denaturation of proteins (202–204). The *statistical* description of a polypeptide chain in a random conformation is that of a chain with completely free rotation about its bonds. In this model each of the angles ϕ and ψ (for definition, see Sect. VII) has the same probability for any of its values.

However, theoretical calculations have shown that the "random" polypeptide chain is not very random in the statistical sense and only a limited range of angles of rotations about the bonds of a polymer backbone are energetically favorable.

Strong restrictions on the randomness are imposed principally by:

a. inherent restriction of rotation about covalent bonds,

b. van der Waals interactions between nonbonded atoms within or directly connected to the backbone,

c. electrostatic interactions of polar groups in the chain backbone,

d. interactions between side chains on two different amino acid units,

e. interactions between an amino acid side chain and the polymer backbone, and

f. interaction between polymer and solvent.

It follows from the preceding considerations that the optical activity and, in general, the spectroscopic properties of a random polypeptide cannot be expected to be that of the isolated peptide chromophore as in the limiting case of the statistical model.

Because of the restrictions in the conformation of the chain, the peptide chromophores of the backbone will, on the average, be in a dissymmetric environment. Small perturbations of the medium modify considerably the array of the chromophores and, consequently, the optical rotation.

In actual cases involving different coil-forming media, polypeptide chains which are random from the overall point of view can have different degrees of local order and, consequently, different spectroscopic properties. It follows that the random chain represents no single structure, but encompasses a range of structural possibilities (205,206).

The ultraviolet absorption spectrum of a polypeptide in the random conformation shows a single broad band centered at about 192 nm (31). This spectrum is essentially similar to that of an N-substituted amide and differs markedly from that of polypeptides in the α-helical conformation. The band at 192 nm is assigned as $\pi \to \pi^*$ on the basis of theoretical calculations and experimental evidence obtained with simple amide model compounds. This result is in agreement with predictions made on the basis that the randomly coiled chain is characterized by a lack of regular geometrical order among the amino acid residues (Fig. 7).

The weak absorption band expected in the 210–230 nm wavelength range, which is present in the uv spectrum of simple amides and corresponds to a low intensity $n \to \pi^*$ electronic transition, is merged with the high-energy, intense band at 192 nm and cannot be seen in the uv spectrum. In addition, the uv absorption of a "random" polypeptide is hyperchromic in comparison with that of a polypeptide in the α-helical conformation.

The ORD spectrum of polypeptides in the "random" conformation is characterized by a negative Cotton effect with a trough and a peak near 205 and 190 nm, respectively, an inflection point near 197 nm, and another much smaller Cotton effect at 232–235 nm. The rotation per residue for this conformation is about -1800 deg. cm^2 decimole^{-1} (52) (see Fig. 17).

The CD spectrum shows a weak and broad positive band at about 218 nm followed by a single negative band at about 200 nm. A third dichroic minimum at about 238 nm is also present in the CD spectrum of charged polypeptides (78,207–209) (see Figs. 18 and 19).

When transitions from the helix to the random coil occur in aqueous solutions for polypeptides such as poly-L-lysine and poly-L-glutamic acid, the negative $n \rightarrow \pi^*$ band is transformed to a weak and broad positive band at about 218 nm and the $\pi \rightarrow \pi^*$ exciton-split bands are replaced by a single negative band at about 200 nm in the CD spectrum.

At present, the interpretation of the CD spectrum of the polypeptide chain in the "random" conformation is equivocal. The presence of the bands at 218 and 200 nm has been attributed to the occurrence of a specific local structure (a polyproline II-type helix) rather than of a random coil (202). The suggestion has been made that the CD spectrum of a random polypeptide chain is characterized by a single negative band centered near 200 nm. As can be inferred from the preceding discussion, the optical characterization of the random coil conformation remains still an unsettled issue (79,206).

Infrared studies on solid films of sodium poly-L-glutamate, poly-L-serine, and poly-L-lysine hydrochloride in the random coil conformation show a broad amide I band at ca. 1655 cm^{-1} and a broad amide II band at ca. 1535 cm^{-1}. The infrared spectrum of sodium poly-α-L-glutamate in the disordered form is shown in Figure 34. The amide II band is overlapped by the band arising from carboxylate groups. It is important to emphasize that the amide I frequency of the random coil conformation is not much different from the amide I frequency of the α-helix, as previously discussed. However, it is possible to distinguish between the two conformations by inspection of the amide V absorption region.

As can be seen in the ir spectrum of a polypeptide in the random coil conformation, the amide V band is observed with the band centered at about 650 cm^{-1}, which is shifted, on N-deuteration, to 510 cm^{-1}. For poly-DL-alanine, which exists also as a random coil, the amide V band has been observed at 665 cm^{-1}.

In general, for random coil conformations the position of the amide V band, which is quite diffuse, is centered at about 650 cm^{-1} and may be readily distinguished from the sharp amide V band of the α-helical form centered at around 600 cm^{-1} (142).

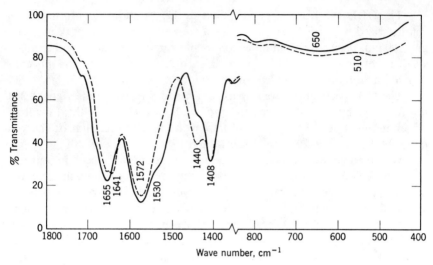

Fig. 34. Infrared spectra of films of sodium poly-L-glutamate (solid line) and its N-deuterated analog (dashed line). These films were cast from water solution and dried (142).

V. β-STRUCTURES FOR POLYPEPTIDES

Many fibrous structural proteins, i.e., keratins, silk, porcupine quill, have been shown by X-ray diffraction studies and infrared spectroscopy to exist in β-conformations (210–212). These structures, as well as the α-helix, are present in the enzyme lysozyme and in other proteins (213). Several synthetic polypeptides of valine, cysteine, serine, isoleucine, and threonine were found to exist in the β-conformation (73). The transformation of synthetic polypeptides from the α-form to the β-form has been achieved in a number of cases by stretching, heating, or treatment with suitable solvents. In this presentation, we will illustrate the structural features common to all β-structures of proteins and polypeptides and present results on optical activity of polypeptides in the β-conformation.

Two β-structures (parallel and antiparallel) consist of extended or nearly extended chains with extensive hydrogen bonding between N—H groups of one chain and C=O groups of adjacent chains. These bonds stabilize the structure and form sheets which accommodate the side-chain groups (214).

The interchain distance is about 4.8 Å in the direction of the hydrogen bond and the lengths of the repeat of patterns (axial translations) are 6.50 and 7.0 Å for the parallel and antiparallel forms, respectively (Figs. 35 and 36).

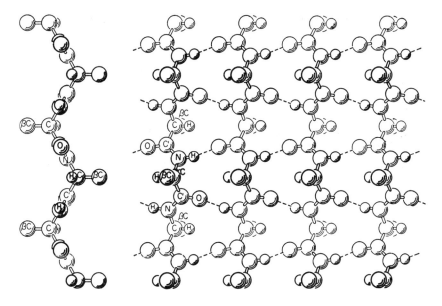

Fig. 35. Parallel-chain pleated sheet (1,2).

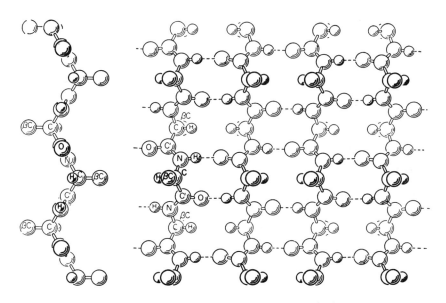

Fig. 36. Antiparallel-chain pleated sheet (1,2).

The reflection originating from the interchain distances may vary with the size of side-chain residues; the strong reflection at 1.15 Å (one-sixth of the repeat of pattern of the peptide backbone which is due to two amino acid residues) appears because the six atoms making up the two residues are nearly equally spaced. Highly oriented samples of poly-L-alanine show a strong meridional reflection corresponding to a spacing of 1.147 Å (215). This result and the striking similarity with the patterns of Tussah silk and β-keratin [for which Meyer and Mark (216) proposed the extended conformation of the polypeptide chain] are related to the occurrence of an antiparallel β-pleated sheet structure in the synthetic polypeptide. In the pleated sheet the atoms of the amide groups are coplanar. Contraction of the chain from the fully extended form occurs by appropriate rotation about backbone single bonds. Hydrogen bonds are formed between adjacent chains in the same plane.

A third β-conformation, the so-called cross-β-structure, with parameters

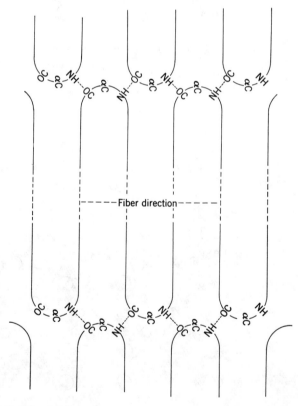

Fig. 37. A schematic representation of a cross-β-structure (6).

similar to those of the parallel and antiparallel β-forms, has also been observed. It differs from the latter structures in that it consists of single extended chains folded back on themselves with the hydrogen bonds parallel to the direction of stretching rather than perpendicular as in the other β-structures (Fig. 37) (217).

Evidence of the cross-β-structure has been obtained by suitable treatment of low molecular weight samples of poly-γ-benzyl L-glutamate and poly-β-benzyl L-aspartate, as well as in samples of poly-O-acetyl-L-serine (218,219) and O-benzyl-L-serine (220,221), poly-O-acetylthreonine and poly-O-acetylallothreonine (222), poly-S-benzyl-L-cysteine and poly-S-carbobenzoxy-L-cysteine (223,224), poly-β-n-propyl L-aspartate (218,225), and sequential polypeptides composed of alanyl and glycyl residues (226,227).

The optical activity of the β-conformation has been qualitatively explained, although there is unexplained disagreement on the relative importance of exciton contributions to the rotatory strength (228,229).

ORD spectra of the antiparallel β-form of poly-L-lysine in solution show a trough at 229–230 nm (with a residue rotation = -6300 deg. cm^2 decimole^{-1}), a peak at 205 nm, and another trough near 190 nm (Fig. 38) (230, 231).

The facts that the antiparallel β-structure ORD spectrum has a trough at 210–230 nm and that a weak trough at 232–235 nm is also present in the ORD spectrum of polypeptides in "random coil" conformation suggest that the use of the 233 nm trough as a diagnostic criterion for the presence of α-helical conformation is not suitable. A proposal has been made to adopt as a better criterion the positive peak at lower wavelengths since its positions in the α-helical, antiparallel β-, and random conformations are sufficiently separated (199, 206, and near 180 nm, respectively) (232).

Films of polypeptides known from ir dichroism to be in the antiparallel β-structure in the solid state give nonidentical ORD spectra. These polypeptides are classified in two families designated as β-I and β-II (233). The β-I type polymers show an ORD spectrum with a trough at 229–230 nm and a peak at 205 nm, very similar to that of β-structured poly-L-lysine in solution; the β-II type has an ORD spectrum composed of a trough close to 240 nm and a peak between 210 and 215 nm.

Besides poly-L-lysine, poly-O-acetyl-L-serine, poly-O-t-butoxy-L-serine, poly-O-benzyl-L-serine, and poly-L-valine are classified as members of the β-I family. Poly-S-benzyl-L-cysteine, poly-S-methyl-L-cysteine, poly-S-carbobenzoxy-L-cysteine, poly-S-carboxymethyl-L-cysteine, and poly-L-serine are classified as members of the β-II family.

The CD spectrum of aqueous solutions of poly-L-lysine in the β-conformation (Fig. 39) shows a negative band at 217–218 nm and a strong positive band at 195–198 nm (234,235). The 195–198 nm band arises from

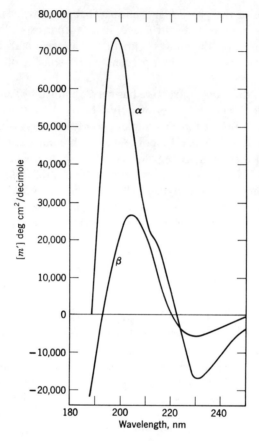

Fig. 38. ORD for poly-L-lysine at pH 11.6. Curve α is obtained by measuring the ORD at 22.5°. Curve β is obtained by measuring the ORD at 22.5° after heating the solution to 51° for 15 min (i.e., until no further decrease in rotation is observed (231).

exciton resonance interactions of the $\pi \rightarrow \pi^*$ transition, while the 217–218 nm band is associated with an $n \rightarrow \pi^*$ electronic transition. Theoretical analyses of the ultraviolet optical properties of polypeptides in the β-conformation are in disagreement as to whether the CD patterns of poly-L-lysine are those of the parallel- or antiparallel-chain β-conformation. Circular dichroism spectra of low molecular weight poly-L-serine in the β-conformation *in solution* show ellipticity extrema at 222 and 197 nm (236). Since this polymer lacks side chains absorbing over the wavelength range studied (260–190 nm), the bands in the CD spectrum can be characterized in terms of rotational strengths and oscillator strengths. These data may be assumed as a reference state for the β-conformation.

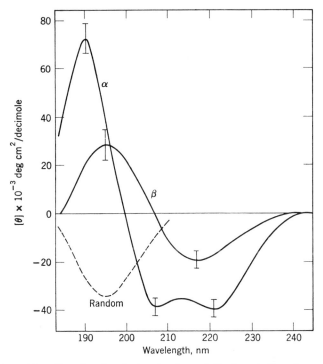

Fig. 39. The CD of poly-L-lysine in the α-helical and β-conformations. The dashed line refers to poly-L-lysine measured at pH 8.0 (i.e., "random coil" conformation).

Figures 40 and 41 show the resolved CD and uv curves of poly-L-serine in 80% trifluoroethanol, respectively. The 222-nm band is due to the $n \to \pi^*$ transition, while the positive band at 198 nm is related to a transition arising from exciton resonance interactions between the monomer $\pi \to \pi^*$ transitions.

CD spectra of films of polypeptides (cast by evaporation of trifluoroacetic acid solution) fall into two categories. In the first (β-I category), the CD spectrum displays a positive band between 216 and 220 nm similar to the CD spectrum of β-structured poly-L-lysine in solution (237). In the second (β-II) category, the CD spectrum shows a very pronounced shift in the high-wavelength negative band which appears at about 228 nm. The positive dichroic band is centered at 198 nm which is similar to its location in the CD spectrum of poly-S-carboxymethyl-L-cysteine in solution (238). CD results obtained with β-structure-forming polypeptides are summarized in Table VII.

From these results it follows that CD spectra of β-structured polypeptides are strongly affected by stabilizing side-chain interactions and steric effects resulting from side-chain bulkiness which cause small variation in the orientation of the transition moments of the β-structured backbone chain.

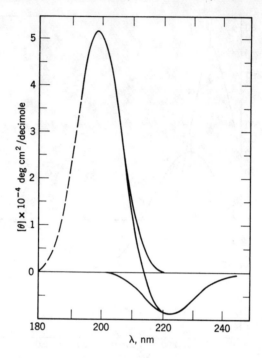

Fig. 40. Resolution of CD curve for poly-L-serine in 80% aqueous trifluoroethanol (236).

TABLE VII

Circular Dichroism Band Positions of β-Structure-Forming Polypeptides

Polypeptide	Band position, nm
Poly-L-lysine	
β-conformation	219 (−); 207 (c-o)[a]; + below 207
α-helix	221 (−); 209 (−); 201 (c-o); 191.5 (+)
Random	202 (−); 220–230 (−) sh
Poly-O-acetyl-L-serine	216 (−); 203 (c-o); 196 (+)
Poly-O-t-butyl-L-serine	220 (−); 212 (c-o); 200 (+)
Polyvaline	217 (−); 203 (c-o); 198 (+)
Poly-S-carbobenzoxymethyl-L-cysteine	228 (−); 217 (c-o); 203 (+)
Poly-S-carboxymethyl-L-cysteine (solution, pH 4.42)	227 (−); 213 (c-o); 198 (+)

[a] Abbreviation used: c-o, crossover point.

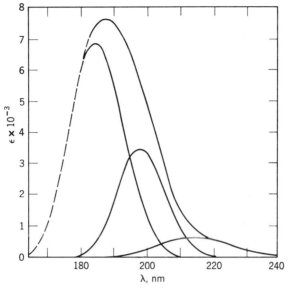

Fig. 41. Resolution of the uv curve for poly-L-serine in 80% aqueous trifluoroethanol (236).

Infrared spectra of polypeptides in the antiparallel-chain β-form are characterized by two amide I bands which have been noted for polyglycine I at 1685 and 1632 cm^{-1} and by one amide II band at 1530 cm^{-1} (239,240). The infrared spectrum of low molecular weight poly-γ-methyl L-glutamate, which exists in the antiparallel-chain β-form, is shown in Figure 42. The amide I bands are located at 1695 and 1629 cm^{-1} and the amide II band is centered at about 1531 cm^{-1}. In the low-frequency region, the strong band at 700 cm^{-1} is shifted, on N-deuteration, to a medium-intensity band at 531 cm^{-1}. From the frequency ratio of 700:531 = 1.31, these bands are identified as the amide V and V' bands of the β-form (142).

The amide I, II, and V frequencies of poly-γ-methyl L-glutamate, poly-γ-benzyl L-glutamate, and poly-L-alanine in the antiparallel β-form and the amide V' frequencies of N-deuterated derivatives are listed for comparison in Table VIII. The amide V bands of the antiparallel-chain β-form are observed in a narrow frequency region of 700–705 cm^{-1} and the amide V' bands are observed in the region of 515–530 cm^{-1}.

Both antiparallel-chain and parallel-chain β-forms exhibit the perpendicular amide I band at 1630 cm^{-1} which is characteristic of the extended conformation. However, it is possible to distinguish between these two structures because the parallel amide I band is observed at ca. 1645 cm^{-1} for the parallel-chain β-form and at ca. 1695 cm^{-1} for the antiparallel-chain

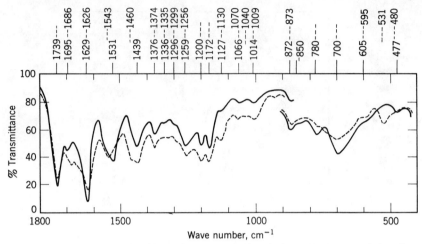

Fig. 42. Infrared spectra for films cast from $CHCl_3$ solutions of poly-γ-methyl L-glutamate ($A/I = 4$) (solid line) and N-deuterated analog (dotted line) (142).

TABLE VIII

Characteristic Amide Frequencies (in cm^{-1}) for the Antiparallel-Chain β-Form of Polypeptides

	I	II	V	V′
Poly-γ-methyl L-glutamate	1629	1531	700	531
Poly-γ-benzyl L-glutamate	1629	1524	a	518
Poly-L-alanine	1632	1548	703	520

[a] Overlapped by the band due to benzyl groups.

β-form. β-Keratin is the only example of a polypeptide chain in the parallel-chain β-form (140).

VI. CONFORMATIONAL TRANSFORMATIONS

A. Helix–Coil and Helix–Helix Transitions

In earlier sections we discussed evidence for the existence of helical and "coil" conformations in solution. Over a decade ago, Doty, Blout, and their collaborators (241) demonstrated that these two conformations interconvert reversibly in solution. They also established the importance of this interconversion in protein denaturation. At this point we will examine some details of helix–coil transformations.

The transition between helix and random coil is sharp, suggesting a cooperative nature of this transformation resembling a phase transition rather than a simple chemical reaction. The cooperative nature of the transition is directly attributable to the interdependence of the conformations of the peptide bonds of neighboring units as discussed in the section on minimum potential energy calculations. The breaking of the first hydrogen bond in a helical sequence does not alter the stereochemistry. It is required that at least three consecutive hydrogen bonds break in the internal structure of the chain to increase entropy. Accordingly, one form will generally be preferred over another, and the relative conformer stabilities depend on thermodynamic variables defining the polymer chain and its environment.

Various attempts have been made to devise a theoretical model in order to explain the experimental observations of the helix–coil transitions. Most of these models are very closely related to each other; hence, we shall not discuss them all in detail (242–253). The major difference in the approaches seems to be in the choice of reference state. For example, Gibbs and Di Marzio choose the helical conformation as a reference state, while Zimm and Bragg choose the coil as the reference state in their model. On the other hand, the treatment of Lifson and Roig is based on the expanded coil as the reference state. For these reasons the notations used by various groups working on the helix–coil transitions do not have the same meaning. Each approach should be considered independently from the standpoint of symbols employed. We shall limit our discussions primarily to the Zimm-Bragg (254) and Lifson-Roig (255) treatments.

1. The Zimm-Bragg Model

This model considers the polypeptide as a chain made up of a combination of helix and coil segments (254). Any state of the polypeptide can be described by a sequence of helix and coil units. For a polypeptide of N units there are 2^N possible states. The relative contributions of each state can be evaluated by assigning the following statistical weight factors to each segment:

a. unity, for any coil unit
b. s, for a helix unit following a helix unit
c. σs, for a helix unit following a coil unit

s is the equilibrium constant for the transformation of a coil segment into a helix segment at the end of a helical sequence. σ is the equilibrium constant for the formation of an interruption in a sequence of helical segments. It should be stressed that the interruption must be such that the number of helical segments remains constant. The factor s is dependent upon external parameters, such as temperature, pH, solvent composition, and pressure.

σ may be considered as a nucleation parameter characteristic of the system which reflects the difficulty of nucleation. In this respect the helix–coil transition is analogous to the equilibrium between a crystal and its liquid. For the helix–coil transition σ is always less than unity, often of the order of 10^{-4}.

Once the two principal parameters, s and σ, are introduced, the partition function Q for a chain of n segments may be obtained from the above assumptions. The partition function for this model is treated in two alternative ways, either as a summation suitable for short chains or in terms of the eigenvalues and eigenvectors of a characteristic matrix:

$$Q = \mathbf{J}^* M^n \mathbf{J} \tag{1}$$

where $\mathbf{J}$ and $\mathbf{J}^*$ are the column and the row vectors, respectively. This approach is more suitable for long chains. The matrix M is of the order $\rho \times \rho$, where $\rho = 2^N$. Therefore, the matrix method can accommodate long-range as well as nearest-neighbor interactions. In the case of the inclusion of long-range interaction ($N = 3$ in the Zimm-Bragg model),

$$M = \begin{vmatrix} 1 & 0 & 0 & 0 & 1 & 0 & 0 & 0 \\ \sigma s & 0 & 0 & 0 & 0 & 0 & 0 & 0 \\ 0 & 1 & 0 & 0 & 0 & 1 & 0 & 0 \\ 0 & s & 0 & 0 & 0 & s & 0 & 0 \\ 0 & 0 & 1 & 0 & 0 & 0 & 1 & 0 \\ 0 & 0 & 0 & 0 & 0 & 0 & 0 & 0 \\ 0 & 0 & 0 & 1 & 0 & 0 & 0 & 1 \\ 0 & 0 & 0 & s & 0 & 0 & 0 & s \end{vmatrix} \tag{2}$$

On the other hand, when one considers nearest-neighbor interactions only ($N = 1$), the matrix of statistical weight, M, is

$$M = \begin{vmatrix} 1 & \sigma s \\ 1 & s \end{vmatrix} \tag{3}$$

and the solution of the characteristic (eq. (3)) gives

$$(1 - \lambda)(s - \lambda) = \sigma s \tag{4}$$

and

$$\lambda = \tfrac{1}{2}\{(1 + s) \pm [(1 - s)^2 + 4\sigma s]^{1/2}\} \tag{5}$$

where λ denotes the eigenvalue of the statistical weight matrix M. Diagonalization of M and substitution in eq. (1) leads to the following:

$$Q = \frac{\lambda_0^{n-2}(\lambda_0 + s) + \lambda_1^{n-2}(s - \lambda_1)}{\lambda_0 - \lambda_1} \tag{6}$$

where λ_0 and λ_1 denote maximum and minimum values of λ as follows:

$$\lambda_0 = \tfrac{1}{2}\{(1 + s) + [(1 - s)^2 + 4\sigma s]^{1/2}\}$$
$$\lambda_1 = \tfrac{1}{2}\{(1 + s) - [(1 - s)^2 + 4s]^{1/2}\}$$

For sufficiently long chains

$$Q \simeq \lambda_c^n \tag{7}$$

The average fraction of units in the helical state is given by

$$\bar{\theta}_h = \frac{1}{n}\frac{\partial \ln Q}{\partial \ln s} \tag{8}$$

The general result applicable to chains of any length is obtained by differentiation of eq. (6) according to eq. (8). In the case of sufficiently long chains, the result can be simplified and given by

$$\bar{\theta}_h = \frac{\partial \ln \lambda_c}{\partial \ln s} = \frac{(\lambda_c - 1)}{(\lambda_c - \lambda_h)} \tag{9}$$

The average fraction $\bar{\theta}_h$ of units in helical form is shown in Figure 43 calculated as a function of s according to eq. (9), with $\sigma = 0.25 \times 10^{-2}$ and $\sigma = 10^{-4}$. It can be seen from these results that the helix–coil transition becomes sharper as σ decreases.

2. Lifson-Roig Model (255)

As in the treatment of Zimm and Bragg, Lifson and Roig apply a matrix representation of a partition function, but differ in choice of matrix elements. The result is the reduction of the order of the matrix of conditional

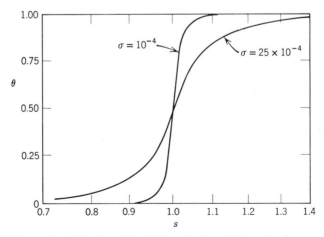

Fig. 43. Hydrogen bond fractions θ for different values of the nucleation parameter σ as a function of the equilibrium constant s (254).

probabilities from eight in the theory of Zimm and Bragg to three in the Lifson-Roig treatment. This simplification in matrix treatment enables them to obtain a solution of the helix–coil transition problem not only in the high molecular weight region, but also in low and intermediate molecular weight ranges. In the Lifson-Roig model the polypeptide chain is defined by specifying the state of each peptide unit as being either in helix or coil form. The contribution of such a state to the partition function can be expressed as a product of n factors, each of which is a statistical factor, u, v, or w, defined as follows:

1. u, a statistical factor for a peptide unit in the coil state;
2. v, a statistical factor for a peptide unit at the beginning or at the end of an uninterrupted sequence of helical states; and
3. w, a statistical factor for a peptide unit at the interior of an uninterrupted sequence of (more than two) helical states.

As an example, a sequence of $n = 15$ may be described by

$$hhhhhccchhchccc$$

and the contribution of this state to the partition function is seen to be

$$vwwwvuuuvvuvuuu$$

The factors u, v, and w represent conditional probabilities of occurrence of their corresponding events. The w corresponds to s in Zimm-Bragg's notation, and v corresponds to $\sigma^{1/2}$. Zimm and Bragg attribute the factor σ to the beginning of a helical sequence, while Lifson and Roig attribute a factor v to both the beginning and the end of a helical sequence.

Once the proper partition function is obtained by summing up all possible values of the state of the peptide units, it can be evaluated by one of the three main approaches described below:

a. The combinatorial approach. In theory, this is the most straightforward approach. It simply involves counting, hoping that some of this counting can be generalized, and as a result, the final expression may represent some of the summations. In practice, however, this approach is the most tedious.

b. The sequence-generating function method. This approach leads to exact solutions of the partition function and requires no more mathematical knowledge than the use of the geometric series. The only drawback is that the results apply only to very long chains.

c. The matrix approach. This approach makes use of the fact that a matrix product generates a sum of terms. Thus, the problem is reduced to finding the proper matrix and performing a product operation. This approach leads to exact expressions for the partition function that apply to chains of any length.

Before we proceed through the discussion of the results of the matrix treatment, we shall touch briefly on the subject of the sequence-generating function method (256).

For the sequence of i helical states, one may write the following sets of expressions which are based on the assignment of statistical weights given by Lifson and Roig (255).

i	v_i
1	v
2	v^2
3	$v^2 w$
⋮	⋮
i	$v^2 w^{i-2}$

We may now assign $v = 1$ to represent one position of rotation. Hence,

$$w = \exp(-\epsilon/RT) \tag{10}$$

where ϵ is the energy of formation of the hydrogen bond.

The expression for the sequence-generating function of the α-helix, $V(x)$, is then

$$V(x) = \sum_{i=1}^{\infty} v_i X^{-i} = \frac{v}{x} + \left(\frac{v}{w}\right)^2 \sum_{i=2}^{\infty} \left(\frac{w}{x}\right)^i = \frac{v}{x} + \frac{v^2}{x(x-w)} \tag{11}$$

Similarly, for the sequence-generating function for a random coil

$$u(x) = u/(x - u) \tag{12}$$

At the point where $x = x_1$, it has also been shown that

$$U(x)V(x) = 0 \tag{13}$$

Substitution of eqs. (11) and (12) into eq. (13) gives

$$\left[\frac{V}{x} + \frac{v^2}{x(x-w)}\right]\left[\frac{U}{x-u}\right] - 1 = 0 \tag{14}$$

which may be expanded to

$$x^3 - x^2(u + w) + x(wu - uv) + uvw - v^2 u = 0 \tag{15}$$

and the fraction of units in the helical state is given by

$$\bar{\theta}_w = \frac{\partial \ln x_1}{\partial \ln w} = \frac{w}{x_1}\frac{\partial x_1}{\partial w} \tag{16}$$

which is the same as the result of Zimm and Bragg (see eq. (7) and Fig. 43).

Turning to the matrix approach, the configurational partition function, Q, of the system may be given by eq. (17),

$$Q = \int_{\phi_1=0}^{2\pi} \cdots \int_{\psi_n=0}^{2\pi} \exp[-\beta V^{(n)}] \, d\phi_1, \ldots, d\psi_n \tag{17}$$

where $\beta = 1/kT$, and $V^{(n)}$ is the configurational energy. The ϕ's and ψ's are the internal rotation angles of N—C$^\alpha$ and C$^\alpha$—C^1, respectively.

The subsequent mathematical procedure leading from the formulation of the configurational partition function, eq. (17), to its final solution is lengthy and will not be reproduced here. It may be viewed as a series of transformations of the state variables which determine the state of the peptide units of the polypeptide chain.

The final expression for the partition function is

$$Q = \sum_{r=1}^{3} Q_r \tag{18}$$

with

$$Q_r = \lambda_r C_r \tag{19}$$

where Q_r is the partition function for the rth state and the λ_r's are the roots of the secular equation. Thus, the whole system can be found, in this representation, in one of the three states $r = 1, 2, 3$, with a probability Q_r/Q for each. The expression $Q_r = \lambda_r n C_r$ suggests that λ_r is the contribution of each unit to Q_r, while C_r is a factor which represents the contribution of end effects to the partition function Q_r. The dependence of these end effects on $\ln w$ is shown in Figure 44.

The relative number or degree of hydrogen bonding, $\theta = \bar{n}_{\text{bond}}/n$, may then be shown by statistical mechanics to be:

$$\theta = \frac{1}{n} \frac{\partial \ln Q}{\partial \ln w} \tag{20}$$

where $\bar{n}_{\text{bond}}$ denotes the average number of intramolecular hydrogen bonds. From the result of eq. (18), eq. (2) may be rewritten

$$\theta = \sum_{r=1}^{3} \frac{\theta_r Q_r}{Q} \tag{21}$$

where θ_r is given by:

$$\theta_r = n^{-1} \frac{\partial \ln Q_r}{\partial \ln w} \tag{22}$$

On the other hand, the number-average length $\bar{l}_n$ of a helical sequence is given by the following:

$$\bar{l}_n = (\bar{n}_{\text{bond}} + 2\bar{n}_{\text{seq}})/(n)_{\text{seq}} = \theta/\eta + 2 \tag{23}$$

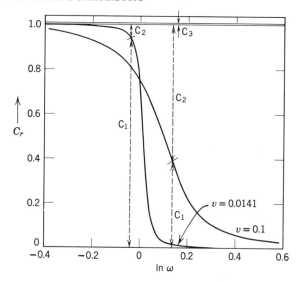

Fig. 44. Dependence of contributions (C_r) of end effects (to the partition function) on the length of uninterrupted sequences of residues in the helical state (ω) for two situations describing the occurrence of residues at the beginning or end of helical arrays (v) (255).

where $\bar{n}_{seq}$ denotes the average number of helical sequences having at least two helical states in a sequence, and $\eta = \bar{n}_{seq}/n$. The term $2\bar{n}_{seq}$ represents the two nonbonded units in the helical state which start and end each sequence. By following the same reasoning which led from eq. (20) to eq. (22),

$$\eta = \sum_{r=1}^{3} \frac{n_r Q_r}{Q} \tag{24}$$

where

$$n_r = n_r^0(1 + n^{-1}\eta_r') \tag{25}$$

and n_r represents the relative number of helical sequences for the rth state.

In summary, the relationship between θ and $\ln \omega$ and the relationship between $\bar{n}_{seq}$ and $\ln \omega$ are shown in Figures 45 and 46, respectively.

B. Kinetics of Transition

Because of the large number of steps involved in a transition and because of its rapid rate, the detailed analysis of the kinetic process is rather difficult. However, it has been possible to obtain the kinetic data for this process with the aid of recently developed fast-reaction techniques, such as the sound-absorption and the temperature-jump methods.

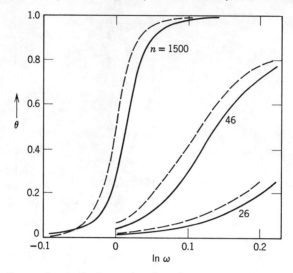

Fig. 45. Intramolecular hydrogen bonding, θ, as a function of $\ln \omega$ for $v = 0.0141$ and $n = 26, 46,$ and 1500. This figure compares the Lifson-Roig approach (solid lines) to that of Zimm and Bragg (dashed lines) (255). (See text for definitions of terms.)

We shall start our discussion from the generalized picture of the growth process (257).

$$\text{cch} \underset{k_B}{\overset{k_F}{\rightleftarrows}} \text{chh}$$

where k_F represents rate constants for the transformation of a residue from the coil to the helical state, k_B is the rate constant for the reverse process, and $k_F/k_B \equiv s$. At the midpoint of transition, $s = 1$ and, therefore, $k_F = k_B$. There are two possible nucleation processes, helix nucleation and coil nucleation.

helix nucleation   ccc $\longrightarrow$ chc

coil nucleation   hhh $\longrightarrow$ hch

It can be shown that the dependence of the mean relaxation time τ^* with respect to s and σ can be derived as follows:

$$1/\tau^* = k_F[(s-1)^2 + 4\sigma] \tag{26}$$

The relaxation time is defined as the time required for the difference between the actual and equilibrium values to be reduced to $1/e$ of its original value. At the midpoint of transition, $s = 1$ and τ^* is at a maximum (258).

$$\tau^*_{\max} = (1/k_F)4\sigma \tag{27}$$

The sharpness of the transition increases with decreasing values of σ as shown

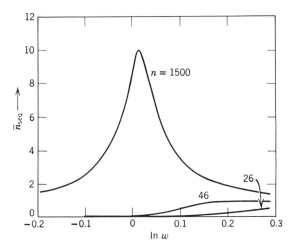

Fig. 46. The dependence of the average number of helical sequences $\bar{n}_{seq}$ on ln w for $v = 0.0141$ and $n = 26$, 46, and 1500 (255). (See text for definitions of terms.)

in Figure 43. Assuming $k_F = 10^{10}$–10^{11} sec^{-1} and using $\sigma = 10^{-4}$ at the midpoint of transition, we find $\tau^*_{max} \simeq 10^{-7}$ sec. The quantities k_F and σ are not easily determined experimentally. The value of τ^* can, however, be obtained experimentally by following the change in fraction of helicity θ with time (258).

$$\frac{1}{\tau^*} = \frac{1}{\Delta\theta}\left(\frac{\partial\theta}{\partial t}\right)_{t=0} \qquad (28)$$

where $\Delta\theta = \theta_\infty - \theta_0$ and $(\partial\theta/\partial t)_{t=0}$ is the instantaneous change in θ at $t = 0$. Since optical activity is rather sensitive to conformational change, the change in θ may be followed by measuring the change in optical activity.

As we have pointed out previously, special techniques are necessary to study kinetics of helix–coil transitions. Relaxation methods may be used for this purpose for reaction times as short as 5×10^{-10} sec (259). In these techniques a system is disturbed from its equilibrium by a sudden change in an external variable of state (temperature, pressure, pH, electric field, etc.). If the rate of change of the variable is fast compared to the rate of the reaction being studied, then the system will be in a state of disequilibrium. The time to reach a new equilibrium can then be measured.

The techniques used in relaxation methods may be classified into two general categories: (*a*) a single displacement method and (*b*) a periodic disturbance method (260). Structural interpretations based on these techniques do not include effects based on molecular weight distribution. It is possible that low molecular weight fractions exhibit significantly different responses from high molecular weight fractions.

Examples of a single displacement technique involve the "temperature-jump method" and "pH-jump method." In an example of the temperature-jump method, the temperature is raised 5–10°C in a few microseconds by a discharge from a high-voltage condensor (260). This method was used to study the helix–coil transition in poly-L-glutamic acid (261). On the other hand, the pH-jump technique utilizes the effect of pH lowering which is caused by a rise in temperature. The lowering of the pH has a much greater effect upon the transition than raising the temperature. This method was tested in the study of side-chain effects on the rates of the helix–coil transition of poly-L-tyrosine (262). In none of the experiments, however, did the authors find a relaxation time long enough to be accessible by these techniques.

More encouraging results are obtained by the utilization of a periodic disturbance. One such method is the ultrasonic absorption method. Chemical relaxation effects from ultrasonic absorption are observed because the relaxations affect the adiabatic compressibility of the system, resulting in a complex contribution to the sound velocity. The velocity dispersion and sound absorption associated with the above effect can be used to measure chemical relaxation times (263). This method was used to measure the relaxation times of poly-L-lysine (264,265). Although a detailed examination of the relaxation times in terms of the kinetics of the elementary steps is difficult because of limitations in the accuracy of the experimental data, it is possible to compute k_F (the rate constant for hydrogen bond formation using the above equations) (263). Values of $k_F = 2.0$–3.0×10^{10} sec^{-1} are obtained for temperatures from 20 to 36°C (using $\sigma = 2 \times 10^{-4}$). These are reasonable values for diffusion-controlled hydrogen bond formation.

Another periodic technique which has been used to study the helix–coil transition is dielectric relaxation (257,263). Schwarz and Seelig studied the effect of an imposed electric field upon the helix–coil transition of poly-γ-benzyl L-glutamate (257). Their results show that $\tau^*_{\max}$ of 5×10^{-7} sec is found by this method, which is in rather good agreement with theory. Using $\sigma = 0.4 \times 10^{-4}$, a value of $k_F = 1.3 \times 10^{10}$ sec^{-1} is obtained (263). This lends support to the idea that helix growth reactions are diffusion controlled. Schwarz also calculated the effect of the electric field on the transition and found that a field of 200 kV/cm causes a change in θ from 0.2 to 0.8. This is a large effect and is indicative of the strong cooperativity of the transition. (See section II, C on nmr of polypeptides.)

C. Thermodynamics of Transitions

Let us now examine the thermodynamic aspect of the transition between helix and coil. The parameter s can be expressed by:

$$s = \exp\left(\Delta F^\circ_{h \to c}/RT\right) \qquad (29)$$

where $\Delta F^\circ_{h \to c}$ is the change in free energy per residue for the conversion of helical to coil form, each in its standard state. It also follows that s should be a function of temperature. Within a narrow temperature range in the vicinity of T_c (transition temperature) at which $s = 1$, s may be considered to be linear in temperature (266).

$$s - 1 = -(\Delta H^\circ_c/RT^2)(T - T_c) \tag{30}$$

or

$$\Delta T = T - T_c = -(RT^2/\Delta H^\circ_c)(s - 1) \tag{31}$$

where standard transition enthalpy, $\Delta H^\circ_c = \Delta H^\circ_{h \to c}$, is the enthalpy associated with the process of "melting" the helix. This will provide, in principle, a direct evaluation of the parameters involved in the transition by calorimetric measurements of ΔH°_m (standard value of calorimetric transition enthalpy).

On the other hand, the maximum value of the transition heat capacity, C_{max}, is defined as (267)

$$C_{max} = \Delta H_m \left(\frac{d\theta}{dT}\right)_{T_c} \tag{32}$$

where θ is the fractional helix content of a polypeptide as we have shown in the previous section. The temperature dependence of θ at T_c can be expressed in terms of the statistical treatments of Zimm-Bragg (254) and of Applequist (268). From an approximate relation between θ and T, as discussed by Applequist (268),

$$C_{max} = (\Delta H_m)^2/4RT_c^2 \sigma^{1/2} \tag{33}$$

or

$$\sigma^{1/2} = [(\Delta H_m)^2/4RT_c^2]C_{max} \tag{34}$$

where R is the gas constant and T_c is the transition temperature (°K). The variability of θ with temperature at T_c can be readily derived from eqs. (32) and (33) as

$$\left(\frac{d\theta}{dT}\right)_{T_c} = \frac{\Delta H_m}{4RT_c^2 \sigma^{1/2}} \tag{35}$$

The so-called van't Hoff heat of transition, ΔH_v, is correlated with the calorimetric transition enthalpy, ΔH_m, by (269):

$$\sigma^{1/2} = \Delta H_v/\Delta H_m \tag{36}$$

At the transition temperature, T_c, $s = 1$ and, therefore, the van't Hoff relation

$$\ln s = \frac{\Delta S_m}{R} - \frac{\Delta H_m}{RT}$$

is reduced to

$$T_c = \Delta H_m/\Delta S_m \tag{37}$$

where ΔS_m is the conformational entropy change related to s.

One of the most significant contributions of the thermodynamic treatment is the determination of σ-values from experimental data.

However, thermodynamic data on helix–coil transitions are even more scarce than kinetic data. This is due to the experimental difficulties associated with delicate measurement of the small changes in enthalpy. The Calvet differential microcalorimeter overcomes these difficulties (270). Differential microcalorimetry was used in the measurements of the heat of solution of poly-L-glutamates as a function of solvent composition; the results are shown in Figures 47 and 48. The abrupt increase in the heat of solution at the solvent composition of the helix–coil transition (as evidenced by optical rotation data) allows the estimation of the transition enthalpy change. The measured value of the heat of transition is in agreement, in sign and in order of magnitude, with previous theoretical estimates based on the model of Zimm, Doty, and Iso (271).

In the past few years, the interest in this area of research has multiplied. For example, Karasz and O'Reilly, as well as Ackermann (267), have made a series of calorimetric determinations on poly-γ-benzyl L-glutamate (272–274) and other polypeptides (269). They demonstrated the dependence of T_c on solvent composition and the critical analysis of σ-values in relation to the thermodynamic findings.

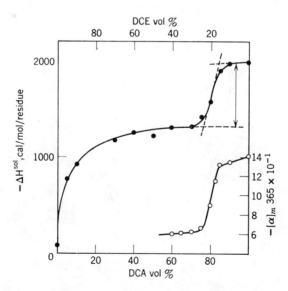

Fig. 47. Values for heats of solution (filled-in circles) and optical rotations (open circles) of poly-γ-benzyl L-glutamate in dichloroethane-dichloroacetic acid (DCE–DCA) mixtures (270).

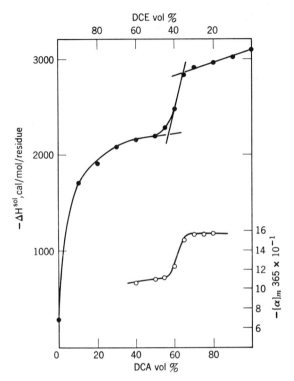

Fig. 48. Values for heats of solution (filled-in circles) and optical rotations (open circles) of poly-γ-ethyl L-glutamate in dichloroethane-dichloroacetic acid (DCE–DCA) mixtures (270).

D. Helix–Helix Transition in Poly-L-proline

As noted earlier (Sect. II), forms I and II of polyproline are interconvertible. In a simple theoretical analysis of the polyproline I–polyproline II system, Engel has applied the Zimm-Bragg theory (275). In this application the propagation step is taken to be the addition of a helix I unit to an existing helix I sequence. The nucleation step involves the formation of a helix I unit between helix II segments. Nucleation is difficult due to the high steric strain at the nucleation site. A σ of approximately 10^{-5} has been estimated by Engel.

We can obtain information about the kinetics of this reaction from a simple relaxation experiment. Since the relaxation times for poly-L-proline are long, there is no need to use the techniques needed for the helix–coil transition. An equilibrated poly-L-proline solution may be disturbed from equilibrium by the addition of a solvent which will shift the equilibrium

(Fig. 49). We can then obtain τ^* (from eq. (28)). Changes in θ_1 are followed by changes in optical rotation. A $\tau^*_{max} = 1000$ min is obtained. Using $\sigma = 10^{-5}$, a value of $k_F = 10^{-1}\,\text{sec}^{-1}$ is found. We see that this reaction is many orders of magnitude slower than the helix–coil transition (Fig. 50). The high degree of cooperativity is indicated by the molecular weight dependence of the transition curve (Fig. 49) and the dependence of the mean relaxation time on degree of conversion and molecular weight.

VII. CALCULATION OF MOST PROBABLE CONFORMATIONS

In the previous section, the dynamic aspects of polypeptide conformation have been discussed. In this section, we shall concern ourselves with the static aspects of polypeptide conformation, that is, the examination of single polypeptide units by both theoretical and instrumental methods.

For a number of years after the evolution of the concept of the α-helix, studies were mainly concerned with elucidation of structures by experimental means. Recently, however, the limits of usefulness of these experimental techniques have begun to be realized and a reexamination has begun of the fundamental aspect of Pauling's concept of α-helix and polypeptide conformation in general from the point of view of the forces governing their molecular architecture.

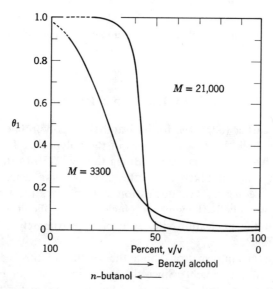

Fig. 49. Mutarotation of polyproline form II into form I as a function of the solvents n-butanol—benzyl alcohol. Transitions for two different molecular weight samples are shown (275).

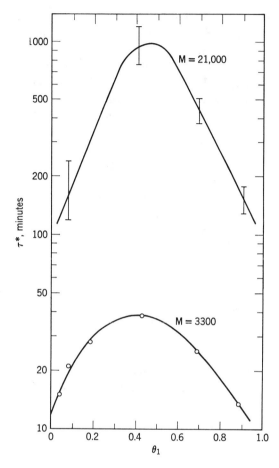

Fig. 50. The average reciprocal relaxation time, τ^*, dependency on the conversion of poly-L-proline I into II, θ_1, at 70° after change in solvent and establishment of new equilibrium (275).

We shall present here a brief picture of the recent advances in some of the theoretical approaches involving potential energy calculations as applied to polypeptide conformations. In doing so, we shall attempt to reproduce experimental data regarding conformational equilibria by *a priori* calculations, both for the purpose of understanding the factors involved in polypeptide conformations and for the better understanding of future discussions. It must be emphasized that this is a fundamentally correct approach, even though the agreement with experiment has not been spectacular, it has been sufficient to encourage the development of better *a priori* methods.

Let us begin our discussion on the theoretical consideration of polypeptide conformations by describing a simplified picture of a section of polypeptide chain and the parameters defining such a model.

The conformation of the backbone of a polypeptide chain can be specified by giving a spatial description of a "dipeptide model" as shown in Figure 51 (276) (see also Table XII for comparison of various notations). Rotation around the N—C^α bond is denoted by ϕ, rotation around the C^α—C' bond is denoted by ψ, and rotation around the peptide (C'—N) bond is denoted by ω. The bond lengths and bond angles of each peptide unit are held fixed, and the amide group is fixed in the planar *trans* conformation (the angle of rotation ω about the C'—N bond being fixed at 0°) because of its partial double bond character. Therefore, the conformation representing the planar *trans* form of the fully extended polypeptide chain is characterized by the rotational angles $\phi_i = \psi_i = \omega_i = 0$. This considerably simplifies the theoretical treatment of polypeptide conformation.

Since each peptide unit has two internal angles of rotation, ϕ_i and ψ_i, the conformation of any polypeptide can be characterized by specifying $2n - 2$ sets of angles. In principle, then, one can calculate the probability of a molecule having a particular conformation if the internal energy of the molecule can be expressed as a function of a coordinate $E(\phi, \psi)$.

$$E(\phi_{i-1}, \psi_{i-1}, \phi_i, \psi_i, \phi_{i+1}, \psi_{i+1})$$

In practice, such calculations have been made satisfactorily only for a few simple systems, such as a random coil or a helix. In these systems, the major interactions are often limited to nearest neighbors or to atoms close to each

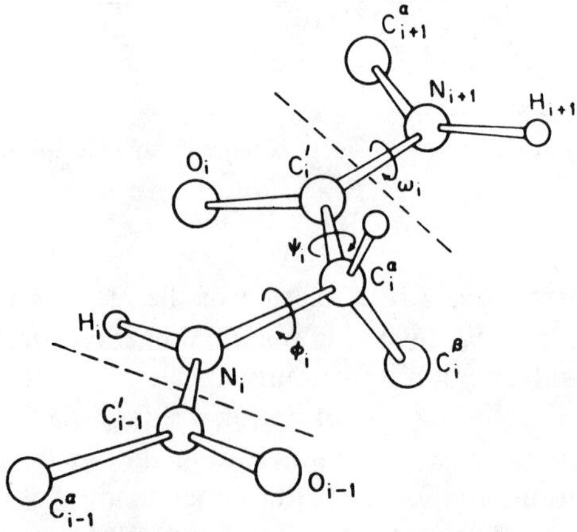

Fig. 51. Drawing of two residues of section of polypeptide chain in perspective. Peptide units are shown by dashed lines. The convention for backbone atoms and internal rotations is indicated (276).

other. Therefore, the mutual orientation of successive bonds can be described in terms of interaction energies between nearest neighbors or between certain residual groups in the chain.

Once the proper model has been designated according to the foregoing presentations, one has to elaborate the possible contributors to the potential functions describing such a system. If the system were an unperturbed random coil of polypeptide chain, one might be able to describe the possible contributors to the potential functions as follows (277): (*a*) inherent torsional potential attributable to the covalent bonding; (*b*) van der Waals interactions between nonbonded atoms within or directly connected to the chain backbone; (*c*) electrostatic interactions of polar groups in the chain backbone; (*d*) interactions between the respective side chains on two different amino acid units; (*e*) interactions between an amino acid side chain and the polymer backbone. Also, the effects of angle deformation may sometimes be considered. But for an unperturbed random coil only the first three factors are major contributors to the potential energy. On the other hand, for helical polypeptides, additional factors, such as contributions from hydrogen bonding, solvent–polymer interactions (including hydrophobic interactions), and dipole–dipole interactions between side chains and polymer backbone, should be included (189,278). Steric factors are among the most important factors in determining polypeptide conformation.

The detailed theoretical treatment of the above-mentioned factors contributing to potential functions may be found in the recent works of Scheraga and co-workers (279,280), Flory (266), Ramachandran and co-workers (20,281–283), Liquori (284), and others (285,286). It should be stressed again that, to evaluate potential minima, both the steric restrictions and the rotational potential functions are taken into account simultaneously.

After these steps in expressing proper potential functions are completed, one may evaluate potential energy minima by following laborious steps of programming, computation, and successive minimization. Computer programs for these steps are well established and are used in various laboratories throughout the world.

The results of these potential energy calculations are usually expressed in the form of energy contour maps. Some typical examples are shown in this section.

A. Construction of Steric Maps

As we have pointed out in the previous section, the importance of steric effects must be considered first before we attempt to evaluate potential functions. The reason is that the stereochemical criteria alone restrict the extent of conformational freedom of a polypeptide chain so severely that one can effectively neglect a large portion of the steric map as not allowed. This will not only give some insight into the problem of a polypeptide conformation, but may also save many steps in computing potential minima.

Pauling and Corey (1) have shown in their original paper that, of the thirty-six theoretically possible "stable conformations," only fifteen are allowed for the formation of regular repeating structures (helices or sheets) because of simple steric reasons. Mizushima and Shimanouchi (287) excluded four more of the above for the same reason.

However, Ramachandran and his associates (282,283) have pursued this subject in more detail from a different point of view. They disregarded the potential minima and evaluated the ranges of ϕ and ψ for which the atoms in two successive peptide residues do not lie closer than the van der Waals contact distances. They selected two sets of these contact distances, corresponding to "normal" and "outer limit" values which they determined by graphical methods (282). It is interesting to note that, of the 36 conformations selected by Pauling and Corey, only five fall inside the fully allowed ("normal") region computed by Ramachandran et al. (282) and twelve fall in the outer limit region.

Scheraga and co-workers (279,280) extended Ramachandran's approach to various di- and tripeptides using mathematical and computer methods. Some of the typical steric maps by these authors are shown in Figures 52–55.

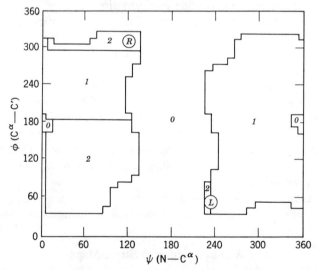

Fig. 52. Energy map showing allowed conformations for gly-gly dipeptides (regions *1* and *2*) and gly-L-ala dipeptides (region *2* only). No dipeptide can assume conformations in regions marked *0* because of overlap of backbone atoms. Right-handed and left-handed α-helices are noted by the letters R and L, respectively. The size of the circles is the approximate range of the angles found experimentally for helix internal rotation angles (280).

POLYPEPTIDE STEREOCHEMISTRY   143

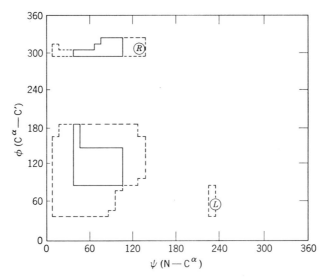

Fig. 53. Map indicating allowed conformations for two dipeptides gly-L-val and gly-L-ileu. The area enclosed by the dotted lines indicates allowed conformations for gly-L-ala (280).

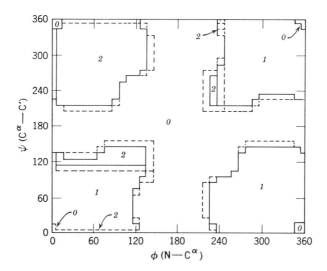

Fig. 54. Steric map drawn according to standard convention (276). The areas enclosed by solid lines denote those internal rotation angles allowed when bond angle N—C$^\alpha$—C′ = 109.5°, and dashed lines show allowed values when the bond angle N—C$^\alpha$—C′ = 112° (280).

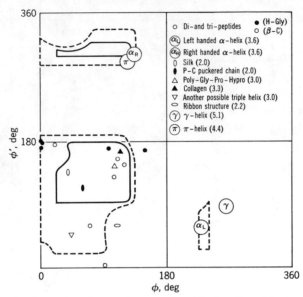

Fig. 55. Steric map based on various peptides. The location of the internal rotation angles ϕ and ϕ' for various biopolymers is listed. All polymers exhibit ϕ and ϕ' values in or close to the allowed regions of the map (282).

The effect of variations in the size and shape of the side-chain groups are discussed by these authors in great detail. It is noteworthy that steric restrictions arising from the backbone atoms alone permit the peptide groups adjacent to glycyl residues to assume only about 50% of all conceivable conformations. An alanine side chain limits these to 16%, and valyl and isoleucyl side chains further reduce them to about 6% of all possible conformations. The rigid ring structures in prolyl peptides provide such severe steric restrictions that a *trans* polyprolyl chain can exist only in a left-handed form of the type observed experimentally for collagen II.

B. Potential Functions

To evaluate potential minima, both the steric restrictions and the rotational potential functions are taken into account simultaneously. Contributions to the potential functions for rotations about the polypeptide chain have been cited previously. In the following, we discuss these contributions briefly.

1. Torsional Potentials

As Brant and Flory (277) have pointed out in their recent work, knowledge about torsional potentials for rotations about the polypeptide bonds C^α—N and C^α—C' is hampered by a scarcity of information, which is largely limited to data obtained by microwave spectroscopy on analogous low molecular weight compounds. For rotation about the bonds C^α—C', the best analogous small molecules are those having the structure CH_3—C—X, where X can be H, F, Cl, Br, CN, OH, CH_3, or C_2H_5. The barriers to rotation about the C—C bond in all of these molecules are known to be threefold, with three minima of equal energy and three maxima of equal energy (288). On the basis of these data, the barrier to rotation about the C^α—C bond is assumed to be threefold with minima at $\psi = 0°$, $120°$, and $240°$. The barrier height ranges from 0.5 to 1.3 kcal/mol. Scott and Scheraga (289) estimate 1.1 kcal/mol for the average value, while Brant and Flory (277) use a barrier of 1.0 kcal/mol [e.g., see the excellent discussion on this subject by Schellman and Schellman (84)]. The torsional potential function for the C^α—C' bond is represented by

$$V(\psi) = \tfrac{1}{2} V^0_\psi (1 - \cos 3\psi)$$

where V^0_ψ is the torsional barrier height (0–0.5 kcal/mol).

The lack of experimental data on the appropriate analogous small molecules makes evaluation of the rotational barrier of the C^α—N bond even more difficult (277,278). Based on the available microwave data of some small molecules and on the assumption of 50% double bond character of the peptide bond, both Flory and Scheraga estimate the torsional contribution to the potential about the C^α—N bond to be threefold, with a barrier height $V°$ of between 1.0 and 2.0 kcal/mol. The favored form of the threefold torsional potential function of the C^α—N bond is given by

$$V(\Phi) = \tfrac{1}{2} V^0_\Phi (1 + \cos 3\Phi)$$

Despite the lack of our present knowledge of torsional potentials, it seems certain that the contribution of this factor to the total rotational potential is rather small or at least not decisive. DeSantis and co-workers (290), for example, neglect C^α—N torsional potentials completely in most of their calculations.

2. Van der Waals Interactions

In view of the importance of steric factors involved in determining polypeptide conformations, careful attention has been given to all the available potential functions by researchers in this field. However, rigorous treatment of this subject is difficult because of the ambiguity inherent in defining the parameters involved and because of the inaccessibility of experimental data applicable to the existing potential functions.

Both Scheraga's and Flory's groups tried the application of two different potential functions: (a) the 6-exponential function (or so-called Buckingham potential) and (*b*) the 6-12 potential function (or Lennard-Jones potential). Modified 6-exponential functions (289,291) are given by

$$V_{ij} = A_{ij} \exp(-b_{ij}r_{ij}) - (C_{ij}/r_{ij}^6)$$

where r_{ij} denotes the separation of the centers of atoms i and j. Values for the parameter C_{ij} have been estimated, using the Slater-Kirkwood equation, from the polarizabilities of the atoms and the effective number of electrons N_{eff} in the outer electronic subshell as suggested by Pitzer (292). Reliable values of b_{ij} are not available for most atom pairs. Brant and Flory (277) use $b_{ij} = 4.60$ based on the limited information available from the molecular-beam scattering studies of Amdur and co-workers (293,294). The parameter a_{ij} is obtained by requiring V_{ij} to be a minimum at a distance r_{ij}° equal to the sum of the van der Waals radii of the interacting species.

On the other hand, the Lennard-Jones 6-12 potential function is given by

$$V_{ij} = d_{ij}/r_{ij}^{12} - e_{ij}/r_{ij}^6$$

where d_{ij} and e_{ij} correspond to a_{ij} and c_{ij}, respectively, as in the 6-exponential function shown previously. In both equations the first term represents a repulsive force due to steric overlap at shorter distances, and the second term represents the attractive force between two atoms at longer distances and is sometimes called the London dispersion force. In his recent paper (295), Flory and co-workers examine the validity of these equations in their application to conformational analysis of polypeptides and point out that there is no real justification for adopting Lennard-Jones 6-12 potential function. However, more and more people have been using the Lennard-Jones potential function in recent studies, including Flory and co-workers (295) and Scheraga and co-workers (189,278). Preferential usage of this potential function is, as Scheraga points out, primarily due to its considerable advantage for writing computer programs rather than to any inherent superiority in expressing van der Waals interactions.

Liquori and his co-workers (290), on the other hand, use both types of equations, depending upon the type of pairwise interaction involved. For example, they prefer Lennard-Jones potential functions for C—C interactions, while they prefer the 6-exponential function for most other pairwise interactions between nonbonded atoms.

3. Electrostatic Interactions

The contribution of dipole–dipole interactions between the polar amide groups of the backbone chain has been neglected until recently. Brant and Flory (277) tested inclusion of this effect in conformational energy calculations and demonstrated its significance by comparing their calculated value of the characteristic dimensionless ratio $r_0^2/n_p l_p^2$ to the experimentally obtained value (296). Here, r_0^2 denotes the mean square end-to-end distance, while n_p is the degree of polymerization and l_p is the fixed distance of 3.80 Å between the α-carbons of the *trans* peptide repeating units in the chain. They used the dipole–dipole type potential function (group dipole) in their earlier calculations (277,297). But more recently, they employed a so-called monopole approximation method which was first introduced by Bradley and co-workers (298,299) [see also Bradley's pioneering work on this subject in relation to the potential energy calculations of the nucleotides (300,301)]. By using partial charges, the dipole–dipole interactions can be calculated with a Coulomb's law type potential function between the partial charges q_i and q_j separated by a distance r_{ij}

$$V_{el}, ij = 332.0 \sum \frac{q_i q_j}{D r_{ij}}$$

where D is the effective dielectric constant. Scheraga and co-workers (189,278) adopted Flory's approach and extended it to peptides containing side-chain dipoles, such as aspartate and glutamate.

4. Hydrogen Bonding

Scheraga and co-workers (189,278,302) used the modified potential function of Lippincott and Schroeder (303,304) to represent the hydrogen bonding between the NH and CO groups of the backbone. They have presented detailed discussions of this subject (189,278). They point out that a reasonable representation of hydrogen bond strength under favorable conditions can be achieved with the modified equation of Lippincott and Schroeder, but it is not well suited for borderline cases, such as poor hydrogen bonding and no bonding at all.

5. Other Interactions

Flory and co-workers (277,295) considered the dependence of electrostatic interaction energy on coil-promoting solvents of relatively high dielectric constant. In a recent study, Scheraga and his associates considered solvent effects with respect to the influence on V_{el}. In a series of articles Scheraga and Poland discuss various aspects of interactions, such as hydrogen bonding of solutes in water, and the binding of water to polypeptides (305), hydrophobic bonding in random coil (306), and other interactions (307–309). A practical application of these results in calculation of polypeptide conformation has appeared in which the conformations of polyalanine in water are predicted and experimentally observed (309a).

C. Conformations of Homopolypeptides

In the foregoing sections we have discussed various factors involved in calculating polypeptide conformation and in constructing a steric map. In the following, we shall discuss some of the results obtained by this theoretical treatment on several isolated (single-stranded) homopolypeptides.

1. Polyglycine

The energy contour diagrams as a function of dihedral angles Φ, ψ have been constructed for this polypeptide by Liquori and co-workers (290), Scott and Scheraga (278), and Flory and co-workers (295). The steric map of this polypeptide by Ramachandram and Scheraga has been mentioned before (see Fig. 52). Figure 56 shows the energy contour map of polyglycine recently published by Flory and co-workers (295). Table IX contains data on dihedral angles and available helical parameters compiled from the literature.

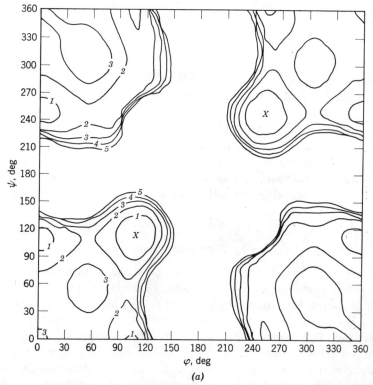

Fig. 56. Contour map of the polyglycine chain according to the treatment by Flory (295). The most stable conformation corresponds to the internal rotation angles denoted by X (295). (a) Neglecting the electrostatic term.

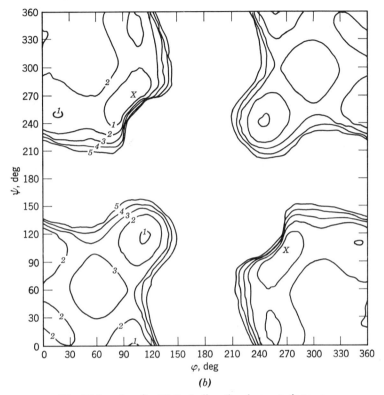

Fig. 56 (*continued*) (*b*) Including the electrostatic term.

Although there is little resemblance among the contour maps of Scheraga, Flory, and Liquori, the dihedral angles corresponding to right- and left-handed α-helix regions obtained from these maps are in good agreement with each other (except for Flory's data). These data are also in close agreement

TABLE IX

Helical Parameters and Angle of Rotation of Polyglycine

Assignment	$\tau(NC^\alpha C')$	$(N—C^\alpha)$	$(C^\alpha—C')$	n	h	Ref.
α-Helix	109.5	125	130	3.63	1.51	290
α-Helix	109.5	125	130	—	—	278
Polyglycine II (threefold helix)	109.5	328	120	3.07	2.70	290
Extended anti-parallel chain	109.5	30	0	2.31	3.60	290

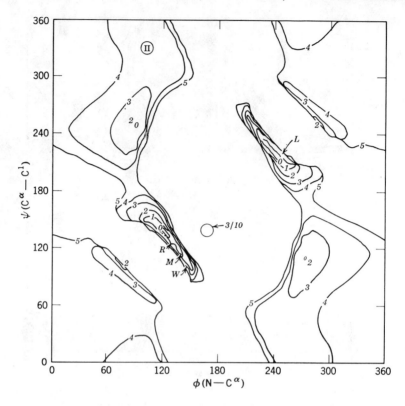

Fig. 56. (*continued*) (*c*) Contour map of the polyglycine chain according to the treatment by Scheraga (280). Included in this map are designations for the internal rotation angles of the right-handed and left-handed α-helix, the helix of myoglobin, the ω-helix, 3/10-helix, and that for the polyglycine II structure (280).

with the crystallographic and other experimental data shown in Table X. One of the most outstanding features of these maps is that there is complete centro-symmetry in these figures.

Two opposing views have been advanced in interpretation of the above results in connection with the stability of the helical conformation of polyglycine. Adopting Schellman's (245) view, Scheraga and co-workers (280,314) interpret the results as follows. When no side chain is present, as in polyglycine, left-handed and right-handed α-helices are equally probable (on steric grounds). Thus, in Figure 56c (and also in Figs. 56a and 56b) the two large areas permitted for glycyl–glycine (or polyglycine) are centro-symmetrical about the planar conformation. This constitutes a destabilizing factor for the helical conformation because each residue has equal intrinsic probability of being in the right-handed or the left-handed helical conformation. On the

POLYPEPTIDE STEREOCHEMISTRY 151

TABLE X
Observed and Postulated Conformations (Φ, ψ) for Polypeptide and Protein (20)

Polypeptide or Protein	$\tau(NC^{\alpha}C')$, deg.	$\Phi(N-C^{\alpha})$, deg.	$\psi(C^{\alpha}-C')$, deg.	n	h	Ref.
α-Helix (3.6$_{13}$)	109.5	133.0	122.8	3.615	1.495	5
γ-Helix (5.1$_{17}$)	110.1	−96.3	258.0	5.143	0.98	2
2$_7$-α-Helix	111.3	105.1	249.5	2.000	2.80	5
2.2$_7$-Helix (ribbon structure)	111.6	10.19	239.2	2.169	2.75	310
3.0$_{10}$-Helix	111.6	130.7	154.3	3.000	2.00	310
4.3$_{14}$-Helix	100.5	−91.9	271.7	4.337	1.20	310
π-Helix (4.4$_{16}$)	114.9	122.9	110.3	4.40	1.15	311
Polyglycine II	109.1	102.0	325.8	−3.00	3.10	312
Poly-L-proline II	110.0	102.8	325.9	−3.00	3.12	159
Poly-L-proline II	108.8	103.7	325.1	−3.00	3.12	158
Poly-L-hydroxyproline A	105.5	103.1	327.6	−3.00	3.05	160
ω-Form of poly-β-benzyl L-aspartate	109.9	115.6	235.4	−4.00	1.325	215
Collagen-type helix	110	116	325.0	−3.28	2.95	281
Silk	110	40	315.0	2.00	3.45	313
P—C puckered chain	110	57	298.0	2.00	3.3	214

other hand, Flory and co-workers (295) have approached this problem from a different angle. They calculate the ratio $r_0^2/n_p l_p^2$ for polyglycine and compare it with that of poly-L-alanine. This ratio is approximately equal to 2 for polyglycine, and for poly-L-amino acids containing the side chains —CH$_2$R the value of this ratio is about 9. They also find that this ratio for polyglycine is rather insensitive to changes in the electrostatic term. In addition, they find that the calculated values of ΔS_{conf} (the configurational entropy) are nearly the same for both polyglycine and poly-L-alanine. Therefore, they conclude that the less extended character of the polyglycine chain results from the chain symmetry rather than from a greater number of accessible chain conformations.

2. Poly-L-alanine

The symmetry of the contour diagram shown for polyglycine is completely lost when a methyl side chain is introduced at the α-carbon. This result is shown in Figure 57. Poor agreement in the general shape of the contour diagram is found among different researchers. However, there are only minor differences in assignment of dihedral angles between Scheraga

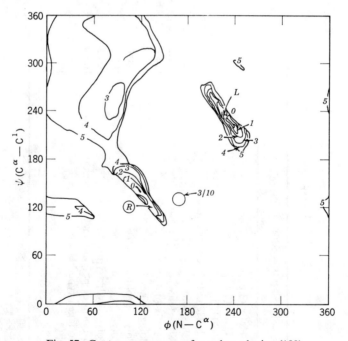

Fig. 57. Contour energy map for poly-L-alanine (189).

and co-workers (189) and Liquori and co-workers (290). Table XI shows the data on dihedral angles compiled from the available literature.

TABLE XI

Helical Parameters and Angle of Rotation of Poly-L-Alanine

Assignment	$\tau(NC^\alpha C')$	$(N—C^\alpha)$	$(C^\alpha—C')$	n	h	Ref.
α-Helix (R)	109.5	120	132	3.70	1.46	290
α-Helix (L)	109.5	232	228	3.45	1.58	290
α-Helix (R)	109.5	132	123	—	—	189
α-Helix (L)	109.5	228	237	—	—	189

The energy contour map indicates that the most stable helical conformations of poly-L-alanine occur in the vicinity of the right- and left-handed α-helices. Furthermore, careful studies by Scheraga and co-workers (189) have shown that the right-handed α-helix is more stable than the left-handed one by 0.4 kcal/mol per residue.

3. Poly-L-valine

Computation of complete potential functions for this polymer involves additional complications chiefly due to the bulkiness of the side-chain isopropyl group attached to the α-carbon and its added rotational freedom. Scheraga and co-workers (189), in their recent report on poly-L-valine, have stated that the right-handed α-helical form of poly-L-valine is the most stable structure. They have argued that the α-helical backbone conformation can accommodate the L-valyl side chain if the latter is rotated about 10–15° away from a position of the torsional potential minimum. Thus, they have stated that the steric strain from the bulky valyl side chain can be relieved at the expense of an increase in the side chain torsional energy. They have found that the increase in this torsional energy due to such rotation is more than compensated by the resulting decrease in nonbonded interaction energy. As additional evidence for their prediction, they have presented their experimental data on a block copolymer of the type $(DL-Lys)_x$-$(L-Val)_x$-$(DL-Lys)_x$ where $x \sim 40$ (315). The optical rotatory dispersion for the helical conformation shows Cotton effects at 233 and 198 nm. The magnitudes of the trough and peak ($m'_{273} = -13{,}500 \pm 500°$ and $m'_{198} = +57{,}000 \pm 6000°$ respectively) of the Cotton effects are consistent with the α-helical structure.

At the present time, this is still a controversial subject, since all the other available experimental data suggest that poly-L-valine cannot exist in the form of an α-helix (316,317).

4. Poly-β-methyl L-aspartate and Poly-γ-methyl L-glutamate

Because of the rotational freedom of the side chains of these peptides, there are many potential energy minima, even if the conformations of the main chain are restricted to fixed values of Φ and ψ in the region of the α-helices. Therefore, calculations by Scheraga and his co-workers (189) have been confined to the regions of the right- and left-handed α-helices to see whether the stability of helices with different screw senses can be predicted. Regardless of the validity of potential functions used in the calculation of these polypeptide conformations, their results are in good agreement with what is known about these polypeptides from experimental data. In the case of poly-β-methyl L-aspartate, the left-handed α-helix is found to be more stable than the right-handed α-helix by 0.1 kcal/residue, but for poly-γ-methyl L-glutamate, the right-handed α-helix is found to be more stable than the left-handed α-helix by 0.4 kcal/residue.

The difference in the screw sense of these two polymers has been studied extensively by Goodman and his co-workers (318,319). They have shown that the mutual interactions of side chains can be the determining factor in the choice between two backbone conformations which differ only slightly in energy (69). One interesting feature of Scheraga's study (189,305) is that the origin of the effect on screw sense lies in the interaction between the dipole of the ester group of the side chain and that of the peptide group of the main chain. This view supports the dipole moment studies of Wada (320) and Hashimoto et al (107). The schematic representation of this view is shown in Figures 58 and 59.

5. Other Polypeptides

Poly-L-tyrosine (189,302), poly-L-proline (280–283,290), and poly-N-methyl-L-alanine (321,322) have also been investigated by the potential energy calculation method.

As we have discussed in the previous section, in the case of poly-L-tyrosine, the ORD curve is complicated by the presence of optically active side-chain transitions which are responsible for an observed positive value of b_0. This would indicate the presence of a left-handed α-helix. Fasman and co-workers (67) were able to resolve this complicated ORD spectrum by careful study of the CD spectrum which consists of three ellipticity bands. However, this view is seriously challenged by Applequist and Mahr (191), who studied the conformation of poly-L-tyrosine in quinoline by dielectric measurements. They have studied the dipole moment effect of bromine substitution at the 3 position of the tyrosine residue (*ortho* to the hydroxyl group) and have observed a decrease in dipole moment of 1.0 ± 0.1 D per residue, which they ascribe to a left-handed helical conformation. This assignment is

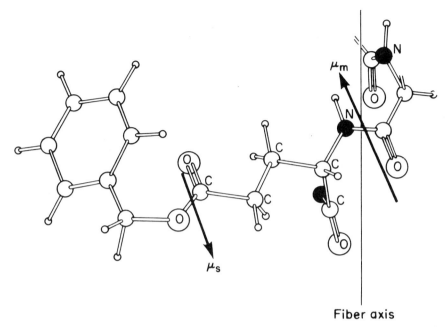

Fig. 58. A schematic representation of the side chain for poly-γ-benzyl L-glutamate and the orientation of its dipole moment (320).

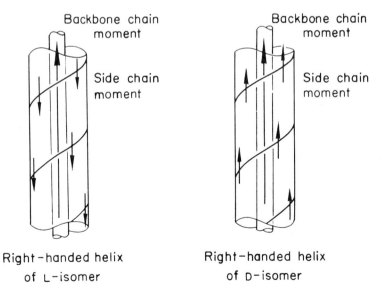

Fig. 59. A schematic representation of the orientation of polar groups in the main chain and side chain for the right-handed helices based on L- and D-amino acids (320).

TABLE XII
Comparison of Various Notations for Rotational Angles

N—C^α	C^α—C'	Positive direction of rotation[a]	Reference
Φ	ψ	Clockwise	Standard convention, (276)
Φ	Φ'	Clockwise	Ramachandran (282,283)
ψ	Φ	Clockwise	Scheraga (279,280)
τ_{NC}	$\tau_{CC'}$	Counterclockwise	Schellman (84)
Φ'''	Φ'	Clockwise	Flory (277)
$\psi_{C_\alpha N}$	$\psi_{C_\alpha C}$	Clockwise	Liquori (290)

[a] looking from N to C^α or C^α to C'.
$\Phi = \Phi \psi = (180° - \tau_{NC}) = \Phi''' = (\psi_{C_\alpha N} \pm 180°)$
$\psi = (\Phi' \pm 180°) - (\Phi \pm 180°) = (180° - \tau_{CC'}) = \Phi' = (\psi_{C_\alpha C} \pm 180°)$

opposite to that made by Fasman and co-workers. Noticing this complication, Scheraga and co-workers (189) have calculated the potential minima around the right- and left-handed α-helical regions of this peptide. They have shown that the right-handed α-helix of poly-L-tyrosine is more stable than the left-handed α-helix by 1.7 kcal/mol.

In the case of poly-L-proline, the rigid ring structures provide such severe steric restrictions that a *trans* polyprolyl chain can exist only in a left-handed helical form (280–283). Liquori and co-workers (290) have calculated the potential minima of both the *cis* and the *trans* conformations of poly-L-proline and were able to assign parameters corresponding to poly-L-proline I and poly-L-proline II.

Other polypeptides of theoretical interest are those of *N*-substituted poly-L-amino acids, since such peptide chains, substituted at the nitrogen atom, are incapable of hydrogen bonding. Conformational energies of poly-*N*-methyl-L-alanine have been calculated by two independent groups, Mark and Goodman (321,323) and Liquori and DeSantis (322). Their results lead to essentially the same conclusion, namely, that both the right-handed threefold helix and the slightly distorted left-handed α-helix are located at potential minima. Liquori and DeSantis, however, suggest that the right-handed threefold helix may be the preferred conformation.

VIII. CONCLUSIONS

Polypeptides have been studied as model systems for the analysis of protein conformations. They have provided a means for understanding the structural implications of spectroscopic findings, conformational transitions, and molecular geometries consistent with conformational potential energy calculations. It is to be hoped that such results with synthetic models will be applicable to problems of protein chemistry.

In writing this report, we have attempted to provide a broad overview of the field. The principles used to deduce conformation have been elaborated and applications of methodology to specific polypeptides have been discussed. Criteria of helicity are numerous. As a consequence, it has been desirable to present a picture of how the many techniques to deduce stereoregularity fit together. It is a more difficult task to discuss nonregular forms. Workers in the peptide conformational field have defined nonhelical forms as "random coils." It is known that polar materials, such as poly-α-amino acids, cannot be truly random. Each specific nonregular polypeptide system can be viewed as being composed of different mixtures of preferred conformations. Since this is so, one cannot obtain direct relationships between structure and spectroscopic properties for random coils.

It appears that many globular proteins are composed of substantially nonregular conformations. Yet each has a specific stereochemistry which is intimately associated with its function. We believe that future successes in research in this field will emanate from studies of stereochemically nonregular polypeptides. With the aid of more refined conformational potential energy calculations and tailor-made synthetic polypeptide systems, it may be possible to unravel the forces leading to given conformations. It is noteworthy that the major void in approaches to polypeptide geometry and protein structure remains solvent–solute interactions. In the near future we may anticipate major advances in solving this problem.

The basic principles of stereochemistry are fully applicable to polymers. Macromolecules are composed of chains, which are the only structural features in which they differ from low molecular weight compounds. It is important to recognize that the same laws govern the stereochemistry of synthetic polymers, of biopolymers, and of low molecular weight compounds. We have striven to stress the unity of the stereochemical approach in this review.

Acknowledgment

The authors wish to thank Dr. Frank Morehouse of our department for his help in composing this manuscript and for his critical comments.

REFERENCES

1. L. Pauling and R. B. Corey, *Proc. Natl. Acad. Sci. U.S.*, **37**, 729 (1951).
2. L. Pauling and R. B. Corey, *Proc. Natl. Acad. Sci. U.S.*, **37**, 235 (1951).
3. M. F. Perutz, *Nature*, **167**, 1053 (1951).

4. C. H. Bamford, L. Brown, A. Elliott, W. E. Hanby, and I. F. Trotter, *Nature*, **169**, 357 (1952).
5. W. Cochran, F. H. C. Crick, and V. Vand, *Acta Cryst.*, **5**, 581 (1952).
6. R. E. Dickerson, in *The Proteins*, Vol. 2, H. Neurath, Ed., Academic Press, New York, 1964, p. 603.
7. L. Brown and I. F. Trotter, *Trans. Faraday Soc.*, **52**, 537 (1956).
8. A. Elliott and B. R. Malcolm, *Proc. Roy. Soc. (London)*, **A249**, 30 (1959).
9. S. Arnott and A. J. Wonacott, *J. Mol. Biol.*, **21**, 371 (1966).
10. S. Arnott and S. D. Dover, *J. Mol. Biol.*, **30**, 209 (1967).
11. H. L. Yakel, *Acta Cryst.*, **6**, 724 (1953).
12. W. Traub, *Acta Cryst.*, **16**, 842 (1963).
13. A. Elliott, in *Poly-α-aminoacids*, G. D. Fasman, Ed., Dekker, New York, 1967, p. 1.
14. C. K. Johnson, Ph.D. Thesis, Massachusetts Institute of Technology, Cambridge (1959).
15. C. H. Bamford, A. Elliott, and W. E. Hanby, *Synthetic Polypeptides*, Academic Press, New York, 1956.
16. E. M. Bradbury, L. Brown, A. R. Downie, A. Elliott, R. D. B. Fraser, and W. E. Hanby, *J. Mol. Biol.*, **5**, 230 (1962).
17. R. D. B. Fraser, T. P. MacRae, and I. W. Stapleton, *Nature*, **193**, 573 (1962).
18. P. Saludjian, C. deLozé, and V. Luzzati, *Compt. Rend.*, **256**, 4297 (1963).
19. G. N. Ramachandran, C. M. Venkatachalam, and S. Krimm, *Biophys. J.*, **6**, 849 (1966).
20. C. Ramakrishnan and G. N. Ramachandran, *Biophys. J.*, **5**, 909 (1965).
21. J. S. Ham and J. R. Platt, *J. Chem. Phys.*, **20**, 335 (1952).
22. H. D. Hunt and W. T. Simpson, *J. Amer. Chem. Soc.*, **75**, 4540 (1953).
23. S. Nagakura, *Mol. Phys.*, **3**, 105 (1960).
24. I. Tinoco, Jr., A. Halpern, and W. T. Simpson, in *Polyamino Acids, Polypeptides, and Proteins*, M. A. Stahman, Ed., University of Wisconsin Press, Madison, 1962, p. 147.
25. E. E. Barnes and W. T. Simpson, *J. Chem. Phys.*, **39**, 670 (1963).
26. D. L. Peterson and W. T. Simpson, *J. Amer. Chem. Soc.*, **79**, 2375 (1957).
27. S. Yomosa, *Biopolymers Symp.*, **1**, 1 (1964).
28. H. Basch, M. B. Robin, and N. A. Kuebler, *J. Chem. Phys.*, **47**, 1201 (1967).
29. D. G. Barnes and W. Rhodes, *J. Chem. Phys.*, **48**, 817 (1968).
30. B. J. Litman and J. A. Schellman, *J. Phys. Chem.*, **69**, 978 (1965).
31. W. B. Gratzer, in *Poly-α-aminoacids*, G. D. Fasman, Ed., Dekker, New York, 1967, p. 177.
32. M. Suard, G. Berthier, and B. Pullman, *Biochim. Biophys. Acta*, **52**, 254 (1961).
33. M. Suard-Sender, *J. Chim. Phys.*, **62**, 79 (1965).
34. E. B. Nielsen and J. A. Schellman, *J. Phys. Chem.*, **71**, 2297 (1967).
35. J. A. Schellman and E. B. Nielsen, in *Conformation of Biopolymers*, Vol. 1, G. N. Ramachandran, Ed., Academic Press, New York, 1967, p. 109.
36. H. Wolf, *Tetrahedron Letters*, **1965**, 1075.
37. D. W. Urry, *J. Phys. Chem.*, **72**, 3035 (1968).
38. S. Feinleib, F. A. Bovey, and J. W. Longworth, *Chem. Commun.*, **1968**, 238.
39. D. Balasubramanian and D. B. Wetlaufer, *Conformation of Biopolymers*, Vol. 1, G. N. Ramachandran, Ed., Academic Press, New York, 1967, p. 147.
40. M. Goodman, C. Toniolo, and J. Falcetta, *J. Amer. Chem. Soc.*, **91**, 1816 (1969).
41. W. Rhodes and D. G. Barnes, *J. Chim. Phys.*, **65**, 78 (1968).
42. K. Rosenheck and P. Doty, *Proc. Natl. Acad. Sci. U.S.*, **47**, 1775 (1961).
43. K. Imahori and J. Tanaka, *J. Mol. Biol.*, **1**, 359 (1959).

44. R. McDiarmid, Ph.D. Thesis, Harvard University, Cambridge, Mass. (1965).
45. R. McDiarmid and P. Doty, *J. Phys. Chem.*, **70**, 2620 (1966).
46. W. B. Gratzer, G. Holzwarth, and P. Doty, *Proc. Natl. Acad. Sci. U.S.*, **47**, 1785 (1961).
47. M. Goodman, I. Listowsky, Y. Masuda, and F. Boardman, *Biopolymers*, **1**, 33 (1963).
48. M. Goodman and I. Listowsky, *J. Amer. Chem. Soc.*, **84**, 3770 (1962).
49. M. Goodman and I. Rosen, *Biopolymers*, **2**, 537 (1964).
50. P. Urnes and P. Doty, *Advan. Protein Chem.*, **16**, 401 (1961).
51. E. Blout, in *Optical Rotatory Dispersion*, C. Djerassi, Ed., McGraw-Hill, New York, 1960, p. 238.
52. J. T. Yang, in *Poly-α-aminoacids*, G. D. Fasman, Ed., Dekker, New York, 1967, p. 239.
53. K. Yamaoka, *Biopolymers*, **2**, 219 (1964).
54. E. Schechter and E. R. Blout, *Proc. Natl. Acad. Sci. U.S.*, **51**, 695 (1964).
55. J. T. Yang, in *Newer Methods of Polymer Characterization*, B. Ke, Ed., Wiley, New York, 1964, p. 103.
56. J. T. Yang, in *Conformation of Biopolymers*, Vol. 1, G. N. Ramachandran, Ed., Academic Press, New York, 1967, p. 157.
57. J. P. Carver, E. Schechter, and E. R. Blout, *J. Amer. Chem. Soc.*, **88**, 2550 (1966).
58. J. T. Yang, *Proc. Natl. Acad. Sci. U.S.*, **53**, 438 (1965).
59. W. Moffitt, *Proc. Natl. Acad. Sci. U.S.*, **42**, 736 (1956).
60. W. Moffitt and J. T. Yang, *Proc. Natl. Acad. Sci. U.S.*, **42**, 596 (1956).
61. W. Moffitt and A. Moscowitz, *J. Chem. Phys.*, **30**, 648 (1959).
62. J. Y. Cassim and E. W. Taylor, *Biophys. J.*, **5**, 553 (1965).
63. E. R. Blout and R. H. Karlson, *J. Amer. Chem. Soc.*, **80**, 1259 (1958).
64. R. H. Karlson, K. S. Norland, G. D. Fasman, and E. R. Blout, *J. Amer. Chem. Soc.*, **82**, 2268 (1960).
65. G. D. Fasman, *Nature*, **193**, 681 (1962).
66. G. D. Fasman, M. Landsberg, and M. Buchwald, *Can. J. Chem.*, **43**, 1588 (1965).
67. G. D. Fasman, E. Bodenheimer, and C. Lindblow, *Biochemistry*, **3**, 1665 (1964).
68. M. Goodman, C. M. Deber, and A. M. Felix, *J. Amer. Chem. Soc.*, **84**, 3773 (1962).
69. M. Goodman, A. M. Felix, C. M. Deber, A. R. Brause, and G. Schwartz, *Biopolymers*, **1**, 371 (1963).
70. M. Goodman, A. M. Felix, C. M. Deber, and A. R. Brause, *Biopolymers Symp.*, **1**, 409 (1964).
71. W. B. Gratzer and D. A. Cowburn, *Nature*, **222**, 426 (1969).
72. D. A. Chignell and W. B. Gratzer, *Nature*, **210**, 262 (1966).
73. G. D. Fasman, in *Poly-α-aminoacids*, G. D. Fasman, Ed., Dekker, New York, 1967, p. 499.
74. A. Moscowitz, *Advan. Chem. Phys.*, **4**, 67 (1962).
75. A. Moscowitz, in *Optical Rotatory Dispersion*, C. Djerassi, Ed., McGraw-Hill, New York, 1960, p. 150.
76. I. Tinoco, Jr., *Advan. Chem. Phys.*, **4**, 113 (1962).
77. J. A. Schellman, *Accounts Chem. Res.*, **1**, 144 (1968).
78. G. Holzwarth and P. Doty, *J. Amer. Chem. Soc.*, **87**, 218 (1965).
79. S. Beychok, in *Poly-α-aminoacids*, G. D. Fasman, Ed., Dekker, New York, 1967, p. 293.
80. L. Velluz, M. Legrand, and M. Grosjean, *Optical Circular Dichroism*, Academic Press, New York, 1965.
81. P. Crabbé, *Optical Rotatory Dispersion and Circular Dichroism in Organic Chemistry*, Holden-Day, San Francisco, 1965.

82. L. Velluz and M. Legrand, *Angew. Chem.*, **77**, 842 (1965); *Angew. Chem., Intern. Ed. Engl.*, **4**, 838 (1965).
83. W. B. Gratzer, *Proc. Roy. Soc. (London)*, **A297** (1448), 163 (1967).
84. J. A. Schellman and C. Schellman, in *The Proteins*, Vol. II, H. Neurath, Ed., Academic Press, New York, 1964, p. 1.
85. W. F. Harrington, R. Josephs, and D. M. Segal, *Ann. Rev. Biochem.*, **35**, 599 (1966).
86. K. Morita, E. R. Simons, and E. R. Blout, *Biopolymers*, **5**, 259 (1967).
87. N. J. Greenfield and G. D. Fasman, *Biopolymers*, **7**, 595 (1969).
88. D. Balasubramanian and D. B. Wetlaufer, *J. Amer. Chem. Soc.*, **88**, 3449 (1966).
89. F. Quadrifoglio and D. W. Urry, *J. Amer. Chem. Soc.*, **90**, 2755 (1968).
90. R. T. Ingwall, H. A. Scheraga, N. Lotan, A. Berger, and E. Katchalski, *Biopolymers*, **6**, 331 (1968).
91. J. Brahms and G. Spach, *Nature*, **200**, 72 (1963).
92. E. M. Bradbury, A. R. Downie, A. Elliott, and W. E. Hanby, *Proc. Roy. Soc. (London)*, **A259**, 110 (1960).
93. E. M. Bradbury, L. Brown, A. R. Downie, A. Elliott, W. E. Hanby, and T. R. R. McDonald, *Nature*, **183**, 1736 (1959).
94. E. M. Bradbury, A. R. Downie, A. Elliott, and W. E. Hanby, *Nature*, **187**, 321 (1960).
95. D. G. H. Ballard, C. H. Bamford, and A. Elliott, *Makromol. Chem.*, **35**, 222 (1960).
96. M. Goodman, A. M. Felix, and G. Schwartz, in *Electronic Aspects of Biochemistry*, B. Pullman, Ed., Academic Press, New York, 1964, p. 365.
97. E. R. Blout, *Biopolymers Symp.*, **1**, 397 (1964).
98. D. F. Bradley, M. Goodman, A. Felix, and R. Records, *Biopolymers*, **4**, 607 (1966).
99. M. Goodman, F. Boardman, and I. Listowski, *J. Amer. Chem. Soc.*, **85**, 2483 (1963).
100. C. Toniolo, M. Falxa, and M. Goodman, *Biopolymers*, **6**, 1579 (1968).
101. M. Goodman and A. M. Felix, unpublished results.
102. M. Goodman and F. R. Prince, Jr., unpublished results.
103. E. M. Bradbury, B. G. Carpenter, and H. Goldman, *Biopolymers*, **6**, 837 (1968).
104. E. M. Bradbury, B. G. Carpenter, and R. M. Stephens, *Biopolymers*, **6**, 905 (1968).
105. M. Hashimoto and J. Aritomi, *Bull. Chem. Soc. Japan*, **39**, 2707 (1966).
106. M. Hashimoto, *Bull. Chem. Soc. Japan*, **39**, 2713 (1966).
107. M. Hashimoto, S. Arakawa, and K. Nakamura, International Symposium on Macromolecular Chemistry, *Tokyo-Kyoto*, Preprints, **9**, 12 (1966).
108. M. Hashimoto and S. Arakawa, *Bull. Chem. Soc. Japan*, **40**, 1698 (1967).
109. D. W. Urry, *Ann. Rev. Phys. Chem.*, **19**, 477 (1968).
110. J. F. Yan, G. Vanderkooi and H. A. Scheraga, *J. Chem. Phys.*, **49**, 2713 (1968).
111. J. N. Vournakis, J. F. Yan, and H. A. Scheraga, *Biopolymers*, **6**, 1531 (1968).
112. F. A. Bovey, *Pure Appl. Chem.*, **16**, 417 (1968).
113. F. A. Bovey, G. V. D. Tiers, and G. Filipovich, *J. Polymer Sci.*, **38**, 73 (1959).
114. M. Goodman and Y. Masuda, *Biopolymers*, **2**, 107 (1964).
115. K. J. Liu, J. S. Lignowski, and R. Ullman, *Biopolymers*, **5**, 375 (1967).
116. D. I. Marlborough, K. G. Orrell, and H. N. Rydon, *Chem. Commun.*, **1965**, 518.
117. D. I. Marlborough and H. N. Rydon, in *Some Newer Physical Methods in Structural Chemistry*, R. Bonnett and J. G. Davis, Eds., United Mechanical Press, London, 1967, p. 211.
118. E. M. Bradbury, C. Crane-Robinson, H. Goldman, and H. W. E. Rattle, *Nature*, **217**, 812 (1968).
119. J. A. Ferretti, *Chem. Commun.*, **1967**, 1030.
120. F. Conti and A. M. Liquori, *J. Mol. Biol.*, **33**, 953 (1968).

121. J. C. Haylock and H. N. Rydon, in *Peptides*, E. Bricas, Ed., North Holland Publ. Co., Amsterdam, 1968, p. 19.
122. H. N. Rydon, *ACS Polymer Preprints*, **10**, No. 1, 25 (1969).
123. J. A. Ferretti, *ACS Polymer Preprints*, **10**, No. 1, 33 (1969).
124. J. A. Ferretti and L. Paolillo, *Biopolymers*, **7**, 155 (1969).
125. S. Hanlon, S. F. Russo, and I. M. Klotz, *J. Amer. Chem. Soc.*, **85**, 2024 (1963).
126. S. Hanlon and I. M. Klotz, *Biochemistry*, **4**, 37 (1965).
127. S. Hanlon, *Biochemistry*, **5**, 2049 (1966).
128. I. M. Klotz, S. F. Russo, S. Hanlon, and M. A. Stake, *J. Amer. Chem. Soc.*, **86**, 4774 (1964).
129. M. A. Stake and I. M. Klotz, *Biochemistry*, **5**, 1726 (1966).
130. J. H. Bradbury and M. D. Fenn, in *Symp. Fibrous Proteins, Australia 1967*, **1968**, p. 69; *Australian J. Chem.*, **22**, 357 (1969).
131. J. H. Bradbury and M. D. Fenn, *J. Mol. Biol.*, **36**, 231 (1968).
132. W. E. Stewart, L. Mandelkern, and R. E. Glick, *Biochemistry*, **6**, 143 (1967).
133. W. E. Stewart, L. Mandelkern, and R. E. Glick, *Biochemistry*, **6**, 150 (1967).
134. F. Quadrifoglio and D. W. Urry, *J. Phys. Chem.*, **71**, 2364 (1967).
135. D. Balasubramanian, *Biochem. Biophys. Res. Comm.*, **29**, 538 (1967).
136. J. Steigman, E. Peggion, and A. Cosani, *J. Amer. Chem. Soc.*, **91**, 1822 (1969).
137. J. Steigman, A. S. Verdini, C. Montagner, and L. Strasorier, *J. Amer. Chem. Soc.*, **91**, 1829 (1969).
138. E. J. Ambrose and A. Elliott, *Proc. Roy. Soc. (London)*, **A205**, 47 (1951).
139. T. Miyazawa and E. R. Blout, *J. Amer. Chem. Soc.*, **83**, 712 (1961).
140. M. Tsuboi, *J. Polymer Sci.*, **59**, 139 (1962).
141. T. Miyazawa, Y. Masuda, and K. Fukushima, *J. Polymer Sci.*, **62**, S62 (1962).
142. Y. Masuda, K. Fukushima, T. Fujii, and T. Miyazawa, *Biopolymers*, **8**, 91 (1969).
143. K. Itoh, T. Nakahara, T. Shimanouchi, and M. Oya, Symposium on Protein Structure, Nagoya, 1967.
144. Y. Masuda and T. Miyazawa, *Makromol. Chem.*, **103**, 261 (1967).
145. T. Miyazawa, K. Fukushima, S. Sugano, and Y. Masuda, in *Conformation of Biopolymers*, G. N. Ramachandran, Ed., Academic Press, New York, 1967, p. 557.
146. T. Miyazawa, in *Polyamino Acids, Polypeptides and Proteins*, M. A. Stahman, Ed., University of Wisconsin Press, Madison, 1962, p. 201.
147. T. Miyazawa, in *Poly-α-aminoacids*, G. D. Fasman, Ed., Dekker, New York, 1967, p. 69.
148. T. Miyazawa, in *Aspects of Protein Structure*, G. N. Ramachandran, Ed., Academic Press, London, 1963, p. 257.
149. H. Susi, in *Structure and Stability of Biological Macromolecules*, S. N. Timasheff and G. D. Fasman, Eds., Dekker, New York, 1969, p. 575.
150. K. Itoh, T. Nakahara, T. Shimanouchi, M. Oya, K. Uno and Y. Iwakura, *Biopolymers*, **6**, 1759 (1968).
151. T. Miyazawa, *Bull. Chem. Soc. Japan*, **34**, 691 (1961).
152. M. Tsuboi, Y. Mitsui, A. Wada, T. Miyazawa, and N. Nagashima, *Biopolymers*, **1**, 297 (1963).
153. Y. Masuda and T. Miyazawa, *Bull. Chem. Soc. Japan*, **42**, 570 (1969).
154. Y. Masuda, T. Miyazawa, and M. Goodman, *Biopolymers*, **8**, 515 (1969).
155. A. Berger, J. Kurtz, and E. Katchalski, *J. Amer. Chem. Soc.*, **76**, 5552 (1954).
156. W. Traub and V. Schmueli, in *Aspects of Protein Structure*, G. N. Ramachandran, Ed., Academic Press, New York, 1963, p. 81.
157. W. Traub and V. Schmueli, *Nature*, **198**, 1165 (1963).
158. P. M. Cowan and S. McGavin, *Nature*, **176**, 501 (1955).

159. V. Sasisekharan, *Acta Cryst.*, **12**, 897 (1959).
160. V. Sasisekharan, *Acta Cryst.*, **12**, 903 (1959).
161. J. P. Carver and E. R. Blout, in *Treatise on Collagen*, Vol. 1, G. N. Ramachandran, Ed., Academic Press, London, 1967, p. 441.
162. W. F. Harrington and M. Sela, *Biochim. Biophys. Acta*, **27**, 24 (1958).
163. W. F. Harrington and P. von Hippel, *Advan. Protein Chem.*, **16**, 1 (1961).
164. E. R. Blout and E. Schechter, *Biopolymers*, **1**, 565 (1963).
165. I. Z. Steinberg, W. F. Harrington, A. Berger, M. Sela, and E. Katchalski, *J. Amer. Chem. Soc.*, **82**, 5263 (1960).
166. A. R. Downie and A. Randall, *Trans. Faraday Soc.*, **55**, 2132 (1959).
167. E. R. Blout, J. P. Carver, and J. Gross, *J. Amer. Chem. Soc.*, **85**, 644 (1963).
168. J. Kurtz, A. Berger, and E. Katchalski, in *Recent Advances in Gelatin and Glue Research*, G. Stainsby, Ed., Pergamon Press, New York, 1958, p. 131.
169. W. B. Gratzer, W. Rhodes, and G. D. Fasman, *Biopolymers*, **1**, 319 (1963).
170. G. D. Fasman and E. R. Blout, *Biopolymers*, **1**, 3 (1963).
171. F. Gornick, L. Mandelkern, A. F. Diorio, and D. E. Roberts, *J. Amer. Chem. Soc.*, **86**, 2549 (1964).
172. E. R. Blout and G. D. Fasman, in *Recent Advances in Gelatin and Glue Research*, **1**, 122 (1957).
173. W. F. Harrington and J. Kurtz, Abstracts, 147th Meeting, American Chemical Society, 10 H (1964).
174. F. A. Bovey and F. P. Hood, *J. Amer. Chem. Soc.*, **88**, 2326 (1966).
175. F. A. Bovey and F. P. Hood, *Biopolymers*, **5**, 325 (1967).
176. F. A. Bovey and F. P. Hood, *Biopolymers*, **5**, 915 (1967).
177. T. Isemura, H. Okabayashi, and S. Sakakibara, *Biopolymers*, **6**, 307 (1968).
178. W. Traub, A. Yonath, and D. M. Segal, *Nature*, **221**, 914 (1969).
179. G. N. Ramachandran and G. Kartha, *Nature*, **176**, 593 (1955).
180. G. N. Ramachandran and V. Sasisekharan, *Biochim. Biophys. Acta*, **109**, 314 (1965).
181. G. N. Ramachandran and R. Chandrasekharan, *Biopolymers*, **6**, 1649 (1968).
182. A. Rich and F. H. C. Crick, *J. Mol. Biol.*, **3**, 483 (1961).
183. P. M. Cowan, S. McGavin, and A. C. T. North, *Nature*, **176**, 1062 (1955).
184. N. S. Andreeva, N. G. Esipova, M. I. Millionova, V. N. Rogulenkova, and V. A. Shibnev, in *Conformation of Biopolymers*, Vol. 2, G. N. Ramachandran, Ed., Academic Press, London, 1967, p. 469.
185. W. Traub and A. Yonath, *J. Mol. Biol.*, **16**, 404 (1966).
186. W. Traub and A. Yonath, *J. Mol. Biol.*, **25**, 351 (1967).
187. A. Scatturin, A. Del Pra, A. M. Tamburro, and E. Scoffone, *Chem. Ind. (London)*, **49**, 970 (1967).
188. S. Beychok and G. D. Fasman, *Biochemistry*, **3**, 1675 (1964).
189. T. Ooi, R. A. Scott, G. Vanderkooi, and H. A. Scheraga, *J. Chem. Phys.*, **46**, 4410 (1967).
190. M. Goodman, G. W. Davis, and E. Benedetti, *Acc. Chem. Res.*, **1**, 1275 (1968).
191. J. Applequist and T. G. Mahr, *J. Amer. Chem. Soc.*, **88**, 5419 (1966).
192. E. Peggion, A. S. Verdini, A. Cosani, and E. Scoffone, *Macromolecules*, **2**, 170 (1969).
193. H. J. Sage and G. D. Fasman, *Biochemistry*, **5**, 286 (1966).
194. H. E. Auer and P. Doty, *Biochemistry*, **5**, 1708 (1966).
195. H. E. Auer and P. Doty, *Biochemistry*, **5**, 1716 (1966).
196. E. Peggion, A. Cosani, A. S. Verdini, A. Del Pra, and M. Mammi, *Biopolymers*, **6**, 1477 (1968).

197. A. Cosani, E. Peggion, A. S. Verdini, and M. Terbojevich, *Biopolymers*, **6**, 963 (1968).
198. M. Goodman and E. Peggion, *Biochemistry*, **6**, 1533 (1967).
199. M. Goodman and A. Kossoy, *J. Amer. Chem. Soc.*, **88**, 5010 (1966).
200. M. Goodman and E. Benedetti, *Biochemistry*, **7**, 4226 (1968).
201. E. Benedetti, A. Kossoy, M. L. Falxa, and M. Goodman, *Biochemistry*, **7**, 4234 (1968).
202. M. L. Tiffany and S. Krimm, *Biopolymers*, **6**, 1379 (1968).
203. M. L. Tiffany and S. Krimm, *Biopolymers*, **6**, 1767 (1968).
204. M. L. Tiffany and S. Krimm, IUPAC International Symposium on Macromolecular Chemistry, Toronto, 1968, B2.5.
205. S. Krimm and J. E. Mark, *Proc. Natl. Acad. Sci. U.S.*, **60**, 1122 (1968).
206. E. R. Blout, J. P. Carver, and E. Schechter, in *Optical Rotatory Dispersion and Circular Dichroism in Organic Chemistry*, G. Snatzke, Ed., Sadtler Research Labs., Philadelphia, Pa., 1967.
207. Y. P. Myer, *ACS Polymer Preprints*, **10**, No. 1, 307 (1969).
208. M. Legrand and R. Viennet, *Compt. Rend.*, **259**, 4277 (1964).
209. A. J. Adler, R. Hoving, J. Potter, M. Wells, and G. D. Fasman, *J. Amer. Chem. Soc.*, **90**, 4736 (1968).
210. K. H. Meyer, in *Natural and Synthetic High Polymers*, Interscience, New York, 1950.
211. W. T. Astbury and A. Street, *Phil. Trans. Roy. Soc. (London)*, **A230**, 75 (1931).
212. W. T. Astbury and H. J. Woods, *Phil. Trans. Roy. Soc. (London)*, **A232**, 333 (1933).
213. C. C. F. Blake, D. F. Koenig, G. A. Mair, A. C. T. North, D. C. Phillips, and V. R. Sarma, *Nature*, **206**, 757 (1965).
214. L. Pauling and R. B. Corey, *Proc. Natl. Acad. Sci. U.S.*, **39**, 253 (1953).
215. C. H. Bamford, L. Brown, A. Elliott, W. E. Hanby, and I. F. Trotter, *Nature*, **173**, 27 (1954).
216. K. H. Meyer and H. Mark, *Ber.*, **61**, 1932 (1928).
217. W. T. Astbury, S. Dickinson, and K. Bayley, *Biochem. J.*, **29**, 2351 (1935).
218. E. M. Bradbury, L. Brown, A. R. Downie, A. Elliott, R. D. B. Fraser, W. E. Hanby, and T. R. R. McDonald, *J. Mol. Biol.*, **2**, 276 (1960).
219. I. Yahara, K. Imahori, Y. Iitaka, and M. Tsuboi, *Polymer Letters*, **1**, 47 (1963).
220. E. M. Bradbury, A. Elliott, and W. E. Hanby, *J. Mol. Biol.*, **5**, 487 (1962).
221. Z. Bohak and E. Katchalski, *Biochemistry*, **2**, 228 (1963).
222. S. Kubota, S. Sugai, and J. Noguchi, *Biopolymers*, **6**, 1311 (1968).
223. A. Elliott, R. D. B. Fraser, T. P. MacRae, I. W. Stapleton, and E. Suzuki, *J. Mol. Biol.*, **9**, 10 (1964).
224. R. D. B. Fraser, B. S. Harrap, T. P. MacRae, F. H. C. Stewart, and E. Suzuki, *J. Mol. Biol.*, **14**, 423 (1965).
225. A. Elliott, E. M. Bradbury, A. R. Downie, and W. E. Hanby, in *Polyamino Acids, Polypeptides and Proteins*, M. A. Stahmann, Ed., University of Wisconsin Press, Madison, 1962, p. 255.
226. R. D. B. Fraser, T. P. MacRae, F. H. C. Stewart, and E. Suzuki, *J. Mol. Biol.*, **11**, 706 (1965).
227. Y. Go, J. Noguchi, M. Asai, and T. Hayakawa, *J. Polymer Sci.*, **21**, 147 (1956).
228. K. Rosenheck and B. Sommer, *J. Chem. Phys.*, **46**, 532 (1967).
229. E. Pysh, *Proc. Natl. Acad. Sci. U.S.*, **56**, 825 (1966).
230. E. Iizuka and J. T. Yang, *Proc. Natl. Acad. Sci. U.S.*, **55**, 1175 (1966).
231. B. Davidson, N. Tooney, and G. D. Fasman, *Biochem. Biophys. Res. Commun.*, **23**, 156 (1966).

232. S. N. Timasheff and M. J. Gorbunoff, *Ann. Rev. Biochem.*, **36**, 13 (1967).
233. G. D. Fasman and J. Potter, *Biochem. Biophys. Res. Commun.*, **27**, 209 (1967).
234. P. K. Sarkar and P. Doty, *Proc. Natl. Acad. Sci. U.S.*, **55**, 981 (1966).
235. R. Townend, T. F. Kumosinski, S. N. Timasheff, G. D. Fasman, and B. Davidson, *Biochem. Biophys. Res. Commun.*, **23**, 163 (1966).
236. F. Quadrifoglio and D. W. Urry, *J. Amer. Chem. Soc.*, **90**, 2760 (1968).
237. L. Stevens, R. Townend, S. N. Timasheff, G. D. Fasman, and J. Potter, *Biochemistry*, **7**, 3717 (1968).
238. S. Ikeda and G. D. Fasman, *J. Mol. Biol.*, **30**, 491 (1967).
239. E. R. Blout and S. G. Linsley, *J. Amer. Chem. Soc.*, **74**, 1946 (1952).
240. A. Elliott and B. R. Malcolm, *Trans. Faraday Soc.*, **52**, 528 (1956).
241. P. Doty, A. M. Holtzer, J. H. Bradbury, and E. R. Blout, *J. Amer. Chem. Soc.*, **76**, 4493 (1954).
242. P. Doty and J. T. Yang, *J. Amer. Chem. Soc.*, **78**, 498 (1956).
243. E. R. Blout and A. Asadourian, *J. Amer. Chem. Soc.*, **78**, 955 (1956).
244. P. Doty and R. D. Lundberg, *J. Amer. Chem. Soc.*, **78**, 4810 (1956).
245. J. A. Schellman, *Compt. Rend. Trav. Lab.*, *Carlsberg*, *Ser. Chim.*, **29**, 230 (1955).
246. J. A. Schellman, *J. Phys. Chem.*, **62**, 1485 (1958).
247. L. Peller, *J. Phys. Chem.*, **63**, 1194 (1959).
248. S. A. Rice, A. Wada, and E. P. Geiduschek, *Discussions Faraday Soc.*, **25**, 130 (1958).
249. J. H. Gibbs and E. A. DiMarzio, *J. Chem. Phys.*, **28**, 1247 (1958).
250. J. H. Gibbs and E. A. DiMarzio, *J. Chem. Phys.*, **30**, 271 (1959).
251. S. A. Rice and A. Wada, *J. Chem. Phys.*, **29**, 233 (1958).
252. T. L. Hill, *J. Chem. Phys.*, **30**, 383 (1959).
253. K. Nagai, *J. Phys. Soc., Japan*, **15**, 407 (1960).
254. B. Zimm and J. Bragg, *J. Chem. Phys.*, **31**, 526 (1959).
255. S. Lifson and A. Roig, *J. Chem. Phys.*, **34**, 1963 (1961).
256. S. Lifson, *J. Chem. Phys.*, **40**, 3705 (1964).
257. G. Schwarz and J. Seelig, *Biopolymers*, **6**, 1263 (1968).
258. G. Schwarz, *J. Mol. Biol.*, **11**, 64 (1965).
259. M. Eigen, in *Techniques in Organic Chemistry*, Vol. 8, Part II, 2nd ed., A. Weissberger, Ed., Interscience, New York, 1963, p. 895.
260. E. F. Caldin, *Fast Reactions in Solutions*, Blackwell Scientific Publications, Oxford, 1964, p. 59.
261. R. Lumry, R. Legare, and W. Miller, *Biopolymers*, **2**, 489 (1964).
262. E. Hamori and H. A. Scheraga, *J. Phys. Chem.*, **70**, 3018 (1966).
263. G. Schwarz, in *Chem. Soc. Publ.* **20**, 191 (1966).
264. R. C. Parker, K. Applegate, and L. J. Slutsky, *J. Phys. Chem.*, **70**, 3018 (1966).
265. R. C. Parker, L. J. Slutsky, and K. Applegate, *J. Phys. Chem.*, **72**, 3177 (1968).
266. P. J. Flory, *Statistical Mechanics of Chain Molecules*, Interscience, New York, 1969, p. 294.
267. Th. Ackermann and H. Ruterjans, *Ber. Bunsenges., Physik. Chem.*, **68**, 850 (1964). Th. Ackermann and E. Neumann, *Biopolymers*, **5**, 649 (1967).
268. J. Applequist, *J. Chem. Phys.*, **38**, 934 (1963).
269. F. E. Karasz, J. M. O'Reilly, and H. E. Bair, *Biopolymers*, **3**, 241 (1965).
270. G. Giacometti, A. Turolla, and R. Boni, *Biopolymers*, **6**, 441 (1968).
271. B. H. Zimm, P. Doty, and K. Iso, *Proc. Natl. Acad. Sci. U.S.*, **45**, 1601 (1959).
272. F. E. Karasz and J. M. O'Reilly, *Biopolymers*, **4**, 1015 (1966).
273. F. E. Karasz and J. M. O'Reilly, *Biopolymers*, **5**, 27 (1967).
274. F. E. Karasz, J. M. O'Reilly, and H. E. Bair, *Nature*, **202**, 693 (1964).

275. J. Engel, *Biopolymers*, **4**, 945 (1966).
276. J. T. Edsall, P. J. Flory, J. C. Kendrew, A. M. Liquori, G. Nemethy, G. N. Ramachandran, and H. A. Scheraga, *Biopolymers*, **4**, 121 (1966); *J. Biol. Chem.*, **241**, 1004 (1966); *J. Mol. Biol.*, **15**, 399 (1966).
277. D. A. Brant and P. J. Flory, *J. Amer. Chem. Soc.*, **87**, 2791 (1965).
278. R. A. Scott and H. A. Scheraga, *J. Chem. Phys.*, **45**, 2091 (1966).
279. G. Nemethy and H. A. Scheraga, *Biopolymers*, **3**, 155 (1965).
280. S. J. Leach, G. Nemethy, and H. A. Scheraga, *Biopolymers*, **4**, 369 (1966).
281. G. N. Ramachandran, C. Ramakrishnan and V. Sasisekharan, in *Aspects of Protein Structure*, G. N. Ramachandran, Ed., Academic Press, New York, 1963, p. 121.
282. G. N. Ramachandran, C. Ramakrishnan, and V. Sasisekharan, *J. Mol. Biol.*, **7**, 95 (1963).
283. C. Ramakrishnan and G. N. Ramachandran, *Proc. Indian Acad. Sci.*, **A59**, 329 (1964).
284. A. M. Liquori, *J. Polymer Sci.*, **C12**, 209 (1966).
285. A. Wada, *Kobunshi, Japan*, **16**, 614 (1967).
286. M. Goodman and N. S. Choi, in *Peptides*, E. Bricas, Ed., North-Holland Publ. Co., Amsterdam, 1968, p. 1.
287. S. I. Mizushima and T. Shimanouchi, *Advances in Enzymology*, Vol. 23, F. F. Nord, Ed., Interscience, New York, 1961.
288. D. R. Herschbach, *Bibliography for Hindered Rotation and Microwave Spectroscopy*, Lawrence Radiation Laboratory, University of California, Berkeley, Calif., 1962.
289. R. A. Scott and H. A. Scheraga, *J. Chem. Phys.*, **42**, 2209 (1965).
290. P. DeSantis, E. Giglio, A. M. Liquori, and A. Ripamonti, *Nature*, **206**, 456 (1965).
291. J. B. Hendrickson, *J. Amer. Chem. Soc.*, **83**, 4537 (1961).
292. K. S. Pitzer, in *Advances in Chemical Physics*, Vol. 2, I. Prigogine, Ed., Interscience, New York, 1959, p. 59.
293. I. Amdur and A. L. Harkness, *J. Chem. Phys.*, **22**, 664 (1954).
294. I. Amdur and E. A. Mason, *J. Chem. Phys.*, **23**, 415 (1955).
295. D. A. Brant, W. G. Miller, and P. J. Flory, *J. Mol. Biol.*, **23**, 47 (1967).
296. D. A. Brant and P. J. Flory, *J. Amer. Chem. Soc.*, **87**, 2788, 2791 (1965).
297. D. A. Brant and P. J. Flory, *J. Amer. Chem. Soc.*, **87**, 663 (1965).
298. D. F. Bradley, I. Tinoco, Jr., and R. W. Woody, *Biopolymers*, **1**, 239 (1963).
299. D. F. Bradley, S. Lifson, and B. Honig, in *Electronic Aspects of Biopolymers*, B. Pullman, Ed., Academic Press, New York, 1964, p. 77.
300. H. A. Nash and D. F. Bradley, *Biopolymers*, **3**, 261 (1965).
301. H. A. Nash and D. F. Bradley, *J. Chem. Phys.*, **45**, 1380 (1966).
302. T. Ooi, R. A. Scott, G. Vanderkooi, R. F. Epand, and H. A. Scheraga, *J. Amer. Chem. Soc.*, **88**, 5680 (1966).
303. E. R. Lippincott and R. Schroeder, *J. Chem. Phys.*, **23**, 1099 (1955).
304. R. Schroeder and E. R. Lippincott, *J. Chem. Phys.*, **61**, 921 (1957).
305. D. C. Poland and H. A. Scheraga, *Biopolymers*, **3**, 275 (1965).
306. D. C. Poland and H. A. Scheraga, *Biopolymers*, **3**, 283 (1965).
307. D. C. Poland and H. A. Scheraga, *Biopolymers*, **3**, 305 (1965).
308. D. C. Poland and H. A. Scheraga, *Biopolymers*, **3**, 315 (1965).
309. D. C. Poland and H. A. Scheraga, *Biopolymers*, **3**, 335 (1965).
309a. R. T. Ingwall, H. A. Scheraga, N. Lotan, A. Berger, and E. Katchalsky, *Biopolymers*, **6**, 331 (1968).
310. J. Donohue, *Proc. Natl. Acad. Sci. U.S.*, **39**, 205 (1953).

311. B. W. Low and H. J. Grenville-Wells, *Proc. Natl. Acad. Sci. U.S.*, **39**, 785 (1953).
312. F. H. C. Crick and A. Rich, *Nature*, **176**, 780 (1955).
313. R. E. Marsh, R. B. Corey, and L. Pauling, *Acta Cryst.*, **8**, 710 (1955).
314. H. A. Scheraga, S. J. Leach, and R. A. Scott, *Discussions Faraday Soc.*, **40**, 268 (1965).
315. N. Lotan, A. Berger, E. Katchalski, R. T. Ingwall, and H. A. Scheraga, *Biopolymers*, **4**, 239 (1966).
316. E. R. Blout, in *Polyamino Acids, Polypeptides and Proteins*, M. Stahmann, Ed., University of Wisconsin Press, Madison, 1962, p. 275.
317. S. M. Bloom, G. D. Fasman, D. de Loze, and E. R. Blout, *J. Amer. Chem. Soc.*, **84**, 458 (1962).
318. M. Goodman, F. Boardman, and I. Listowsky, *J. Amer. Chem. Soc.*, **85**, 2491 (1963).
319. M. Goodman, E. E. Schmitt, and D. A. Yphantis, *J. Amer. Chem. Soc.*, **84**, 1288 (1962).
320. A. Wada, in *Poly-α-aminoacids*, G. D. Fasman, Ed., Dekker, New York, 1967, p. 369.
321. J. E. Mark and M. Goodman, *Biopolymers*, **5**, 809 (1967).
322. A. M. Liquori and P. DeSantis, *Biopolymers*, **5**, 815 (1967).
323. J. E. Mark and M. Goodman, *J. Amer. Chem. Soc.*, **89**, 1267 (1967).

Rotational Isomerism about sp^2-sp^3 Carbon–Carbon Single Bonds

GERASIMOS J. KARABATSOS AND DAVID J. FENOGLIO

Department of Chemistry, Michigan State University, East Lansing, Michigan

I.	Introduction	167
II.	Carbonyl Compounds	171
	A. Aldehydes	172
	1. Methodology	172
	2. Factors Affecting Rotamer Stabilities	175
	B. Ketones	179
	1. Aliphatic	180
	2. α-Heteroatom Substituted	182
	C. Acyl Halides	183
	D. Carboxylic Acids, Esters, and Related Compounds	185
III.	Olefins	187
	A. 3-Substituted	187
	B. 2-Substituted	190
	C. 1-Substituted	191
IV.	Imino Compounds	193
	A. *syn* Isomers	193
	B. *anti* Isomers	193
V.	Carbonium Ions and Free Radicals	195
VI.	General Comments	198
	References	200

I. INTRODUCTION

Rotational isomerism about single bonds has received considerable attention in the last 15 years. A major factor for the increased activity in this area has been the application of nuclear magnetic resonance spectroscopy, which has, at least in quantity of data, supplanted infrared and Raman spectroscopy, electron diffraction, and microwave spectroscopy. We shall confine this review to problems associated with rotational isomerism about sp^2-sp^3 hybridized carbon–carbon single bonds, with particular emphasis on the factors that influence the relative stabilities of rotamers **1–4**.

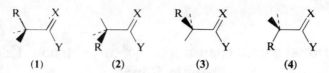

Rotamers **1** and **2** have a single bond eclipsing the C=X double bond, whereas **3** and **4** have a single bond eclipsing the C—Y single bond. We will refer to **1** and **2** as the eclipsing or eclipsed rotamers or conformations and to **3** and **4** as the bisecting, the latter terminology being used to denote the fact that the C=X double bond bisects the angle made by two of the single bonds of the sp^3-hybridized or *alpha* carbon atoms.

In view of the fact that applications of nuclear magnetic resonance (nmr) spectroscopy have provided the bulk of the data discussed in this review, we shall briefly outline the method used to treat the nmr data and point out the limitations that the method places on quantitative interpretations of such data.

The application of nmr spectroscopy to the conformational analysis about the sp^2-sp^3 carbon–carbon bond is based, primarily and most conveniently, on the resonance of a proton bonded to the sp^3-hybridized carbon atom. Let us consider conformations **1–4** and assume that R and Y are protons H_α and H_1, respectively. Since barriers to rotation about the relevant sp^2-sp^3 carbon–carbon bonds are less than 5 kcal/mol, only time-averaged couplings and chemical shifts for H_α are observed. The averaged vicinal spin–spin coupling $J_{H_\alpha H_1}$ may then be expressed by eq. (1), where P_1, P_2, P_3, and P_4 are the fractional populations (mole fractions) of rotamers **1**, **2**, **3**, and **4**, respectively.

$$J_{H_\alpha H_1} = J_{H_\alpha H_1}(\mathbf{1}) \cdot P_1 + J_{H_\alpha H_1}(\mathbf{2}) \cdot P_2 + J_{H_\alpha H_1}(\mathbf{3}) \cdot P_3 + J_{H_\alpha H_1}(\mathbf{4}) \cdot P_4 \quad (1)$$

Equation (1) may be rewritten as eq. (2), where J_t is the *anti* or *trans* vicinal

$$J_{H_\alpha H_1} = J_t \cdot P_1 + J_g \cdot P_2 + J_{120} \cdot P_3 + J_c \cdot P_4 \quad (2)$$

coupling (180° dihedral angle), J_g is the *gauche* (60° dihedral angle), J_{120} is the coupling for the 120° dihedral angle, and J_c is the *syn* or *cis* coupling (0° dihedral angle). The average chemical shift ν_{H_α} of H_α is analogously expressed by eq. (3).

$$\nu_{H_\alpha} = \nu_{H_\alpha}(\mathbf{1}) \cdot P_1 + \nu_{H_\alpha}(\mathbf{2}) \cdot P_2 + \nu_{H_\alpha}(\mathbf{3}) \cdot P_3 + \nu_{H_\alpha}(\mathbf{4}) \cdot P_4 \quad (3)$$

If the system were to be perturbed, either by temperature or solvent changes, so as to change the relative populations of the various rotamers, then the average vicinal coupling and chemical shifts would change. It is this change of $J_{H_\alpha H_1}$ and ν_{H_α} that can be used to deduce the relative stabilities of the various rotamers. Since the chemical shift of H_α in the various conformations is neither known nor even understood very well as yet, the changes in

ν_{H_α} cannot presently be used for quantitative conformational analysis about the sp^2-sp^3 carbon–carbon bond. For this reason, the nmr technique is best suited to the analysis of compounds where Y is proton, or some other convenient nucleus with a spin, such as fluorine or carbon-13, so that the coupling of Y with H_α may be observed.

Let us then focus attention on eq. (2). From the Karplus equation (1), which relates the vicinal coupling $J_{H_\alpha H_1}$ to the dihedral angle, J_t and J_c are considerably larger than J_g and J_{120}, with J_t being larger than J_c. Consider the case where $J_{H_\alpha H_1}$ increases with temperature. In this case the more stable rotamers are those with the small vicinal couplings, i.e., **2** and/or **3**. The reverse is true, i.e., **1** and/or **4** are more stable if the average coupling decreases with temperature. To extract any meaningful quantitative conclusions regarding the relative stabilities of the various rotamers from this treatment of the system, however, is futile and almost impossible.

The treatment of the data becomes more manageable if one considers only two conformations at a time, e.g., **1** versus **2**, **3** versus **4**, or **1** versus **4**. Let us consider the eclipsed conformations **1** versus **2** and discuss the problem in some detail by using specific examples, such as **5** versus **6** (monosubstitution at the sp^3-hybridized carbon atom) and **7** versus **8** (disubstitution at the sp^3-hybridized carbon). Rotamer populations and free energy differences, $\Delta G°$, between one of the two energetically degenerate forms of **5** and **6** are

expressed by eqs. (4) and (5), where p is the fractional population of **5**, $1 - p$ that of **6**, and J_{av} is

$$J_{av} = p(J_t + J_g)/2 + (1-p)J_g \tag{4}$$

$$\Delta G° = -RT \ln \frac{1-p}{p/2} = -RT \ln \frac{(J_t + J_g - 2J_{av})}{(J_{av} - J_g)} \tag{5}$$

the observed $J_{H_\alpha H_1}$ time-averaged vicinal coupling. The corresponding equations for **7** and **8** (again, **7** versus one of the two energetically degenerate forms of **8**) are eqs. (6) and (7), where p is the fractional population of **7** and $1 - p$ that of **8**. Solution of eqs. (5) and (7), which require measurements of

$$J_{av} = pJ_t + (1-p)J_g \tag{6}$$

$$\Delta G° = -RT \ln \frac{1-p}{2p} = -RT \ln \frac{\frac{1}{2}(J_t - J_{av})}{(J_{av} - J_g)} \tag{7}$$

J_{av} at different temperatures, may lead* to J_t, J_g, and enthalpy differences, $\Delta H°$, between the various rotamers (2). A simpler method of calculating $\Delta H°$ is to estimate from the data (3) reasonable values for J_t and J_g and then calculate $\Delta H°$ from plots of log K_{eq} versus $1/T$, where K_{eq} for monosubstituted and disubstituted compounds, respectively, is expressed by eqs. (8) and (9). As pointed out (4), both methods of treating the data suffer from one

$$K_{eq} = 2(1 - p)/p \tag{8}$$

$$K_{eq} = (1 - p)/2p \tag{9}$$

limitation whose seriousness cannot be evaluated, namely, the implicit assumption that the potential minima are sharp enough to make contributions by torsional oscillations to the averaged coupling constants insignificant. Thus, changes in the coupling constant with changes in temperature are attributed solely to changes in the relative populations of the various rotamers, whereas, if the potential minimum is broad—as it is in one of the rotamers of fluoroacetyl fluoride (5)—significant changes in the vicinal coupling may be brought about by populating higher energy levels of one minimum energy conformation without significantly altering the relative populations of the various rotamers.

Let us now consider the bisecting conformations **9–12** as the minimum

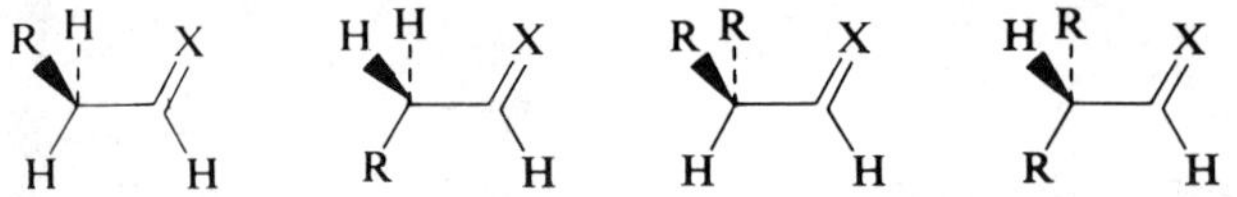

| (9) | (10) | (11) | (12) |

energy conformations. Equations (4) and (6) now take the forms of eqs. (10)

$$J_{av} = p(J_c + J_{120})/2 + (1 - p)J_{120} \tag{10}$$

$$J_{av} = pJ_c + (1 - p)J_{120} \tag{11}$$

and (11), where p is the fractional population of **9** and of **11** and $(1 - p)$ that of **10** and **12**. Similarly, eqs. (5) and (7) become eqs. (12) and (13). The similarities between the two sets of equations are quite obvious, especially

$$\Delta G° = -RT \ln \frac{1 - p}{p/2} = -RT \ln \frac{(J_c + J_{120} - 2J_{av})}{(J_{av} - J_{120})} \tag{12}$$

$$\Delta G° = -RT \ln \frac{1 - p}{2p} = -RT \ln \frac{\frac{1}{2}(J_c - J_{av})}{(J_{av} - J_{120})} \tag{13}$$

since $J_t > J_g$ and $J_c > J_{120}$. Indeed, the same set of data, i.e., the dependence

* By measuring J_{av} at four temperatures at least and by setting $\Delta G° = \Delta H° - T\Delta S°$, one can, in principle, solve for the four unknowns J_t, J_g, $\Delta H°$, and $\Delta S°$.

of J_{av} on temperature, may be treated by either set of equations and be consistent with either a pair of eclipsing or a pair of bisecting conformations. A way out of this dilemma, i.e., a choice between a set of bisecting and one of eclipsing conformations, is usually provided by the dependence of J_{av} on the dielectric constant of the solvent (6). To illustrate this point, let us consider dichloroacetaldehyde. If the minimum energy conformations are eclipsing (**13** and **14**), then increase in the dielectric constant of the solvent will decrease

(13)   (14)   (15)   (16)

J_{av}, as a result of increasing the concentration of the more polar rotamer **14** that has the smaller vicinal coupling constant (J_g). If the minimum energy conformations are bisecting, **15** and **16**, then the reverse will be observed, as the more polar rotamer **15** has the higher coupling constant ($J_c > J_{120}$).

The limitations of the nmr method become quite serious if one must choose between a set of two eclipsing conformations and a set of one eclipsing and one bisecting, e.g., **13** and **14** versus **13** and **15**. Since either **14** or **15** is more polar than **13** and both have vicinal coupling constants smaller than that of **13**, it becomes virtually impossible to make a choice between the two sets. The only good recourse that one has available to solve this problem is to know the values J_t/J_g and J_t/J_c for the system under examination. Unfortunately, in most cases all one can do is to guess these values by using arguments of analogy. This problem has been discussed in detail (7) and cogently illustrates the major weakness of the nmr method in rendering an unambiguous verdict in such cases of rotational isomerism. It is, therefore, important to realize that many conclusions drawn from nmr studies of this type on the nature of minimum energy conformations about sp^2-sp^3 carbon–carbon bonds need corroborative evidence from electron diffraction and microwave spectroscopic methods. It should also be made clear at this point that the designation throughout this text of the various conformations as "perfectly" eclipsing or bisecting no doubt represents an oversimplification. Deviations from these conformations are probably the rule rather than the exception.

II. CARBONYL COMPOUNDS

Carbonyl compounds, as a result of their easy accessibility, variety of forms in which they exist, and importance in chemistry, constitute the major class of compounds whose conformational analysis has been intensively investigated by practically all modern methods.

A. Aldehydes

1. Methodology

In 1957, Kilb, Lin, and Wilson (8) showed by microwave spectroscopy that acetaldehyde has a threefold barrier to rotation about the carbon–carbon bond, the minimum energy conformation of the compound being **17**. The barrier to rotation is small (9), 1100 ± 60 cal/mol. The conformation is not perfectly eclipsed as written in **17**, but somewhat distorted, as the C—H

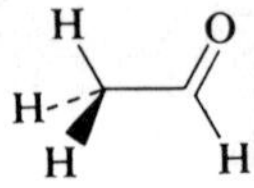

(17)

bond is out of the plane of the carbonyl by about 9°.

Similar microwave studies on propionaldehyde (10) established that the minimum energy conformations of this compound are also eclipsed (**18** and **19**), with **18** favored over **19** by about 900 ± 100 cal/mol. It was further

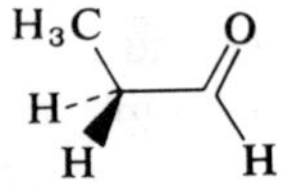

(18) 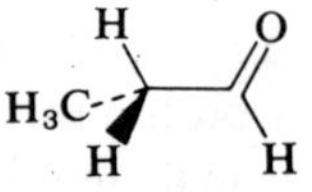   (19)

shown that **19** is not perfectly eclipsed, but again slightly distorted, as the C—H bond is out of the plane of the carbonyl by about 11 ± 6°. The electron diffraction studies of Bartell, Carroll, and Guillory (11) on isobutyraldehyde have also shown that the minimum energy conformations of this compound are **20** and **21** in a ratio of 9:1 in favor of **20**. The only aldehyde

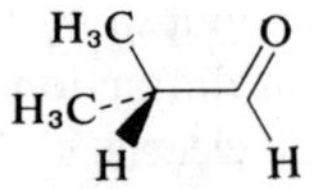

(20) 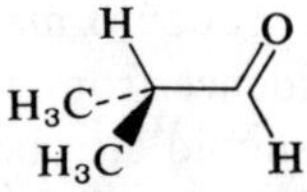   (21)

that has been shown by electron diffraction (11) and microwave spectroscopy (12) to exhibit a twofold barrier to rotation (one eclipsed and one staggered minimum energy conformation) is cyclopropanecarboxaldehyde, the two conformations **22** and **23** being of equal stability in the gas phase.

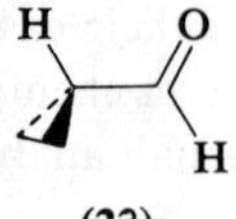

(22) 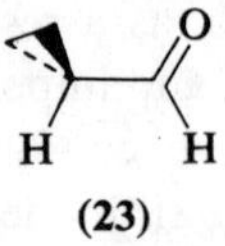   (23)

Cyclopropanecarboxaldehyde, however, should be considered as a special case, as the hybridization of the relevant carbon atomic orbital is sp^2 rather than sp^3. In this sense, it is more appropriate to place it in the group of α,β-unsaturated carbonyl compounds, rather than in the group of compounds under discussion in this section. The extended Hückel MO calculations of Hoffmann (13) on various aldehydes show that cyclopropanecarboxaldehyde falls in the same group with acrolein and benzaldehyde in exhibiting twofold barriers to rotation, rather than in the group of aliphatic aldehydes. Karabatsos and Hsi (6) interpreted their nmr data on this compound as consistent with either a twofold or a threefold barrier to rotation, but preferred a threefold barrier because of the rapid decrease of the vicinal coupling constant with temperature. As pointed out elsewhere (3) and in the preceding section of this review, nmr studies cannot render an unambiguous verdict in such cases and must yield to the evidence from electron diffraction and microwave studies.

Propionaldehyde was studied by nuclear magnetic resonance spectroscopy by Abraham and Pople (2), and, in good agreement with the results obtained from microwave spectroscopic studies (10), it was found that **18** is favored, enthalpywise, over **19** by about 1000 cal/mol. Several monosubstituted and disubstituted acetaldehydes have been studied by Karabatsos and his co-workers (3,4,6,7) by nmr spectroscopy and the enthalpy differences between the various rotamers have been summarized in Table I. In addition to cyclopropanecarboxaldehyde, it appears that glycidaldehyde also exhibits a twofold barrier to rotation (3). Analysis of the microwave spectrum of glycidaldehyde (14) shows this compound to exist essentially as rotamer **24** in the gas phase, with no other rotamer having been detected. In the liquid state, the less stable rotamer, presumably **25**, is also detectable (3).

(24)   (25)

There are very few results available in the literature from other studies with which to compare those summarized in Table I. The -800 cal/mol enthalpy value for propionaldehyde agrees with those already mentioned (2,10). The qualitative conclusions of Buchanan, Stothers, and Wu (15) on cyclohexanecarboxaldehyde and the results of Bartell and his co-workers (11) on isobutyraldehyde are also consonant with the values given in Table I. The previous conclusion from infrared studies (16) that chloroacetaldehyde exists essentially in conformation **26** conflicts with that drawn from the nmr studies and was shown to be incorrect (4).

TABLE I
Enthalpy Differences, $\Delta H°$, between Rotamers of Aldehydes

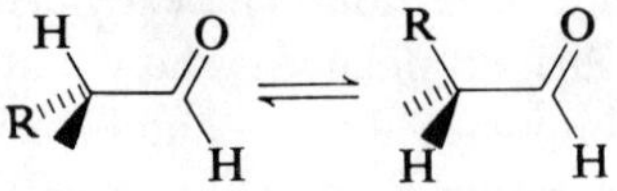

Aldehyde	Solvent	$\Delta H°$ (cal/mol)	Method	Ref.
MeCH$_2$CHO	Gas	− 900	Microwave	10
MeCH$_2$CHO	Neat	−1000	nmr	2
MeCH$_2$CHO	Neat	− 800	nmr	6
EtCH$_2$CHO	Neat	− 700	nmr	6
n-PrCH$_2$CHO	Neat	− 600	nmr	6
n-AmCH$_2$CHO	Neat	− 500	nmr	6
i-PrCH$_2$CHO	Neat	− 400	nmr	6
t-BuCH$_2$CHO	Neat	+ 250	nmr	6
ϕCH$_2$CHO	Neat	− 300	nmr	6
Me$_2$CHCHO	Neat	− 500	nmr	6
Et$_2$CHCHO	Neat	+ 250	nmr	6
(t-Bu)$_2$CHCHO	CCl$_4$	+1100	nmr	6
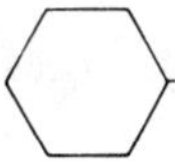 cyclohexyl-CHO	Neat	− 400	nmr	6
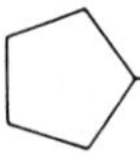 cyclopentyl-CHO	Neat	ca. 0	nmr	6
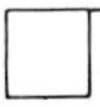 cyclobutyl-CHO	Neat	− 150	nmr	6
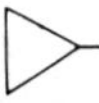 cyclopropyl-CHO	Gas[a]	ca. 0	Electron diffraction	11
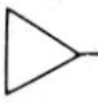 cyclopropyl-CHO	Gas[a]	ca. 0	Microwave	12
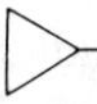 cyclopropyl-CHO	Neat	ca. +1000	nmr	6
ClCH$_2$CHO	trans-Decalin	− 300	nmr	4
ClCH$_2$CHO	(CH$_3$)$_2$NCHO	−2100	nmr	4
BrCH$_2$CHO	trans-Decalin	ca. 0	nmr	4
BrCH$_2$CHO	(CH$_3$)$_2$NCHO	−1500	nmr	4
CH$_3$OCH$_2$CHO	trans-Decalin	ca. − 400	nmr	3
CH$_3$OCH$_2$CHO	(CH$_3$)$_2$NCHO	ca. −1500	nmr	3
ϕOCH$_2$CHO	trans-Decalin	ca. − 400	nmr	3
ϕOCH$_2$CHO	(CH$_3$)$_2$NCHO	ca. −1500	nmr	3

TABLE I (*Continued*)

Aldehyde	Solvent	$\Delta H°$ (cal/mol) 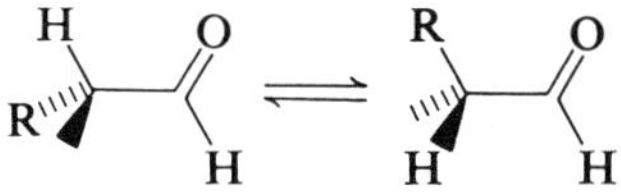	Method	Ref.
CH₃SCH₂CHO	*trans*-Decalin	+1000	nmr	3
CH₃SCH₂CHO	(CH₃)₂NCHO	+ 500	nmr	3
Cl₂CHCHO	*trans*-Decalin	+ 300	nmr	7
Cl₂CHCHO	(CH₃)₂NCHO	−1400	nmr	7
Br₂CHCHO	*trans*-Decalin	+ 500	nmr	7
Br₂CHCHO	(CH₃)₂NCHO	− 500	nmr	7
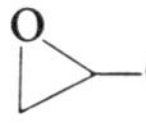 cyclopropyl-CHO	Neat	> 0	nmr	7
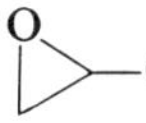 cyclopropyl-CHO	Gas	> +2000	Microwave	14

ᵃ Twofold barrier to rotation.

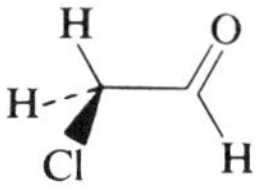

(26)

2. Factors Affecting Rotamer Stabilities

Since one of the major objectives of the various conformational analysis studies is to understand the factors that influence rotamer stabilities, we shall discuss this point in some detail.

a. Solvent Dielectric Constant. If the dipole moments of individual rotamers are different, the rotamer populations will depend on the dielectric constant of the solvent. This dependence will be particularly pronounced whenever the difference between the dipole moments of individual rotamers is large. For aliphatic aldehydes the effect of solvent on rotamer stability is, as expected, small. The nmr data were interpreted (6) in terms of **27** and **28** with the concentration of **27** increasing at the expense of that of **28** as the

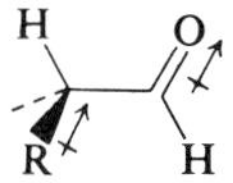

(27) 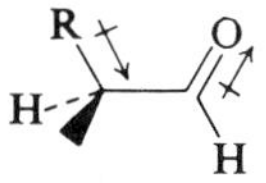   (28)

dielectric constant of the medium increase. In energy terms this increase corresponded to about 100 cal/mol in going from cyclohexane ($\epsilon \sim 2$) to acetonitrile ($\epsilon \sim 36$) and was sufficient to reverse the stabilities of the rotamers of cyclopentanecarboxaldehyde (6). The difference was considerably higher, about 400 cal/mol, in the case of cyclopropanecarboxaldehyde. This observation is certainly consonant with the view that cyclopropanecarboxaldehyde is more similar to α,β-unsaturated aldehydes, whose rotamers ought to have relatively large dipole moment differences, than to alkyl-substituted acetaldehydes.

The trend was reversed with phenylacetaldehyde as the concentration of **30** increased at the expense of **29** with increase of the dielectric constant of

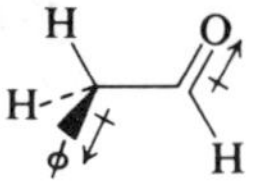

(29) 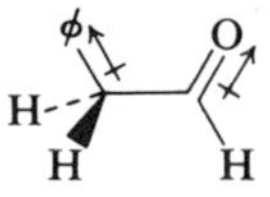   (30)

the medium. In terms of energy this increase corresponded to about 350 cal/mol in going from cyclohexane to acetonitrile (6).

It was not surprising to find that the relative stability of the rotamers of aldehydes bearing a halogen atom or an oxygen function at the sp^3-hybridized carbon changed enormously with large changes in the dielectric constant of the medium. Thus, in going from *trans*-decalin to N,N-dimethylformamide the enthalpy changes in the direction of the more polar rotamer **32** were about 1800 cal/mol for chloroacetaldehyde (4), 1500 cal/mol for bromoacetaldehyde

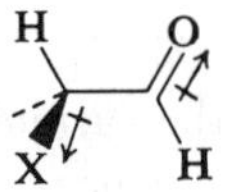

(31) 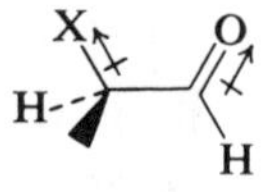   (32)

(4), 1100 cal/mol for methoxyacetaldehyde and phenoxyacetaldehyde (3), 1700 cal/mol for dichloroacetaldehyde (7), and 1000 cal/mol for dibromoacetaldehyde (7). In the case of dibromoacetaldehyde, the relative stability of **31** and **32** was inverted, **31** being more stable by 500 cal/mol in *trans*-decalin and less stable by 500 cal/mol in N,N-dimethylformamide. It was pointed out (7) that the inversion occurs in a medium of a dielectric constant of about 9, provided the medium is nonaromatic. In aromatic solvents, specific interactions between solute and solvent often overshadow the relationship between rotamer stabilities and the dielectric constant of the medium (7). Since the carbon–chlorine and carbon–oxygen bonds are more polar, respectively, than the carbon–bromine and the carbon–sulfur bonds, it was understandable to find more pronounced changes in the rotamer populations of chloro- than of bromo- and of oxygen than of sulfur compounds. Indeed, the least pronounced variation with solvent dielectric was observed with methylmercaptoacetaldehyde, the compound with the least polar bond (C—S).

It was pointed out (3) that, whereas the cyclopropane ring of cyclopropanecarboxaldehyde donates electronic charge (π–π electron orbital overlap), the oxirane ring of glycidaldehyde withdraws it. The oxirane ring, however, acts as a much weaker electron-withdrawing group than might be expected for an alkoxy group. In this sense, an epoxy substituent acts as a hybrid between a cyclopropyl and an alkoxy group, the extent of its electron-withdrawing effect placing it closer to cyclopropyl than to alkoxy.

b. Substituent on sp^3-Hybridized Carbon. *1. Alkyl Substituted Acetaldehydes.* Part of the effect of a single alkyl α-substituent on the relative stabilities of **27** and **28** has been explained by Karabatsos and Hsi (6) in

(33)   (34)   (35)   (36)

terms of formulations **33–36**. Let us take as a starting point of our argument propionaldehyde, whose rotamer with the methyl group eclipsing the carbonyl (**33**) is more stable than the rotamer with a hydrogen eclipsing the carbonyl by 800 cal/mol. When the methyl is substituted by an *n*-alkyl group, the corresponding rotamer (**34**) is not destabilized appreciably, as the R group may be positioned away from the carbonyl oxygen. Thus, in going from methyl to ethyl, to *n*-propyl, to *n*-amyl, the enthalpy favoring this rotamer changes only from 800 to 700, to 600, to 500 cal/mol. This effect is analogous to the one that an *n*-alkyl substituent has on the relative stabilities of the equatorial and axial alkylcyclohexanes. When the β-carbon is disubstituted, as in **35**, then one of the alkyl groups finds itself close to the carbonyl oxygen, and the energy of this rotamer increases. Finally, when the β-carbon is trisubstituted (**36**), the alkyl–oxygen nonbonded interactions are sufficiently repulsive to make the *gauche* rotamer (**27**) more stable. Thus, rotamer **27** of *t*-butylacetaldehyde is now favored over **28** by about 250 cal/mol.

An analogous explanation to that just given has also been suggested (6) to account for the facts that, whereas $\Delta H°$ for ethylacetaldehyde is -700 cal/mol, $\Delta H°$ for diethylacetaldehyde is $+250$ cal/mol and whereas $\Delta H°$ for *t*-butylacetaldehyde is $+250$ cal/mol, that for *di-t*-butylacetaldehyde is $+1100$ cal/mol.

Although the above explanations account reasonably well for the effects that changes in the alkyl groups produce on rotamer stabilities, they do not answer the basic question, namely, why the *cis* rotamer of propionaldehyde is

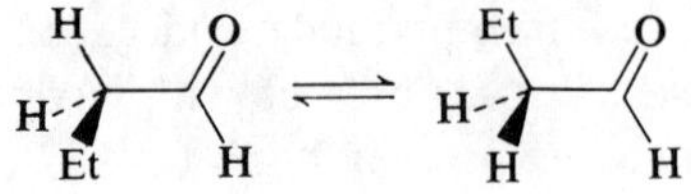

$$\text{(H,H,Et,Et,H,O)} \rightleftharpoons \text{(H,Et,H,H,H,O)} \qquad \Delta H° = -700 \text{ cal/mol}$$

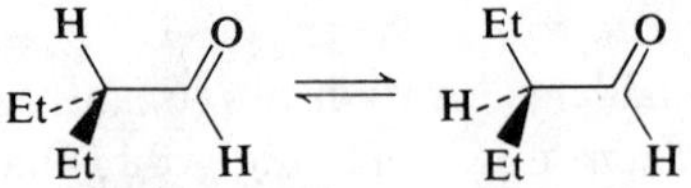

$$\text{(H,Et,Et,Et,H,O)} \rightleftharpoons \text{(Et,H,H,Et,H,O)} \qquad \Delta H° = +250 \text{ cal/mol}$$

favored over the *gauche* by about 900 cal/mol. On the basis of nonbonded interactions one might intuitively have expected the reverse to be true. To gauge the contributions of nonbonded interactions to the relative stabilities of the two rotamers of propionaldehyde, Sonnichsen and Karabatsos (17) calculated the nonbonded interactions associated with the two rotamers. They obtained a value of 170 cal/mol in favor of **18**, the *cis* rotamer. Although this value is in the opposite direction of what chemical intuition might have led to believe,* it is nevertheless considerably smaller than the 900 cal/mol experimental value. The nature of the factor responsible for the considerably greater stability of **18** over **19** therefore, remains to be uncovered.

2. *Heteroatom-Substituted Acetaldehydes.* Whereas the rotamer stabilities of alkyl-substituted acetaldehydes do not depend to any great extent on dipole–dipole or dipole-induced dipole interactions, those of acetaldehydes substituted with a heteroatom on the sp^3-hybridized carbon atom ought to depend strongly on such factors. Electrostatic dipole–dipole interactions probably are responsible for the fact that, whereas $\Delta H°$ for chloroacetaldehyde and bromoacetaldehyde in *trans*-decalin ($\epsilon \sim 2$) are -300 and 0 cal/mol, respectively, they are $+300$ and $+500$ cal/mol for dichloro- and dibromoacetaldehyde. Karabatsos and Fenoglio, (4) have argued that such polar interactions, which favor **37** over **38** and **39** over **40**, lead to greater energy differences between the rotamers of the disubstituted than between those of the monosubstituted compounds, on account of larger differences in the dipole moments of the rotamers of the disubstituted compounds. As pointed out

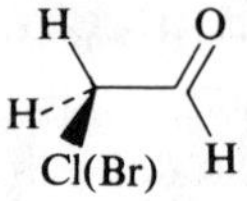

(37) H, H, H, Cl(Br), O, H

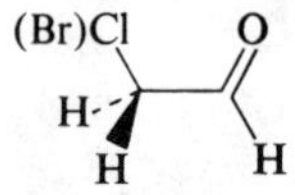

(38) (Br)Cl, H, H, H, O, H

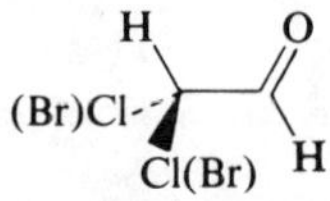

(39) H, (Br)Cl, Cl(Br), H, O, H

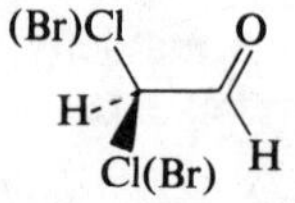

(40) (Br)Cl, H, Cl(Br), H, O, H

*It must, however, be recalled that nonbonded interactions may be attractive as well as repulsive.

(4), however, these interactions cannot be the principal factor controlling rotamer stabilities, as **38** is more stable than **37** by 300 cal/mol in the case of chloroacetaldehyde, and by 400 cal/mol in the cases of methoxy- and phenoxyacetaldehyde.

On the basis that in the low dielectric constant solvent *trans*-decalin the enthalpy for bromoacetaldehyde (0 cal/mol) is more positive than that for chloroacetaldehyde (-300 cal/mol), that of methylmercaptoacetaldehyde ($+1000$ cal/mol) more positive than that of methoxyacetaldehyde (-400 cal/mol), and that of dibromoacetaldehyde ($+500$ cal/mol) more positive than that of dichloroacetaldehyde ($+300$ cal/mol), it has been argued (3) that the dipole-induced dipole interactions are not significant enough to control the relative stabilities of the relevant rotamers. Had they been so, the reverse would have been true.

Karabatsos and Fenoglio (3) have summarized the relative stabilities of

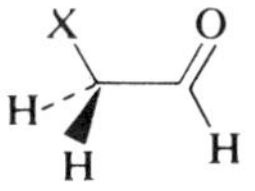

(41) 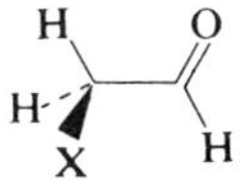   (42)

41 and **42** as follows:

$X = CH_3 > CH_2CH_3 \sim CH_3O \sim C_6H_5O > CH(CH_3)_2$

$> C_6H_5 \sim Cl > Br > C(CH_3)_3 > CH_3S$

Increased stability ⟵⟶ Increased stability
of **41** of **42**

The sequence indicated is valid only in solvents of low dielectric constant, such as saturated hydrocarbons. In solvents of high dielectric constant, the methoxy, phenoxy, chloro, and bromo groups move ahead of the methyl group. For groups preceding bromine in the indicated order, $\Delta H°$ favors **41**, whereas for those following bromine, it favors **42**. On the basis of what has been presented up to now, we would predict that the fluoro group will be as good as methoxy in favoring **41**, and that the iodo group will be as bad as or worse than methylmercapto.

B. Ketones

The conformational analysis of ketones has been confined primarily to infrared and Raman spectroscopy studies—the majority of which are qualitative—and to a few investigations by electron diffraction, microwave, and nmr spectroscopy. The paucity of quantitative studies on ketones, when compared to those on the aldehydes, is understandable, as—spectroscopically speaking—most ketones are quite complex molecules.

1. Aliphatic

The microwave spectrum of acetone has been interpreted (18,19) in terms of **43**, the eclipsed conformation of C_{2v} symmetry. The two methyl

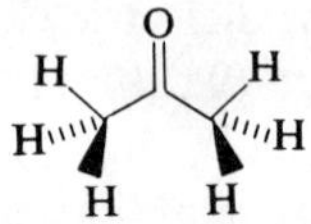

(43)

groups were found to be "tilted" by about 1.3° toward the double bond, a phenomenon that appears to be general with methyl groups bonded to double bonds (18). The barrier to rotation was found to be relatively small, about 778 cal/mol.

There is a dearth of information regarding the effect that alkyl substitution on the α-carbon has on rotamer populations. Electron diffraction studies (20) on diethyl ketone have shown that the most stable rotamer of this compound is **44**, a conclusion that is certainly reasonable and is in accord with

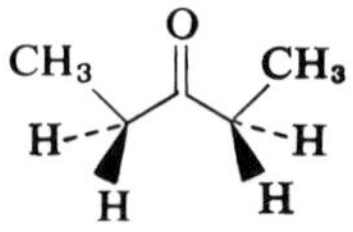

(44) 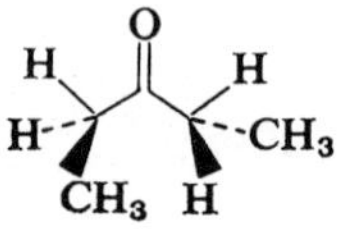   (45) 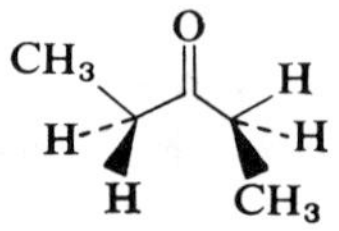   (46)

what was found to be the major rotamer of propionaldehyde. On the basis of the infrared and Raman spectra of this compound, Jones and Noack (21) suggested that **45** might be the major rotamer of diethyl ketone. Their conclusion, however, was speculative and based on the intuitive assumption that nonbonded interactions would be quite repulsive in **44**. By applying the more recent model (22) for the anisotropic effects of the carbonyl group, Karabatsos and his co-workers interpreted the temperature dependence of the chemical shifts (21) of diethyl ketone in favor of **44**, not **45**. In fact, **46** would be expected to be more stable than **45**.

It should be recognized at this point that the relative populations of **47** and **48** ought to depend on R′, with the concentration of **48** increasing at

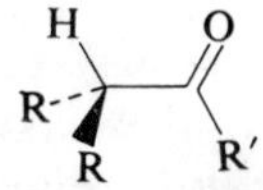

(47) 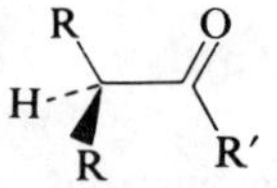   (48)

the expense of **47** as R′ increases in size. From studies on the chemical shifts of the α-hydrogens, it was concluded (22) that this is indeed the case. Thus,

ROTATIONAL ISOMERISM ABOUT sp^2-sp^3 BONDS

t-butyl *n*-alkyl ketones exist essentially in conformation **49**. If the minimum energy conformation of pivaldehyde is **50**, i.e., eclipsing, then the *t*-butyl

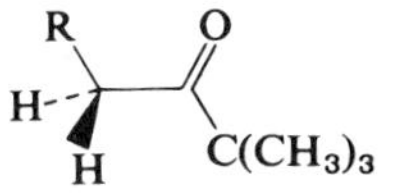

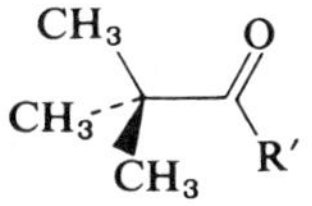

group of all other *t*-butyl alkyl ketones (provided R' is not trisubstituted) also exists in this conformation (22). It was further concluded that the conformation of the R' group in **50**, when R' is a disubstituted alkyl group, is bisecting (**51**) rather than eclipsing (**52**). Thus, the extremely large downfield

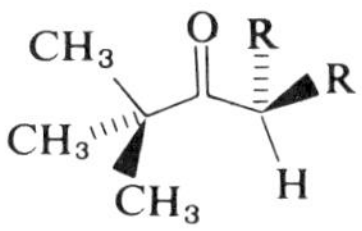

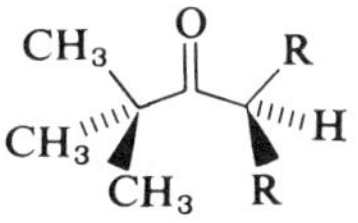

shift ($\tau \sim 7$) of the resonance of the α-methine proton of *t*-butyl diethylcarbinyl ketone was interpreted in support of **51** rather than in support of any of the other conformations. Such a conclusion is not only consistent with the model, but also reasonable if one considers the energetically unfavorable 1,3-eclipsing interactions between methyl and alkyl in **52**. It should be reiterated, however, that although conformations are written for simplicity as perfectly eclipsing or bisecting, they may deviate considerably from such ideal models.

From nmr studies, Gough, Lin, and Woolford (23) also concluded that the major rotamers of various 2-substituted 3-pentanones (X = C_6H_5, NO_2, OCH_3, F, Cl, Br, I, SH) are those with the methyl groups eclipsing the carbonyl (**53**).

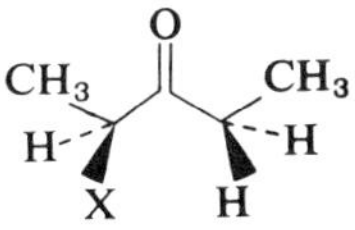

As was the case with aldehydes, the cyclopropyl group behaves more as a vinyl than as an alkyl group. Thus, electron diffraction studies of methyl cyclopropyl ketone (24) have been interpreted in terms of a twofold barrier to rotation, with **54** and **55** as the minimum energy rotamers of the compound.

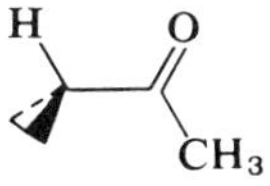

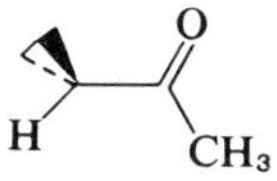

It is not clear from the data, however, how much distorted rotamer **54** is from the perfectly eclipsed conformation. Whereas in cyclopropanecarboxaldehyde the analogous rotamers were equally populated, in the present case **55** is more stable than **54** by a factor of 4. This difference has been attributed to nonbonded repulsions between methyl and cyclopropyl in **54** that destabilize it in favor of **55**. In microwave spectroscopic studies on this compound, Schwendeman and Lin (25) have been able to detect only **55**.

2. α-Heteroatom Substituted

In 1953, Mizushima and his co-workers (26) published a series of infrared and Raman spectroscopic studies on chloroacetone. They concluded that the minimum energy rotamers of the compound were **56** and **57**, with rotamer

(56)   (57) $\theta = 30°$

57 being considerably more stable than **56** in the gas phase and of equal stability with **56** in the neat liquid ($\epsilon \sim 30$). If one were to compare the relative stabilities of **56** and **57** in the neat liquid with those of the corresponding rotamers of chloroacetaldehyde, where the rotamer with the chlorine group eclipsing the carbonyl is favored (neat liquid) by at least 1000 cal/mol (4), one would conclude that the nonbonded interactions between methyl and chlorine in **57** are not repulsive or that, if repulsive, they are completely overshadowed by some other factor. We prefer to view these results as another example where attractive, rather than repulsive, nonbonded interactions govern the stabilities of the rotamers.

More recent infrared studies on fluoro- (27), bromo- (28), and iodoacetone (28) have been interpreted in terms of two rotamers, presumably analogs of **56** and **57**. The conclusions reached were as follows. In all cases, the less polar rotamer, presumably **57**, is more abundant than **56** in the gaseous state. In the liquid state, the two rotamers of fluoroacetone and iodoacetone are present in equal concentrations. Those of bromoacetone are present in unequal concentrations, with the more polar rotamer being favored over the less polar. In the solid state, the more polar rotamer is the more stable of the two; in fact, it is the only detectable rotamer in the case of bromoacetone. Similar studies on *sym*-difluoroacetone (29) and *sym*-dichloroacetone (30) have led to the conclusion that, whereas in the gas phase the less polar rotamers are more stable than the more polar ones, in the liquid phase the reverse is true. From infrared studies on α-chloroacetophenone (31) and various acetophenones (32) monosubstituted (Cl, Br, I, CN, OH, C_6H_5) and disubstituted (Cl, Br) at the α-carbon atom, conclusions have been drawn that parallel all those cited earlier, namely, that the more stable rotamer in solution is the one where the substituent eclipses the carbonyl group. The qualitative nature of all the above-mentioned studies, however, precludes any meaningful comparisons in terms of enthalpy differences that might have been useful in understanding some of the factors influencing rotamer stabilities in this series of compounds.

C. Acyl Halides

In addition to acetyl fluoride (33), chloride (34), bromide (35), and iodide (36), whose microwave spectra have been interpreted in terms of a threefold barrier to rotation with 58 as the minimum energy conformation,

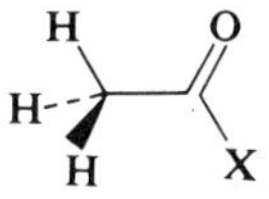

(58)

haloacetyl halides constitute the major class of acyl halides whose conformation has been studied. From infrared and Raman spectroscopic studies, Mizushima and his co-workers (37) concluded that 59 and 60 represent the minimum energy conformations of chloroacetyl chloride, bromoacetyl chloride, and bromoacetyl bromide. In the gas phase, 59 is more stable than 60 by about 1000 cal/mol for bromoacetyl chloride and by about 1900 cal/mol for bromoacetyl bromide. A value of about 1200 cal/mol favoring 59 over 60

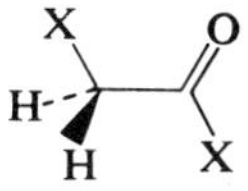

(59) 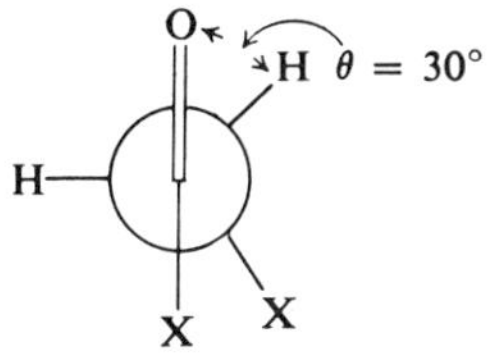 (60)

for chloroacetyl chloride was estimated (7) from the data of Morino, Kuchitsu, and Sugiura (38). In the liquid state the concentration of 60, the more polar rotamer, increases at the expense of 59 to the point where the two rotamers are equienergetic in the case of chloroacetyl chloride, although not so in the case of the other two halides (59 is still more stable than 60). On the basis of their infrared studies, Bellamy and Williams (39) reached similar conclusions on chloroacetyl chloride. In the solid state these compounds were found to exist essentially in conformation **59**. In contrast to these halides, fluoroacetyl fluoride exhibits a twofold barrier to rotation in the gas state (5), the minimum energy conformations being **59** (eclipsing) and **61** (bisecting). This change has been rationalized in terms of differences in nonbonded interactions

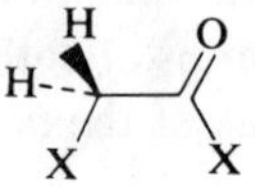

(61)

between the two halogens. Thus, when the X groups are the small fluorine atoms, their repulsion is very small, whereas when they are the larger chlorine and bromine atoms, their repulsion is sufficiently large to change the equilibrium conformation from **61** to **60**. In agreement with this explanation is the fact that the potential minimum of **61** of fluoroacetyl fluoride is quite broad (5). The more stable rotamer of fluoroacetyl fluoride, by about 900 cal/mol, is again **59**.

Dihalo substitution at the α-carbon reverses the relative stabilities of the rotamers. Thus, $\Delta H°$ favors **62** over **63** by 200 cal/mol (40). The reason for

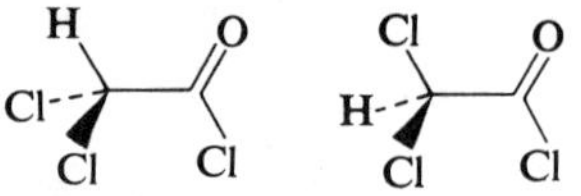

(62)   (63)

this reversal is presumably the one already given for mono- and dihalosubstituted acetaldehydes, namely, the larger difference between the dipole moments of **62** and **63** than between those of **59** and **60**.

The chloride and fluoride of cyclopropanecarboxylic acid have been studied by electron diffraction (24) and microwave spectroscopy (12), respectively. For both compounds the barrier to rotation was found to be twofold. In both cases **65** is more stable than **64**, the ratio **65:64** being 74:26 in the case of the fluoride.

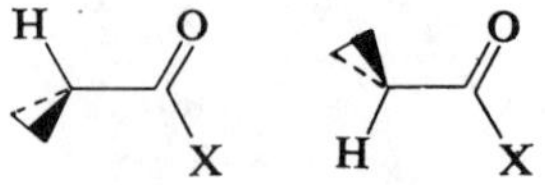

(64)   (65)

Two α-trisubstituted acyl halides have been investigated by infrared spectroscopy. They are trichloroacetyl chloride (39) and 2-bromo-2-methylpropionyl bromide (40). Since a single carbonyl stretching vibration was observed for the former, it was concluded that the compound exists in a single minimum energy conformation whose structure remains unknown and presumably is the one with a carbon–chlorine bond eclipsing the carbonyl. Crowder and Northam (41) concluded that 2-bromo-2-methylpropionyl bromide exists as two rotamers, with the less polar rotamer being more stable in both the liquid and the vapor states. They suggested structures **66** and **67**, respectively, for the less and more polar rotamer. The choice of **67** over **68**,

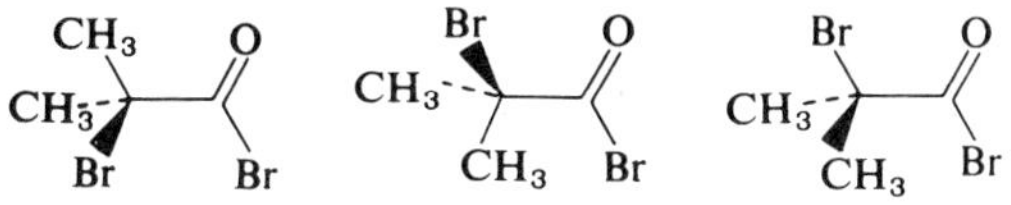

(66)   (67)   (68)

however, was based on the reasoning that "inspection of molecular models shows that for the *trans** form of the haloacetyl halides, the halogen of the CH_2X group probably overlaps the oxygen and it does not seem reasonable that this would be a stable form since there would be considerable electrostatic interaction. What seems more reasonable is that in one of the two forms, a hydrogen is *cis* to the halogen of the COX group (this would be the more polar form), and in the other form, a hydrogen is *cis* to the oxygen...." The weakness of this intuitive argument is obvious, as microwave spectroscopy has shown that there is nothing wrong with **59** in fluoroacetyl fluoride. Our own choice of the two minimum energy rotamers would be **66** and **68** rather than **66** and **67**.

D. Carboxylic Acids, Esters, and Related Compounds

In consonance with the previously mentioned microwave spectroscopic results for carbonyl compounds, analogous findings concerning the structure of acetic acid (42), methyl acetate (43), and acetyl cyanide (44) have been interpreted in terms of **58** as the minimum energy conformation of these compounds. Conformational studies of α-substituted compounds of this type have been confined to a few infrared and Raman spectroscopic investigations.

From infrared spectroscopic studies T. L. Brown concluded (45) that the minimum energy conformations of ethyl α-haloacetates are **69** and **70**. In chloroform or carbon tetrachloride solution, enthalpy differences in favor of

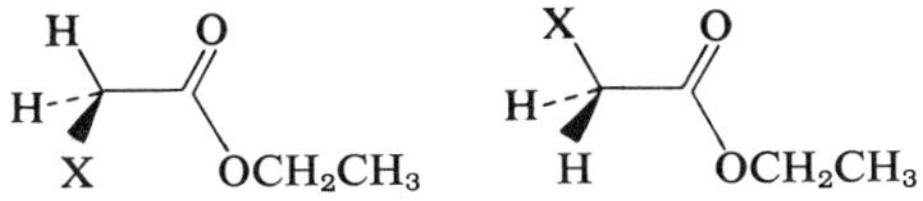

(69)   (70)

*That is, the form with the halogen atoms *trans* or *anti* to each other.

70 were obtained of 560 ± 100 and 500 ± 100 cal/mol for ethyl α-fluoro- and α-chloroacetate, respectively. The enthalpy difference for ethyl α-bromoacetate was found to be 0 ± 100 cal/mol. This trend, i.e., destabilization of **70** with respect to **69** in going from the α-chloro to the α-bromo compound, is analogous to that observed for the α-haloacetaldehydes (7).

The most interesting conclusion drawn by Brown was that the minimum energy conformations of ethyl α,α-difluoro- and α,α-dichloroacetate are now **71** and **72**. Thus, in contrast to the monohalo compounds, the dihalo-

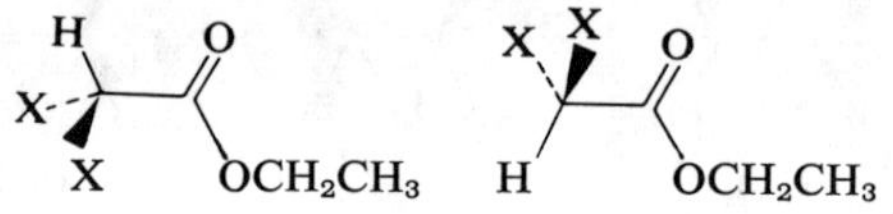

(71)   (72)

substituted compounds exhibit a twofold barrier to rotation about the carbon–carbon bond. The enthalpy differences for **71** ⇌ **72** were found to be +25 ± 100 cal/mol and 0 ± 100 cal/mol for the ethyl α,α-difluoro- and α,α-dichloroacetate, respectively. Whereas the finding that $\Delta H°$ for conversion of the less polar to the more polar conformer is more positive for the dihalo compounds (**71** ⇌ **72**) than for the monohalides (**69** ⇌ **70**) parallels that observed for the corresponding aldehydes (4,7) and acyl halides (37,40) and may be interpreted in terms of differences in the dipole moments of the rotamers, the change in the nature of one of the minimum energy conformations (**70** versus **72**) is novel. We have no good explanation for this change, but we would like to point out that these results serve as a cogent example of how dangerous conclusions drawn from arguments of analogy may be.

Mizushima and his co-workers (46) interpreted their data from infrared and Raman spectroscopy and dipole moment studies on N-methylchloroacetamide in terms of **73** and **74** as the minimum energy conformations of this

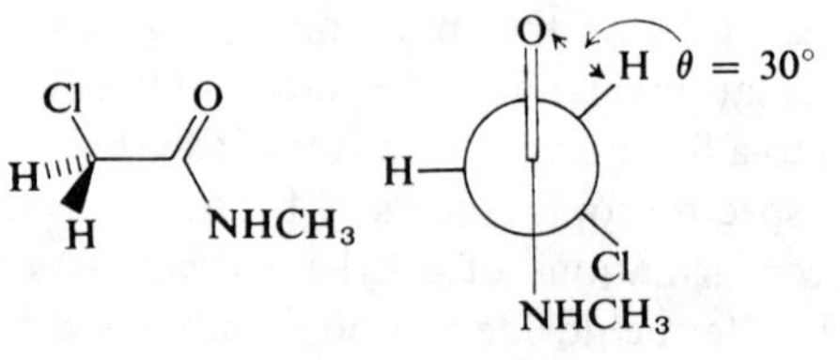

(73)   (74) $\theta = 30°$

compound. In the solid and gaseous states and in nonpolar solvents, only **74** was found to be present. The two rotamers were judged to be of approximate stability in the neat liquid, whose dielectric constant is about 100. The authors (46) have given an excellent explanation in terms of electrostatic dipole–dipole interactions of why **74** is so much more stable than **73**, when the reverse is true with haloacetyl halides and with chloroacetone, haloaldehydes (4), and ethyl haloacetates (44). Their explanation may be better understood by considering resonance structures **75** and **76**, which focus

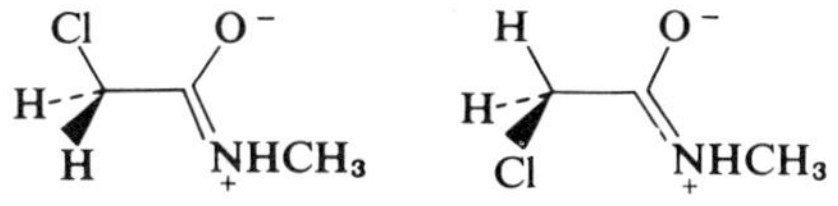

(75)   (76)

attention on the large difference between the dipole moments of the two rotamers. The difference in electrostatic dipole–dipole interactions, which favors **74** over **73**, is much larger for this compound than is the analogous difference between all the other halo compounds.

III. OLEFINS

A. 3-Substituted

From analysis of the microwave spectra of propene (47) and 3-fluoropropene (48) it was concluded that structures **77–79** are the minimum energy

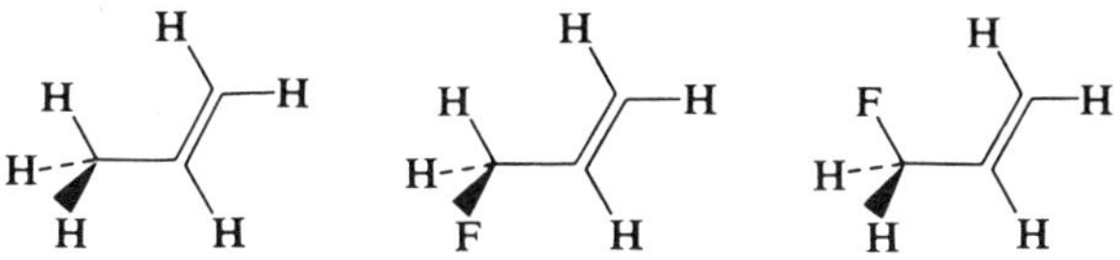

(77)   (78)   (79)

conformations of these compounds. For the monofluoro compound, rotamer **79** was favored over **78** by 167 ± 67 cal/mol. Bothner-By and his co-workers (49–53) have addressed themselves to the problem of the relative stabilities of **80** and **81** by studying the vicinal proton–proton couplings. Before embarking

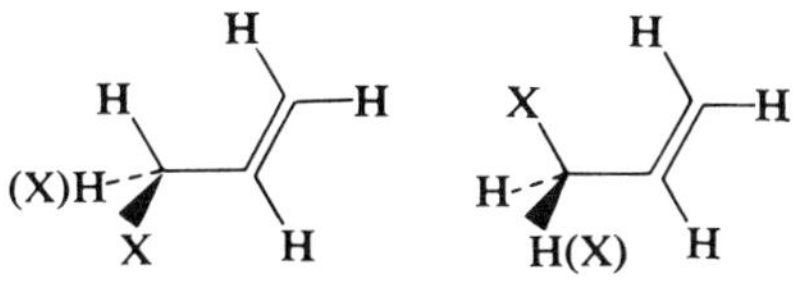

(80)   (81)

on a discussion of their results, we wish to reiterate that the nmr data are consistent with **80** and **81** as the minimum energy conformations, but they do not necessarily prove that these are the only conformations of these compounds.

Various enthalpy differences for **80** ⇌ **81** have been summarized in Table II. The values of about +400 cal/mol for 1-butene and 3-methyl-1-butene were estimated by Karabatsos and Taller (54) from the relative magnitudes of the vicinal coupling constants of these compounds (49,50) and that of propene was estimated (49) by taking into account the effect that an alkyl group has on the average vicinal coupling constant. For this reason, we consider this estimate more realistic than that of Bothner-By and his co-workers ($\Delta H°$ of about zero). It is worth pointing out that the recent calculations of Allinger and his group (56) led to a value of +690 cal/mol.

TABLE II

Enthalpy Differences, $\Delta H°$, between Rotamers of 3-Substituted Propenes

Propene	$\Delta H°$ (cal/mol) for **80** ⇌ **81**	Method	Ref.
$MeCH_2CH=CH_2$	ca. + 400	nmr	49[a]
$n\text{-}PrCH_2CH=CH_2$	ca. > + 400	nmr	49[a]
$i\text{-}PrCH_2CH=CH_2$	≫ + 400	nmr	50[a]
$t\text{-}BuCH_2CH=CH_2$	≫ + 1000	nmr	50
$Me_2CHCH=CH_2$	> + 400	nmr	50[a]
$(t\text{-}Bu)_2CHCH=CH_2$	≫ + 1000	nmr	51
$\phi CH_2CH=CH_2$	+ 100	nmr	51
$CNCH_2CH=CH_2$	< 0	nmr	55
$FCH_2CH=CH_2$	− 170	Microwave	48
$FCH_2CH=CH_2$	− 100	nmr	52
$ClCH_2CH=CH_2$	+ 100	nmr	53
$BrCH_2CH=CH_2$	> + 500	nmr	51
$ICH_2CH=CH_2$	≫ + 1000	nmr	51
$CH_3OCH_2CH=CH_2$	− 115	nmr	53
$F_2CHCH=CH_2$	+ 500 to + 1500	nmr	52
$Cl_2CHCH=CH_2$	+ 800	nmr	53
$(CH_3O)_2CHCH=CH_2$	− 110	nmr	53

[a] The values in this table are recalculated (54); see text.

For most compounds, except 3-fluoropropene, 3-methoxypropene, 3-cyanopropene (55), and 3,3-dimethoxypropene, rotamer **80** is favored over **81**. This fact should be contrasted with that pertaining to the corresponding aldehydes, where the rotamer with the group X eclipsing the carbonyl was more stable. We ascribe this reversal primarily to larger nonbonded repulsions (or smaller nonbonded attractions) in rotamer **81** than in the corresponding rotamer of aldehydes. Aside from this difference, the trends observed with 3-substituted propenes and aldehydes are quite similar. For example, as the size of the alkyl group increases, the stability of **80** increases to the point where 3-*t*-butylpropene exists essentially in conformation **80**. As the size of the halogen increases from fluorine to iodine, the stability of **80** over that of **81** also increases. Furthermore, in going from the monohalo to the dihalo compounds the stability of **80** over **81** also increases, as was the case with various carbonyl compounds. Bothner-By and his co-workers have rationalized this change (52) in terms of decreased attractive interactions between X and the olefinic proton in rotamer **81** as being due to the C—X bond being less polar in the dihalo than in the monohalo compounds. The only disturbing exception to this trend involves the 3-methoxypropene and the 3,3-dimethoxypropene, where $\Delta H°$ for **80** $\rightleftharpoons$ **81** is the same for both compounds.

Analysis of the infrared spectra of 3-chloro, 3-bromo-, and 3-iodopropene led Radcliffe and Wood (57) to conclusions similar to those drawn by Bothner-By and his co-workers, namely, that **80** is more stable than **81** in all cases, with the iodo compound existing essentially in conformation **80** in the gaseous, liquid, and solid states.

The cyclopropyl problem encountered in the nmr studies of cyclopropanecarboxaldehyde has also shown up in the interpretation of the nmr spectra of vinylcyclopropanes. Lüttke and de Meijere (58) have interpreted the temperature dependence of the relevant vicinal coupling of hexadeuteriovinylcyclopropane in terms of **82** and **83** (twofold barrier to rotation) as the

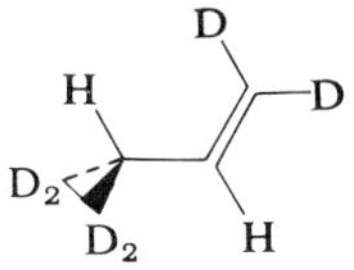

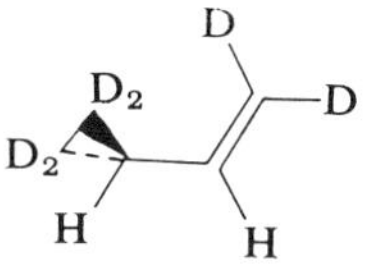

minimum energy conformations of the compounds. They concluded that **82** is favored over **83** by about 1100 ± 200 cal/mol. On the other hand, Günther and Wendisch (59) have interpreted their nmr data in terms of **84** and **85** (threefold barrier to rotation) as the minimum energy conformations of these

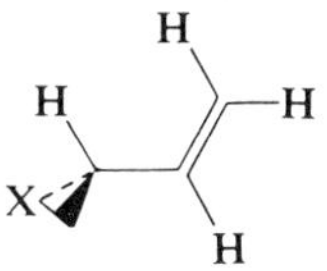

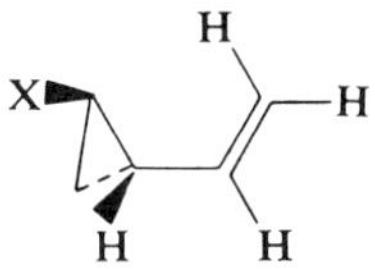

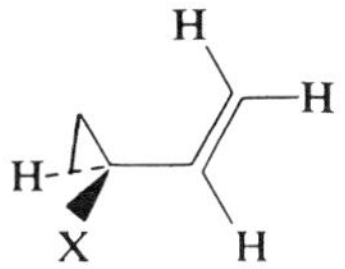

compounds. They concluded that enthalpy favors **84** by about 600–700 cal/mol. De Mare and Martin (60) examined both the relevant coupling constants and chemical shifts of vinylcyclopropane as a function of temperature and concluded that a threefold barrier to rotation best fits the data. The energy favoring rotamer **84** (X = H) was estimated to be about 800 ± 150 cal/mol. From analysis of the microwave spectrum of vinylcyclopropane, Schwendeman and Codding (61) have been able to detect only rotamer **84**. More recently, Lüttke and de Meijere (62) have concluded that their electron diffraction data best fit a threefold, rather than a twofold, barrier to rotation. It appears, therefore, that the V_3 term may be contributing more than the V_2 term to the torsional barrier of vinylcyclopropane. The question, however, is not as yet satisfactorily answered.

Not surprisingly, the chemical shifts of phenylcyclopropane have been interpreted (63) in terms of **86** as the minimum energy conformation of the compound.

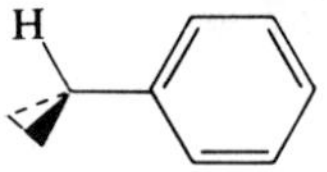

(86)

B. 2-Substituted

Studies on rotational isomerism of 2-substituted olefins have been confined to only a few compounds. The microwave spectrum of 2-chloropropene has been interpreted (64) on the assumption that **87** is the minimum

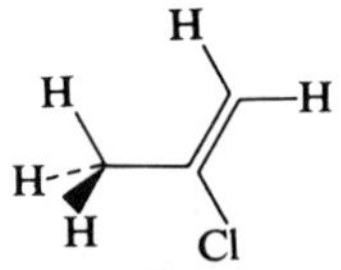

(87)

energy conformation of the compound. The infrared spectrum of methallyl chloride (65) was interpreted in terms of **88–90**. The contention of the

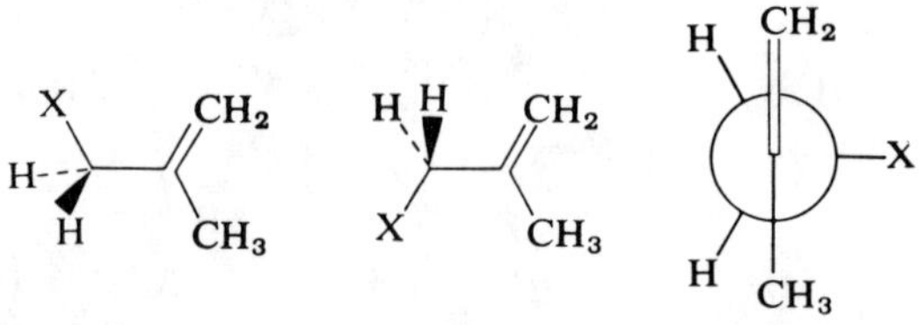

(88)   (89)   (90)

existence of three different minimum energy rotamers for this compound has been disputed by Northam, Oliver, and Crowder (66), who prefer only two on the basis that methallyl iodide exists in only one conformation in the liquid state (89), with a second rotamer becoming detectable in the vapor state at elevated temperatures. The stable rotamer (89) is favored over the less stable, presumably 88, by about 4000 ± 500 cal/mol. The choice of 89 as the stable rotamer, however, is quite intuitive and arbitrary and its structure might very well be 91, or one closely related to it.

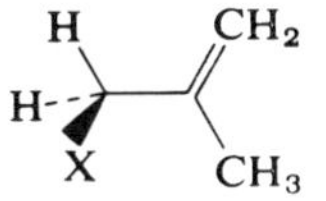

(91)

C. 1-Substituted

When the substituent at C-1 of the olefin is *cis* to the bond whose internal rotation is under investigation, it is expected that not only will the relative stabilities of 92 and 93 change in the direction favoring 92 over 93, but also the energy differences between eclipsing and bisecting (94) conformations

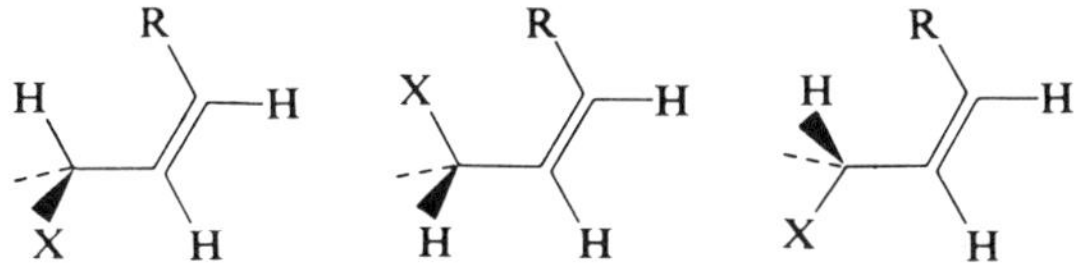

(92)   (93)   (94)

will decrease (67), perhaps to the point where 94 is more stable than 92. There are no data available that prove the first point, but if one were to judge from the data obtained with the analogous aldehydic derivatives (*vide infra*), this point is certainly reasonable. The second point, namely, the decrease in energy difference between 92 and 94, is supported by the fact that the barriers to rotation of 1-substituted propenes are lower for the *cis* than for the *trans* compounds. For example, the barrier to methyl rotation of *cis*-1-fluoropropene is 1057 cal/mol (68), whereas that of the *trans* isomer is 2150 cal/mol (69); that of *cis*-1-chloropropene is 620 cal/mol (70) and that of the *trans* isomer 2170 cal/mol (71); that of *cis*-1-cyanopropene is 1400 cal/mol (72) and that of the *trans* isomer >2100 cal/mol (73). Finally, the barrier to rotation of the methyl groups of *trans*-butene is 1950 cal/mol (74), whereas that of *cis*-butene is only 730 cal/mol (75). The decrease in the barrier to internal rotation of the methyl groups of *cis*-butene has been attributed by Dauben and Pitzer (67) to destabilization of the minimum energy conformation without any appreciable effect on the energy of the maximum. It is quite conceivable that the minimum energy conformation of *cis*-butene is 95, i.e., the conformation where both methyl groups are staggered with respect to the double bond.

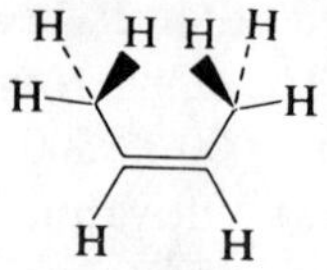

(95)

Karabatsos and Lande (76) calculated the hydrogen–hydrogen nonbonded interactions for **95**, **96**, and **97** and concluded that **95** is favored over **96** by

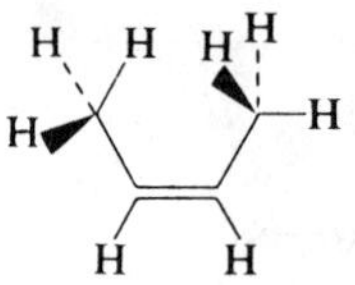

(96) 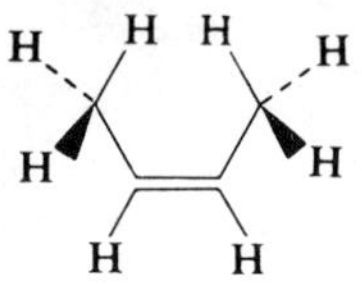   (97)

about 250 cal/mol and over **97** by at least 2500 cal/mol. These results, viz. the great instability of **97** and the low barrier to rotation, further suggest that the two internally rotating groups may be interacting in some form of a gearing or cogwheel process. Similarly, Grant and Cheney (77) calculated **98** to be the minimum energy conformation of *ortho*-xylene.

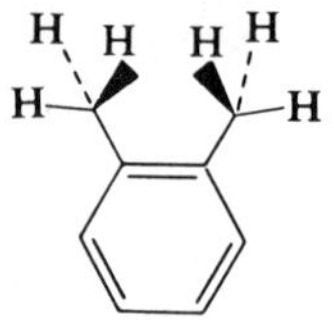

(98)

The microwave spectra of 1-halopropenes have been interpreted on the assumption that **99** is the minimum energy conformation of these compounds.

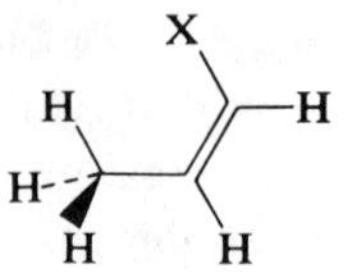

(99)

Boden, Emsley, Feeney, and Sutcliffe (78) have interpreted the fluorine and proton nmr spectra of *cis*-3,3-ditrifluoromethyl-1,3-difluoropropene in terms of a twofold barrier to internal rotation about the sp^3-sp^2 bond, with a $\Delta H°$ value of 900 ± 200 cal/mol favoring **100** over **101**. The analogous rotamer of the *trans* compound was found to be favored by 1700 ± 100 cal/mol.

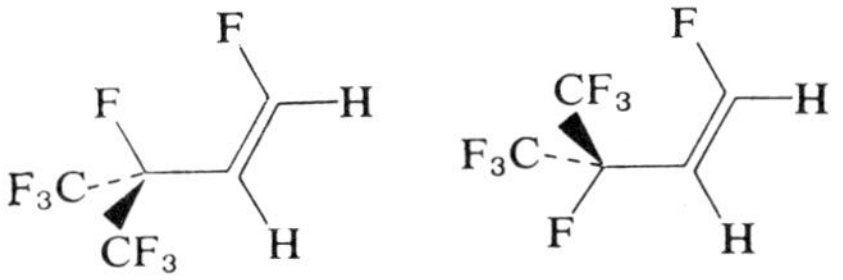

(100)   (101)

IV. IMINO COMPOUNDS

A. *syn* Isomers

Karabatsos and his co-workers have investigated, by nmr spectroscopy, the relative stabilities of **102** and **103**, not only as a function of R, but also as

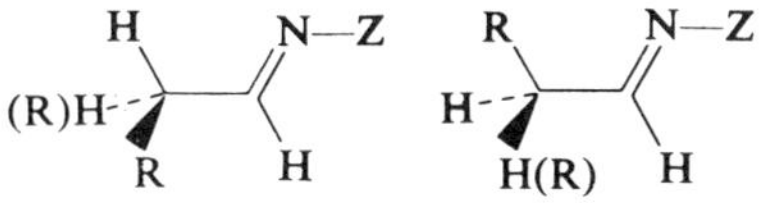

(102)   (103)

a function of Z, where Z is methyl (76), amino (79), *N*-methylamino (80), *N,N*-dimethylamino (54), *N*-methylanilino (81), hydroxy (82), and methoxy (83). Enthalpy differences for several of these compounds are summarized in Table III.

N-Methylpropionaldimine, as well as any other *N*-alkylpropionaldimine (78), is the only compound in this class whose rotamer **103**, i.e., the one where the alkyl group eclipses the carbon–nitrogen double bond, is favored over rotamer **102**. Altogether, *N*-alkylaldimines have $\Delta H°$ values for **102** $\rightleftharpoons$ **103** that are less positive than those of the other imino compounds. They fall between those of aldehydes and alkenes. For all other compounds the relative stabilities of **102** and **103** are similar to those of alkenes rather than those of aldehydes. From the $\Delta H°$ values of *N*-methylpropionaldimine (-200 cal/mol), propionaldehyde hydrazones ($+100$ to $+250$ cal/mol), and propionaldoxime and its *O*-methyl ether ($+400$ to $+500$ cal/mol), it appears that $\Delta H°$ for **102** $\rightleftharpoons$ **103** becomes more positive as the electronegativity of Z increases. Nonbonded interactions between the methyl group and the lone electron pair on the nitrogen may, therefore, be ruled out as the major factor controlling the relative stabilities of these rotamers, unless these interactions are strongly dependent on the polarizabilities of the lone electron pairs. This argument is based on the assumption that Z does not substantially alter the structures of these compounds.

B. *anti* Isomers

A variety of imino compounds have served as suitable models to study the relative stabilities of the various rotamers about the C—C=NZ single

TABLE III

Enthalpy Differences, $\Delta H°$, between Rotamers of Various Imino Compounds

$R_1R_2CHCH=NZ$

$\Delta H°$ (cal/mol) for

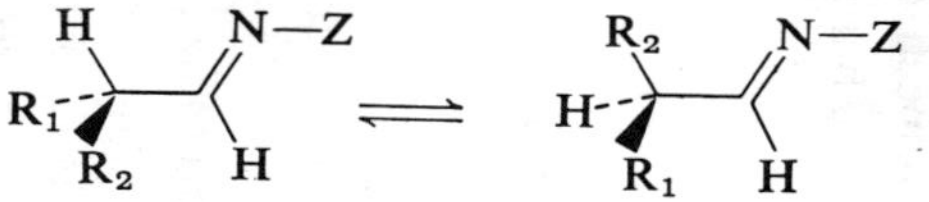

$$\begin{array}{c}H\\R_1\overset{\displaystyle\diagup}{}\overset{N-Z}{\diagdown}\\R_2\quad H\end{array} \rightleftharpoons \begin{array}{c}R_2\overset{\displaystyle\diagup}{}\overset{N-Z}{\diagdown}\\H\overset{}{}\quad\overset{}{}R_1\quad H\end{array}$$

R_1	R_2	$Z = CH_3$ [a]	$Z = NHCH_3$ [b]	$Z = N(CH_3)_2$ [c]	$Z = NCH_3C_6H_5$ [d]	$Z = OH$ [e]	$Z = OCH_3$ [f]
H	CH_3	-200	$+200$	$+250$	$+100$	$+500$	$+400$
H	CH_2CH_3	0			$+300$	$+700$	$+600$
H	$(CH_2)_4CH_3$	0					$+600$
H	$CH(CH_3)_2$	$+350$			$+600$	$+700$	$+650$
H	$C(CH_3)_3$	$>+1000$	$>+2000$	$>+2000$	$>+2000$	$>+2500$	$>+2500$
H	C_6H_5				$+700$	$+700$	$+1200$
CH_3	CH_3	$+100$	$+300$	$+300$	$+100$	$+500$	$+300$
CH_2CH_3	CH_2CH_3	$+450$	$+500$	$+600$	$+400$	$+800$	$+700$
$CH(CH_3)_2$	$CH(CH_3)_2$				$+700$		
$C(CH_3)_3$	$C(CH_3)_3$						$+3000$
△							$+1000$
⬠					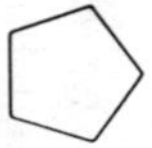 $+700$	$+800$
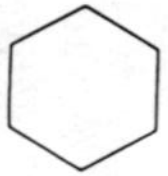 ⬡		0	$+400$		$+50$	$+400$	$+450$

[a] Ref. 76. [b] Ref. 80. [c] Ref. 54. [d] Ref. 81. [e] Ref. 82. [f] Ref. 83.

bond when the tetrahedral carbon is *cis* to Z. *A priori*, one would expect the energy differences between eclipsed and staggered conformations to decrease, as was the case with 1-substituted *cis*-propenes. That this is indeed so has been shown by Schwendeman and Rogowski (84), who found, by microwave spectroscopy, that the barrier to internal rotation of the methyl group is 1840 cal/mol for **104**, but only 350 cal/mol for **105**. Karabatsos and his group

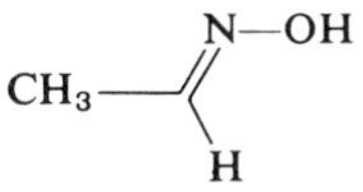

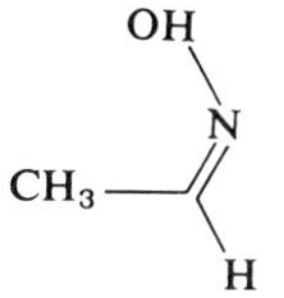

(54,76,79–83) have concluded from the behavior of the average vicinal coupling constants of these compounds that **106** and **107** must be of comparable energy, and that **108**, or some structure closely related to it, i.e., one

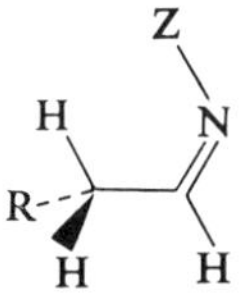

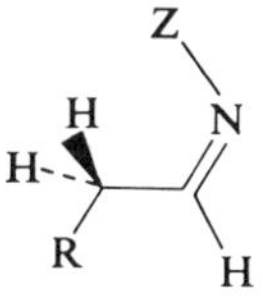

where the dihedral angle relating the two vicinal hydrogens is not smaller than 150°, is the major conformation of compounds disubstituted at the sp^3-hybridized carbon atom. The problem of whether **106** or **107** represents the minimum energy conformation of the monosubstituted compounds has not been solved. Karabatsos and Lande (76) have calculated the relative stabilities of **109–111** and concluded that **109** is favored over **110** by about 250 cal/mol and over **111** by considerably more than 5 kcal/mol. We would like to

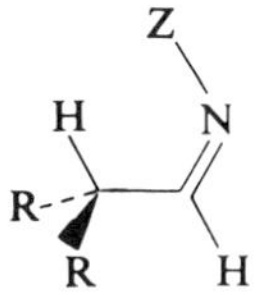

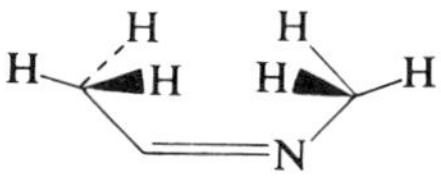

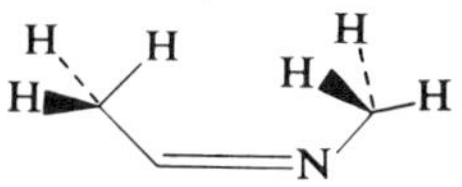

suggest, therefore, that the minimum energy conformations of such compounds that are either unsubstituted or monosubstituted at the tetrahedral carbon atom are staggered, not eclipsed.

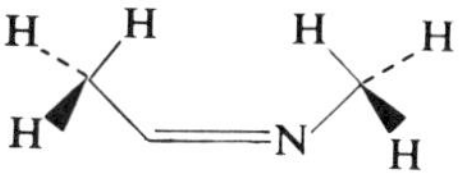

V. CARBONIUM IONS AND FREE RADICALS

It is customary to write carbonium ions and free radicals in conformation **112**, whereby the plane defined by cCC is perpendicular to that defined by

$$\text{(112)} \qquad \text{(113)}$$

RRC, rather than in the eclipsed conformation **113**. If one were to judge from what has already been discussed in this review, one might conclude that **113** may be more stable than **112**, at least in those cases where the R groups are hydrogens. No data are available in the literature on the basis of which an unambiguous decision can be made as to which of these two conformations is more stable. The magnetic nonequivalence of the methyl groups of dimethyl cyclopropylcarbinyl cation in FSO_3H—SO_2—SbF_5 at about $-60°$ to $-65°$ was interpreted by Pittman and Olah (85) in terms of **113**, where a is the hydrogen and c and b are the methylene groups of the cyclopropyl ring. The data are equally compatible, however, with **112** as the minimum energy conformation of the species, provided the hydrogen does not occupy position c. The choice of **113**, a reasonable one, is based on the fact that cyclopropanecarboxaldehyde and various other related compounds exist in the conformation analogous to **113**. As was already pointed out, however, the cyclopropyl compounds are special cases and may best be treated as sp^2-sp^2 rather than sp^2-sp^3 cases. Thus, even if the conformation of the dimethylcyclopropylcarbinyl cation were to be **113**, the minimum energy conformations of the simple cations may very well be **112**. At the present time the choice between the two is a matter of conjecture.

Free radicals of analogous cyclopropyl compounds, such as those of cyclopropyl semidiones, have been shown to be best interpreted in terms of a conformation resembling **113** rather than **112**. For example, Russell and Malkus (86) have calculated from the small a_β (0.57 gauss) and by using eq. (14)—which relates the splitting of β-hydrogens to a constant characteristic

$$a_H = B\rho \cos^2 \theta \qquad (14)$$

of a system (B), to the electron density at the α-carbon(ρ), and to the dihedral angle θ between the β-hydrogen and the p-orbital of the free radical electron—that **114**, with the hydrogen 8° out of the nodal plane, is the minimum energy

$$\text{(114)} \quad COR \ (R = CH_3, \Delta)$$

conformation of the two compounds shown. In this sense, **114** is closer to **113** (8° apart) than to **112** (22° apart).

The electron spin resonance spectra of several simple alkyl radicals have

ROTATIONAL ISOMERISM ABOUT sp^2-sp^3 BONDS

been studied by Fessenden and Schuler (87) and, more recently, by Krusic and Kochi (88). Although these authors have not interpreted their data in terms of different rotamers, we will venture to do so in terms of conformation **113**. The a_β of the ethyl radical, which may be considered from **115** as the time

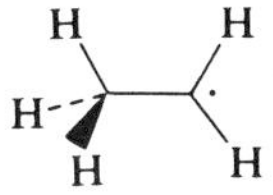

(115)

average of three splittings, a very small one (about zero) involving the hydrogen eclipsing the —CH$_2$· group ($\theta = 90°$) and two larger ones involving the two *gauche* hydrogens ($\theta = 30°$), was found to be 26.87 gauss. The a_β of the 1-propyl radical, whose barrier to rotation was estimated (87) as 400–500 cal/mol, was found to be 33.2 gauss at $-180°$ and 29.4 gauss at $-50°$. Since this coupling is larger than that of the ethyl radical and decreases with increase in temperature, one may conclude that $\Delta H°$ favors **117** over **116**. This conclusion

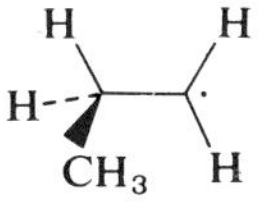

(116) 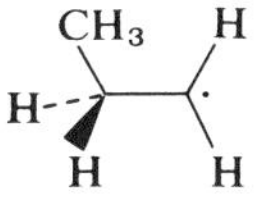   (117)

is by no means unreasonable, as these stabilities parallel those of the analogous rotamers of propionaldehyde. The a_β of the isobutyl radical, 35.1 gauss, was also found to be larger than that of the ethyl. Thus, the enthalpy difference between **118** and **119** must again be in favor of **119**.

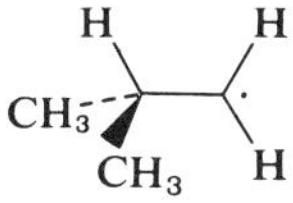

(118) 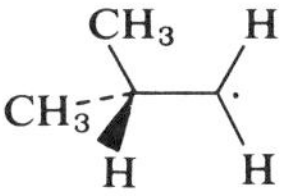   (119)

When the radical is secondary, twice as many eclipsing conformations are possible. From the a_β of 24.69 gauss for the isopropyl radical and the 27.9 gauss of the methylene of the 2-butyl radical, the conclusion may be drawn that **120** is the most stable conformation of this species.

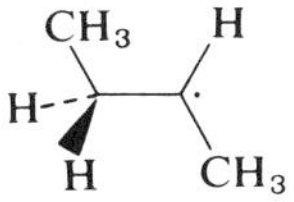

(120)

The relative stabilities of these rotamers are reversed when the radical is trisubstituted. Thus, a_β of the *t*-butyl radical was found to be 22.72 gauss and the a_β of the methyl groups of the *t*-amyl radical was found to be 22.8 gauss at $-125°$ and 22.7 at $-5°$. On the other hand, the a_β of the methylene hydrogens of this radical was found to be only 17.6 gauss at $-125°$ and considerably larger, 18.70 gauss, at the higher temperature of $-5°$. Thus, rotamer **121** is more stable than **122**. This conclusion is certainly reasonable

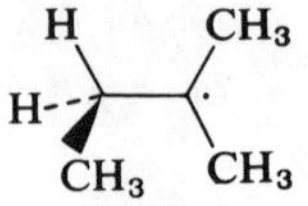

(121) 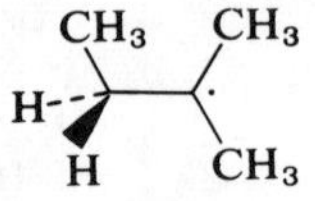   (122)

and consistent with the results obtained with various 3-substituted alkylpropenes, which should serve as the appropriate model for this radical.

Fessenden and Schuler (87) have reported some interesting splittings regarding the ethyl-d_5 and the ethyl-1,1,2,2-d_4 radicals. In both cases the a_β for the deuterium is about the same, 3.47 gauss and 3.46 gauss, respectively. These values correspond to about 22.60 gauss for a hydrogen. The a_β for the deuterium of the ethyl-d_5 radical at $-175°$ was found to be 4.08 gauss, which corresponds to 26.57 gauss for a hydrogen and compares well with the a_β of 26.87 gauss for the ethyl radical. However, the a_β for the deuterium of the ethyl-1,1,2,2-d_4 radical at $-175°$ was only 3.88 gauss, corresponding to 25.24 gauss for a hydrogen, whereas the a_β for the hydrogen of this radical was considerably higher, 29.77 gauss at $-175°$ and 28.39 gauss at $-85°$. These results may be interpreted in terms of **123** and **124** with an energy difference that favors **123** over **124** by more than the statistical factor of two.

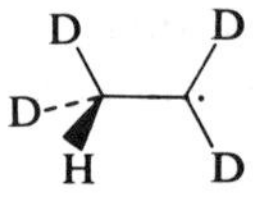

(123) 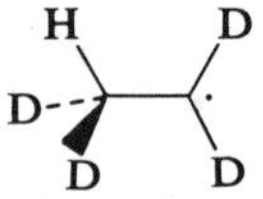   (124)

Although we have interpreted the above-mentioned results in terms of **113** as the minimum energy conformation of these radicals, it should be made clear that they may also be adequately interpreted in terms of **112** or other conformations intermediate between these two. Our discussion of these radicals in terms of **113** was primarily aimed at pointing out that these radicals have detectable conformational preferences, rather than at deciding whether **112** or **113** best represents the minimum energy conformation of radicals. As with the carbonium ions, the choice between **112** and **113** is a matter of personal choice.

VI. GENERAL COMMENTS

Although the methods used to study rotational isomerism about sp^2-sp^3 carbon–carbon bonds have been refined and extended in the last 10 to 15 years and the amount of data available has accordingly increased considerably, several basic problems associated with this type of isomerism remain; for example:

1. Why is the eclipsed conformation of such simple compounds as acetaldehyde, propene, and acetic acid and its derivatives favored over the staggered conformation? One may view the electron distribution in a double bond in terms of two bent orbitals (89) and argue (57) that the electron configuration is now staggered, as shown in **125**. Thus, spin–spin interactions

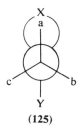

(125)

between electrons, which may control the barrier to rotation, are minimized. Although this explanation is attractive, no experimental results are available that support it. In fact, Flygare and his group (90) have concluded that the electron density of the double bond of formaldehyde is concentrated along the geometric internuclear line of the carbon–oxygen double bond. This conclusion certainly mitigates against bent bonds in this compound.

2. Why is the rotamer with two large groups eclipsing each other so often favored over the one where a small and a large group eclipse each other? As pointed out, nonbonded interaction energies calculated from available nonbonded potential functions often fail to account for this result. It is conceivable that such functions are too "hard," especially in the case of atoms with polarizable electrons, and perhaps new ones are needed. The tendency among several investigators has been to explain such relative rotamer stabilities in terms of nonbonded attractions rather than repulsions (6,53). Whatever the reason may be, it has become clear that, if either one or both of the two nonbonded interacting groups have unshared or polarizable electrons, conformations with the two groups closer together are frequently more stable than those with the two groups farther apart. Only when both groups are alkyls can one safely apply the intuitive rule that the more stable conformations are those with the two groups farthest apart. This observation applies not only to rotational isomerism about sp^2-sp^3 single bonds, but also about sp^3-sp^3 single bonds and about double bonds. For example, to cite only a few cases, the enthalpy favoring the *gauche* rotamer (**127**) over the *trans* rotamer (**126**) is 500 cal/mol (microwave) for 1-fluoropropane (91), about 0 cal/mol (microwave) for 1-chloropropane (92), 100–500 cal/mol

(126) (127)

(infrared) for 1-bromopropane (93), and 300 cal/mol (electron diffraction) for 1-chlorobutane (94). It is about 60 cal/mol and 0 cal/mol, respectively, for 1-fluoro-2-chloroethane (95) and 1,2-difluoroethane (96) in the gaseous state, despite the fact that dipole–dipole interactions would favor the *trans* rotamer by about 800 cal/mol (97). Similarly, the enthalpy difference in favor of the *cis* isomer (128) over the *trans* (129), is 928 cal/mol for 1,2-

(128) (129)

difluoroethylene (97), 170 ± 120 cal/mol for crotononitrile (98), and 620 cal/mol for 1-chloropropene (70). Finally, in the liquid state, 128 is more stable than 129 for acetaldoxime (82) and for its *O*-methyl ether (83), and 128 is favored over 129 by a factor of 4 for 1-bromopropene (99).

References

1. M. Karplus, *J. Amer. Chem. Soc.*, **85**, 2870 (1963).
2. R. J. Abraham and J. A. Pople, *Mol. Phys.*, **3**, 609 (1960).
3. G. J. Karabatsos and D. J. Fenoglio, *J. Amer. Chem. Soc.*, **91**, 3577 (1969) and previous references cited therein.
4. G. J. Karabatsos and D. J. Fenoglio, *J. Amer. Chem. Soc.*, **91**, 1124 (1969).
5. E. Saegebarth and E. B. Wilson, Jr., *J. Chem. Phys.*, **46**, 3088 (1967).
6. G. J. Karabatsos and N. Hsi, *J. Amer. Chem. Soc.*, **87**, 2864 (1965).
7. G. J. Karabatsos, D. J. Fenoglio, and S. S. Lande, *J. Amer. Chem. Soc.*, **91**, 3572 (1969).
8. R. W. Kilb, C. C. Lin, and E. B. Wilson, Jr., *J. Chem. Phys.*, **26**, 1695 (1957).
9. P. H. Verdier and E. B. Wilson, Jr., *J. Chem. Phys.*, **29**, 340 (1958).
10. S. S. Butcher and E. B. Wilson, Jr., *J. Chem. Phys.*, **40**, 1671 (1964).
11. (a) L. S. Bartell, B. L. Carroll, and J. P. Guillory, *Tetrahedron Letters*, **1964**, 705; (b) L. S. Bartell and J. P. Guillory, *J. Chem. Phys.*, **43**, 647 (1965).
12. R. N. Schwendeman and H. N. Volltrauer, Symposium on Molecular Structure and Spectroscopy, Columbus, Ohio, September 1966.
13. R. Hoffmann, *Tetrahedron Letters*, **1965**, 3819.
14. R. N. Schwendeman and H. N. Volltrauer, private communication.
15. G. W. Buchanan, J. B. Stothers, and S. Wu, *Can. J. Chem.*, **45**, 2955 (1967).

16. L. J. Bellamy and R. L. Williams, *J. Chem. Soc.*, 3465 (1958).
17. G. C. Sonnichsen and G. J. Karabatsos, unpublished results.
18. R. Nelson and L. Pierce, *J. Mol. Spectry*, **18**, 344 (1965).
19. J. D. Swalen and C. C. Costain, *J. Chem. Phys.*, **31**, 1562 (1959).
20. C. Romers and J. E. G. Creutzberg, *Rec. Trav. Chim.*, **75**, 331 (1956).
21. R. N. Jones and K. Noack, *Can. J. Chem.*, **39**, 2214 (1961).
22. G. J. Karabatsos, G. C. Sonnichsen, N. Hsi, and D. J. Fenoglio, *J. Amer. Chem. Soc.*, **89**, 5067 (1967).
23. T. E. Gough, W. S. Lin, and R. G. Woolford, *Can. J. Chem.*, **45**, 2529 (1967).
24. L. S. Bartell, J. P. Guillory, and A. T. Parks, *J. Phys. Chem.*, **69**, 3043 (1965).
25. R. N. Schwendeman and P. L. Lee, private communication.
26. S. Mizushima, T. Shimanouchi, T. Miyazawa, I. Ichishima, J. Kuratani, I. Nakagawa, and N. Shido, *J. Chem. Phys.*, **21**, 815 (1953).
27. G. A. Crowder and B. R. Cook, *J. Chem. Phys.*, **47**, 367 (1967).
28. B. R. Cook and G. A. Crowder, *J. Chem. Phys.*, **47**, 1700 (1967).
29. G. A. Crowder and B. R. Cook, *J. Mol. Spectry.*, **25**, 133 (1968).
30. L. W. Daasch and R. E. Kagarise, *J. Amer. Chem. Soc.*, **77**, 6156 (1955).
31. L. J. Bellamy, L. C. Thomas, and R. L. Williams, *J. Chem. Soc.*, 3704 (1956).
32. R. N. Jones and E. Spinner, *Can. J. Chem.*, **36**, 1020 (1958).
33. L. Pierce and L. C. Krisher, *J. Chem. Phys.*, **31**, 875 (1959).
34. K. M. Sinnott, *J. Chem. Phys.*, **34**, 851 (1961).
35. L. C. Krisher, *J. Chem. Phys.*, **33**, 1237 (1960).
36. M. J. Moloney and L. C. Krisher, *J. Chem. Phys.*, **45**, 3277 (1966).
37. I. Nakagawa, I. Ichishima, K. Kuratani, T. Miyazawa, T. Shimanouchi, and S. Mizushima, *J. Chem. Phys.*, **20**, 1720 (1952).
38. Y. Morino, K. Kuchitsu, and M. Sugiura, *J. Chem. Soc. Japan*, **75**, 721 (1954).
39. L. J. Bellamy and R. L. Williams, *J. Chem. Soc.*, 3465 (1958).
40. A. Miyake, I. Nakagawa, T. Miyazawa, I. Ichishima, T. Shimanouchi, and S. Mizushima, *Spectrochim. Acta*, **13**, 161 (1958).
41. G. A. Crowder and F. Northam, *J. Mol. Spectry.*, **26**, 98 (1968).
42. W. J. Tabor, *J. Chem. Phys.*, **27**, 974 (1957).
43. E. B. Wilson, Jr., "The Problems of Barriers to Internal Rotation in Molecules," in *Advances in Chemical Physics*, Vol. 2, I. Prigogine, Ed., Interscience, New York, 1959, pp. 367–396.
44. L. C. Krisher and E. B. Wilson, Jr., *J. Chem. Phys.*, **31**, 882 (1959).
45. T. L. Brown, *Spectrochim. Acta*, **18**, 1615 (1962).
46. S. Mizushima, T. Shimanouchi, I. Ichishima, T. Miyazawa, I. Nakagawa, and T. Araki, *J. Amer. Chem. Soc.*, **78**, 2038 (1956).
47. D. R. Herschbach and L. C. Krisher, *J. Chem. Phys.*, **28**, 728 (1958).
48. E. Hirota, *J. Chem. Phys.*, **42**, 2071 (1965).
49. A. A. Bothner-By and C. Naar-Colin, *J. Amer. Chem. Soc.*, **83**, 231 (1961).
50. A. A. Bothner-By, C. Naar-Colin, and H. Günther, *J. Amer. Chem. Soc.*, **84**, 2748 (1962).
51. A. A. Bothner-By and H. Günther, *Discussions Faraday Soc.*, **34**, 127, (1962).
52. A. A. Bothner-By, S. Castellano, and H. Günther, *J. Amer. Chem. Soc.*, **87**, 2439 (1965).
53. A. A. Bothner-By, S. Castellano, S. J. Ebersole, and H. Günther, *J. Amer. Chem. Soc.*, **88**, 2466 (1966).
54. G. J. Karabatsos and R. A. Taller, *Tetrahedron*, **24**, 3923 (1968).
55. R. C. Hirst, quoted in ref. 30 and by M. Barfield and D. M. Grant, *J. Amer. Chem. Soc.*, **85**, 1899 (1963).

56. N. L. Allinger, J. A. Hirsch, M. A. Miller, and I. J. Tyminski, *J. Amer. Chem. Soc.*, **90**, 5773 (1968).
57. K. Radcliffe and J. L. Wood, *Trans. Faraday Soc.*, **62**, 2038 (1966).
58. W. Lüttke and A. de Meijere, *Angew. Chem.*, **78**, 544 (1966); *Angew. Chem. Intern. Ed Engl.*, **5**, 512 (1966).
59. H. Günther and D. Wendisch, *Angew. Chem.*, **78**, 266 (1966); *Angew. Chem. Intern. Ed. Engl.*, **5**, 251 (1966).
60. G. R. De Mare and J. S. Martin, *J. Amer. Chem. Soc.*, **88**, 5033 (1966).
61. R. N. Schwendeman and E. G. Codding, Symposium on Molecular Structure and Spectroscopy, Columbus, Ohio, September 1968.
62. A. de Meijere and W. Lüttke, *Tetrahedron*, **25**, 2047 (1969).
63. G. L. Closs and H. B. Klinger, *J. Amer. Chem. Soc.*, **87**, 3265 (1965).
64. M. L. Unland, V. Weiss, and W. H. Flygare, *J. Chem. Phys.*, **42**, 2138 (1965).
65. Y. A. Pentin, E. V. Morozov. *Opt. Spectry (USSR) English Transl.*, **20**, 357 (1966).
66. F. Northam, J. Oliver, and G. A. Crowder, *J. Mol. Spectry.*, **25**, 436 (1968).
67. W. G. Dauben and K. S. Pitzer, in *Steric Effects in Organic Chemistry*, M. S. Newman, Ed., Wiley, New York, 1956, pp. 58–59.
68. R. A. Beaudet and E. B. Wilson, Jr., *J. Chem. Phys.*, **37**, 1133 (1962).
69. S. Siegel, *J. Chem. Phys.*, **27**, 989 (1957).
70. R. A. Beaudet, *J. Chem. Phys.*, **40**, 2705 (1964).
71. R. A. Beaudet, *J. Chem. Phys.*, **37**, 2398 (1962).
72. R. A. Beaudet, *J. Chem. Phys.*, **38**, 2548 (1963).
73. V. W. Laurie, *J. Chem. Phys.*, **32**, 1588 (1960).
74. J. E. Kilpatrick and K. S. Pitzer, *J. Res. Natl. Bur. Std.*, **37**, 163 (1946).
75. T. N. Sarachman, *J. Chem. Phys.*, **49**, 3146 (1968).
76. G. J. Karabatsos and S. S. Lande, *Tetrahedron*, **24**, 3907 (1968).
77. D. M. Grant and B. V. Cheney, *J. Amer. Chem. Soc.*, **89**, 5315 (1967).
78. N. Boden, J. W. Emsley, J. Feeney, and L. H. Sutcliffe, *Proc. Roy. Soc. (London)* **A282**, 559 (1964).
79. G. J. Karabatsos and C. E. Osborne, *Tetrahedron*, **24**, 3361 (1968).
80. G. J. Karabatsos and R. A. Taller, *Tetrahedron*, **24**, 3557 (1968).
81. G. J. Karabatsos and K. L. Krumel, *Tetrahedron*, **23**, 1097 (1967).
82. G. J. Karabatsos and R. A. Taller, *Tetrahedron*, **24**, 3347 (1968).
83. G. J. Karabatsos and N. Hsi, *Tetrahedron*, **23**, 1079 (1967).
84. R. H. Schwendeman and R. S. Rogowski, private communication.
85. C. U. Pittman, Jr., and G. A. Olah, *J. Amer. Chem. Soc.*, **87**, 2998 (1965). See also, N. C. Deno, J. S. Liu, J. O. Turner, D. N. Lincoln, and R. E. Fruit, Jr., *J. Amer. Chem. Soc.*, **87**, 3000 (1965).
86. G. A. Russell and H. Malkus, *J. Amer. Chem. Soc.*, **89**, 160 (1967).
87. R. W. Fessenden and R. H. Schuler, *J. Chem. Phys.*, **39**, 2147 (1963). See also, R. W. Fessenden, *J. Chim. Phys.*, **61**, 1570 (1964).
88. P. J. Krusic and J. K. Kochi, *J. Amer. Chem. Soc.*, **90**, 7155 (1968).
89. J. A. Pople, *Quart. Rev.*, **11**, 273 (1957).
90. W. H. Flygare, *J. Chem. Phys.*, **41**, 206 (1964); W. H. Flygare and J. T. Lowe, *ibid.*, **43**, 3645 (1965); W. H. Flygare and V. W. Weiss, *ibid.*, **45**, 2785 (1966).
91. E. Hirota, *J. Chem. Phys.*, **37**, 283 (1962).
92. T. N. Sarachman, *J. Chem. Phys.*, **39**, 469 (1963).
93. N. Sheppard, *Advan. Spectry.*, **1**, 288 (1959).
94. F. A. Momany, R. A. Bonham, and W. H. McCoy, *J. Amer. Chem. Soc.*, **85**, 3077 (1963).

95. P. A. Bazhulin and L. P. Osipova, *Opt. Spektroskopiya*, **6**, 406 (1959).
96. P. Klaboe and J. R. Nielsen, *J. Chem. Phys.*, **33**, 1764 (1960).
97. N. C. Craig and A. E. Entemann, *J. Amer. Chem. Soc.*, **83**, 3047 (1961).
98. J. N. Butler and R. D. McAlpine, *Can. J. Chem.*, **41**, 2487 (1963).
99. P. S. Skell and R. G. Allen, *J. Amer. Chem. Soc.*, **80**, 5997 (1958).

The Use of Ultrasonic Absorption and Vibrational Spectroscopy to Determine the Energies Associated with Conformational Changes

E. WYN-JONES AND R. A. PETHRICK

Department of Chemistry and Applied Chemistry, University of Salford, Salford, Lancashire, England

I.	Introduction	206
II.	Ultrasonic Absorption	208
	A. Introduction	208
	B. Theory	208
	1. Plane Wave Propagation and Attenuation in Liquids	208
	2. Basic Concepts of Relaxation	212
	3. The Kinetics of a Two-State Equilibrium Perturbed by an Ultrasonic Wave	215
	4. Thermodynamics of a Two-State Equilibrium Perturbed by an Ultrasonic Wave	219
	C. Experimental	225
	1. Apparatus and Evaluation of Relaxation Parameters	225
	2. Determination of Energy Parameters	228
	D. Substituted Ethanes	230
	1. Barriers	230
	2. Enthalpy Differences	240
	E. Tertiary Amines	242
	F. Esters	245
	G. Acyclic and Heterocyclic Ring Inversions	249
	H. α,β-Unsaturated Aldehydes and Ketones	253
	I. Other Systems	255
III.	Vibrational Spectroscopy	257
	A. Introduction	257
	B. Enthalpy Differences	258
	1. Theory	258
	2. Results	260
	C. Potential Barriers	265
	1. Introduction	265
	2. Experimental Methods	267
	3. Results	267
	D. The Use of Infrared Spectroscopy as a Probe to Determine Potential Barriers	269
	Addendum in Proof	270
	References	271

I. INTRODUCTION

The rotation of one part of a molecule with respect to another part about an internal chemical bond is usually accompanied by a change in the potential energy of the molecule. In some molecules internal rotation of this kind gives rise to the phenomenon of rotational isomerism where the molecules exist as an equilibrium mixture of two or more conformers. One of the classical examples of a molecule exhibiting rotational isomerism is 1,2-dichloroethane, where the variation of potential energy with azimuthal angle is as shown in Figure 1. Also included in this diagram are the conformations of the molecule corresponding to the various maxima and minima of the potential energy curve. The minima in these potential energy curves correspond to the stable staggered conformations of the molecules which are the so-called "rotational isomers." The difference in energy between the minima is known as the enthalpy difference between the isomers, and the difference between neighboring maxima and minima is called the potential barrier hindering rotation.

The potential energy curve associated with internal rotation in furan-2-aldehyde is shown in Figure 2. The equilibrium between the stable forms differing in energy in both 1,2-dichloroethane and furan-2-aldehyde is equivalent to a two-state process which can be represented by a simple equation of the type:

$$A \rightleftharpoons B$$

In the case of 1,2-dichloroethane, the upper state is degenerate owing to the statistical weights of the optically active *gauche* isomers.

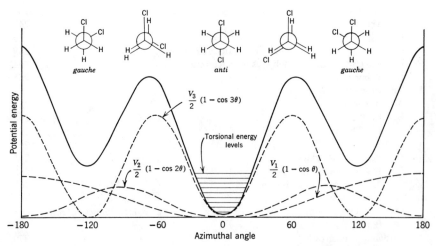

Fig. 1. Potential energy function for 1,2-dichloroethane.

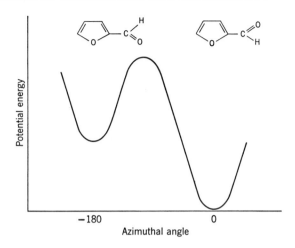

Fig. 2. Internal rotation in Furan-2-aldehyde.

In some asymmetrically substituted ethanes, such as $CF_2BrCFClBr$, the potential energy diagram is similar to that shown in Figure 1, but with all maxima and minima having different energies.

Rotational isomerism of this kind is now basic in theories covering large areas of chemistry, including the conformational analysis of alicyclic structures and the physical properties of polymers. This phenomenon also has considerable importance in protein chemistry, where the coiling and uncoiling of polypeptide chains play an important role in biological reactions.

There are several articles (1–6) available on this subject which deal with the basic principles of internal rotation in molecules as well as covering other interesting aspects, such as the historical approach and the types of physical methods that can be used to study this phenomenon. However, one of the most surprising aspects of the work described in the above references is the apparent lack of quantitative information about the rotation, especially the energy values associated with the potential energy of the molecule. Up to a few years ago most of the quantitative data available consisted of the enthalpy differences between the rotamers. Recent developments (7) in the fields of ultrasonic absorption, nmr spectroscopy, far infrared spectroscopy, microwave spectroscopy, and neutron scattering have provided the experimentalist with a variety of techniques which can lead to a wealth of information about the energy parameters, particularly the potential barrier associated with internal rotation.

This article describes the use of ultrasonic absorption and vibrational spectroscopy to determine the energies associated with conformational changes in molecules.

II. ULTRASONIC ABSORPTION

A. Introduction

When the potential barrier to a conformational change is in the range of 3 to 12 kcal/mol, the rate constants of the isomeric exchange are of the order of 10^6 to 10^9 sec^{-1} [$k = A(\sim 10^{13})e^{-E/RT}$] and the kinetics of the conformational change are fast. In the liquid phase the kinetic and thermodynamic properties of fast equilibria can be investigated with the ultrasonic relaxation technique (8). The principle involved in relaxation studies of fast reactions (9) is to impose a sudden change in the external conditions (e.g., temperature, pressure) of the system which is in equilibrium. This sudden impulse will obviously affect the position of the equilibrium, which in turn will respond to this change. The response to this sudden change will not be instantaneous but the equilibrium will adjust itself over a finite time interval, reaching its new position at an exponential rate with a time constant, τ, known as the relaxation time, which is a function of the forward and reverse rate constants of the equilibrium. In relaxation studies the position of the equilibrium is monitored after the application of the change in external parameter. In the case of ultrasonic relaxation in conformational equilibria, the external parameter that is changed is temperature, which alters periodically according to the frequency of the sound wave. The position of the equilibrium is monitored indirectly by observation of the change in sound absorption and velocity with frequency.

B. Theory

1. Plane Wave Propagation and Attenuation in Liquids

It is possible, experimentally, to produce sound waves in the frequency range of a few cycles to over 10 GHz (10^{10} Hz). This acoustic spectrum can be divided into four ranges (10): the infra-audible (10^{-2} to 10 Hz), the audible (10 Hz to 14 kHz), the ultrasonic (14 kHz to 800 MHz), and the hypersonic range ($\gtrsim 10^9$ Hz).

Sound waves are propagated in a liquid as longitudinal waves. A longitudinal wave may be considered as being the superposition of a pure compressional and a pure shear wave (11). A compressional wave may be considered as creating an alternating periodic pressure variation, which results in a given volume being successively compressed and decompressed. In all pure liquids, except water at 4°C, the specific heat at constant pressure C_p exceeds that at constant volume C_v and, hence, $\gamma(=C_p/C_v) > 1$. It therefore follows that the alternating periodic variation in the pressure will also be manifest as a corresponding alternation in the temperature. The pressure amplitude,

p, of a one-dimensional sound wave traveling in the positive x-direction is given by:

$$p = p_0 \exp(-\alpha x) \exp i\omega\left(t - \frac{x}{c}\right) \qquad (1)$$

where p_0 is the pressure amplitude at $x = 0$ and time $t = 0$, α is the absorption coefficient related to the attenuation of the sound wave, c is the phase velocity, and $\omega = 2\pi f$ where f is the frequency of the sound wave. In eq. (1) the quantity $p_0 \exp(-\alpha x)$ represents the amplitude of the sound wave at a distance x from the source. This real part of eq. (1) can be seen to be analogous to the Beer-Lambert law governing light absorption. The imaginary part of eq. (1) represents the particle velocity or density variation.

Herzfeld and Litovitz (11) have reviewed the description of the "classical" contribution to the absorption coefficient, α, based on the Stokes-Navier equation. When a sound wave passes through a liquid, the regions under compression will have a temperature above the average, the regions being decompressed will have a temperature below the average, and heat conduction will tend to equalize the temperature. This in turn means that a certain loss in sound energy will occur because of the tendency for heat to flow from the hotter to the colder regions. Kirchoff (12) has shown that α is related to the thermal conductivity, Q, of the liquid by the expression:

$$\alpha_{\text{cond}} = \left(\frac{2\pi^2}{\rho c^3}\right)(\gamma - 1)\left(\frac{Qf^2}{C_p}\right) \qquad (2)$$

where ρ is the density of the liquid. For all liquids except metals, the contribution to the absorption from thermal conductivity given by eq. (2) may be taken to be negligible.

At very low frequencies the pressure variation takes place so slowly that the temperature can be regarded as following the pressure variation exactly. It follows that since $\gamma > 1$, the sound wave is propagated essentially adiabatically. However, the adiabatic approximation fails at very high frequencies (above 10^3 MHz in gases) because the mean free path becomes comparable with the wavelength of the sound wave, so that enough energy can be exchanged by direct transfer of molecular momentum (heat conduction) to make the propagation isothermal (10). Just as energy is lost from the acoustic compression by heat conduction, so also can it be lost by radiation. Since the regions of compression are at a slightly higher temperature than their surroundings, their rate of emission of energy according to Stefan's law is greater than the average. In ultrasonic relaxation experiments we are only concerned with small temperature perturbations of the order of ± 0.01–$0.001°C$ and so this cause of energy loss may be neglected as not contributing to sound attenuation in normal liquids.

The action of the sound wave in successively compressing and decompressing a given volume of a liquid will cause the density and, hence, the free volume occupied by the molecules to alternate about a given value. During compression a loss in the free volume which the molecule may occupy leads to a more compact and highly ordered structure of the liquid. To achieve the more compact structure, movement of the molecules in three dimensions will occur. Viscosity is defined as resistance to flow and it thus follows that a liquid can have a "volume" or compressional viscosity in addition to the more familiar shear viscosity. Truesdell (13) has shown that the concept of "volume" viscosity can be accommodated within hydrodynamic theory and leads to an expression for the absorption coefficient of the sound wave of the form:

$$\alpha_{\text{visc}} = \left(\frac{2\pi^2}{\rho c^3}\right)(\eta_v + \tfrac{4}{3}\eta_s)f^2 \tag{3}$$

where η_v is the volume viscosity and η_s is the shear viscosity of the liquid. It is normal to term the combined effects of viscosity and thermal conductivity as the "classical" absorption. The classical absorption can thus be seen to be given by eq. (4).

$$\left(\frac{\alpha}{f^2}\right)_{\text{classsical}} = \left(\frac{2\pi^2}{\rho c^3}\right)\left(\eta_v + \tfrac{4}{3}\eta_s + \frac{(\gamma - 1)\,Q}{C_p}\right) \tag{4}$$

The quantity (α/f^2) is important in the analysis of results from measurements of acoustic absorption and for a given liquid at a stated pressure and temperature.

$$\left(\frac{\alpha}{f^2}\right)_{\text{classical}} = \text{constant} = B \text{ (for example)} \tag{5}$$

Strictly speaking, eq. (5) is not valid since both the viscosity and the thermal conductivity may be considered to be relaxational in behavior and will be frequency dependent. This type of relaxational behavior would, however, only be expected to be observed at very high frequency, and eq. (5) can be considered to be valid for the experimental range used in this work.

The difference between the measured absorption α and the classical coefficient $\alpha_{\text{classical}}$ represents all absorption in excess of that due to the additive effects of shear viscosity, volume viscosity, and thermal conductivity and is usually referred to as the "excess" absorption. The volume viscosity, η_v, can be shown to be expressed by eq. (6).

$$\eta_v = \frac{4}{3}\eta_s\left(\frac{\alpha - \alpha_{\text{classical}}}{\alpha_{\text{classical}}}\right) \tag{6}$$

When this volume viscosity arises from a perturbation of a molecular equilibrium due to the temperature or pressure changes imposed by a compressional wave, ultrasonic relaxation will occur. Several different kinds of mechanisms are responsible for volume viscosity, all being relaxational in behavior and giving rise to excess absorption.

The velocity of the sound wave can similarly be considered in terms of the density variations of the liquid accompanying its propagation. A relationship has been proposed between the speed of sound in the liquid and that in the vapor phase based on a molecular model proposed by Kincaid and Eyring (14). They assume molecules to be hard elastic spheres (internal compressibility of molecules being zero) and picture the sound wave as traveling in the space between the molecules with velocity identical with that in the gas, while inside the molecule it travels with infinite velocity. They arrive at the result:

$$c_{\text{liq}} = c_{\text{gas}} \left[1 - \left(1 - \frac{V_a}{V}\right)^{1/3}\right]^{-1} \tag{7}$$

where V_a is the available volume, which is the difference between the volume corresponding to the close packed structure for the molecules and the total volume. Derivation of an unambiguous theoretical relationship for V_a/V is not possible as this requires a knowledge of the structure of the liquid. The velocity of sound in a liquid has been shown to be given by the Laplace formula:

$$c^2 = \frac{1}{\rho\beta} \tag{8}$$

and alternatively can be expressed in the form:

$$c^2 = (\gamma - 1)\frac{JC_p}{\theta^2 T \rho V} \tag{9}$$

where ρ is the density, β the adiabatic compressibility, θ the coefficient of thermal expansion, and V the molar volume. The cubic coefficient of thermal expansion, θ, for a liquid is given by eq. (10). Kittel (15) has shown that the

$$\theta = V^{-1}\left(\frac{dV}{dT}\right)_p = \frac{V_a}{V}\frac{1}{T} \tag{10}$$

temperature dependence of the velocity of sound can be approximately represented by combination of eqs. (7), (8), and (9). Semiquantitative agreement between experiment and theory has been observed for a large number of organic liquids. Since application of pressure decreases the available volume, the pressure coefficient of the sound velocity is expected to be positive, as is actually observed; numerically the calculated values are 30–100% lower than the observed values. Abnormalities in the velocity as a function of frequency would be expected at high frequency due to the presence of viscous forces, the shear forces becoming predominant over compressional forces in the consideration of propagation of the sound wave.

2. Basic Concepts of Relaxation

In order to explain the principles involved in ultrasonic relaxation, Litovitz (16) has described the energy content of a liquid in terms of the energy box shown diagrammatically in Figure 3. In brief, the total energy content of the liquid is the sum of many components, including translational, rotational, and vibrational components and the less familiar contributions from the existence of conformational and structural equilibria. Conformational equilibria arise when a molecule can exist as a mixture of conformers differing in energy and able to undergo interconversion. Structural energy originates from the existence of a short range order associated with the quasi-crystalline structure of liquids. In the energy box, the content of each different energy is allocated to a segment as shown in the diagram. Energy can be transferred from one segment to another and this may be described by a first-order exponential rate equation with an appropriate time constant, τ the relaxation time, being defined as the time required for the concentration

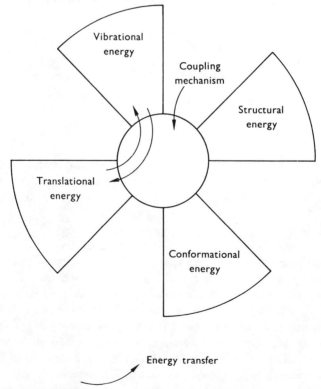

Fig. 3. A modified version of the energy box introduced by Litovitz. (From E. Wyn-Jones, *Roy. Inst. Chem. Rev.* **2**, No. 1, 61 (1969), by courtesy of the Editor.)

of a species to have been reduced to 1/eth of its original value. This relaxation time depends upon the nature of the coupling mechanism which governs the energy transfer between boxes. It is the relaxation time of the change in energy content in the conformational equilibrium segment that we are concerned with in this article. This change will be reflected in an adjustment of the position of equilibrium.

When a sound wave is passed through a liquid, the periodic compression and decompression accompanying the propagation of the pressure wave will produce a corresponding alternation in the temperature and, hence, energy content of the translational box. Since the energies of all the boxes are coupled, an increase in the translational energy content will lead to some of this energy being transferred to all other modes until a new equilibrium is reached. The energy content of the translational box conversely will be lowered during the decompression cycle and energy will then flow in the reverse direction from the coupled boxes into the translational box. Consider the case of energy exchange only occurring between the translational and vibrational boxes; this energy will be exchanged via the coupling mechanism of molecular collision. Obviously not every collision will be efficient in the exchange of energy, which results in the process taking a finite time to occur. It can be easily seen that if the time of the compression–decompression cycle for the sound wave is long compared with the energy exchange time (relaxation time), some of the extra energy in the translational box will be transferred to the vibrational box and return again during this same half cycle and no net loss per cycle in the sound energy will be observed. If the period of the sound wave is decreased (i.e., the frequency is increased), a time will be reached when the energy perturbation of the translational box occurs so fast that effectively no change can occur with other boxes. In between these two extreme cases will occur a condition when the period of the sound wave is approximately equal to the relaxation time for the exchange. At these acoustic frequencies some of the energy lost by the translational box during compression will be returned to the sound wave during the decompression cycle. Energy is then returned out of phase, resulting in a net loss of energy from the sound wave during that cycle, this being observed as attenuation of the sound wave and over the frequency range where $f < 1/2\pi\tau < f$ this behavior is known as ultrasonic relaxation.

Consider the passage of a sound wave through a system in which a conformational equilibrium of the type:

$$A \rightleftharpoons B \quad (\Delta H^\circ \neq 0)$$

is present. Because there is an enthalpy difference between the two forms, the temperature fluctuations accompanying a compressional sound wave will impose a periodic change in the chemical equilibrium about a mean value. At very low acoustic frequencies the periodic changes in both the sound wave and the equilibrium will be in phase. This corresponds to the situation where enthalpy is taken and returned to the sound wave during the same cycle. If the acoustic frequency is progressively increased, the equilibrium will not be able to respond instantaneously to the temperature fluctuations and will become out of phase with the sound wave. This lag enables the equilibrium to accept enthalpy during the temperature crest of the sound wave and to give up enthalpy during the temperature trough with an inevitable attenuation of the sound wave intensity as the wave travels through the liquid. When the acoustic frequency is very high, the equilibrium is unable to follow the temperature fluctuations imposed by the sound wave, which again results in a zero loss of sound energy per cycle. The maximum loss in sound energy will occur at an acoustic frequency f_c given by $f_c = 1/2\pi\tau$ where τ is the relaxation time of the chemical equilibrium. The simple processes described above are the basic principles involved in the relaxation techniques to study the kinetics of fast reactions.

Two factors determine the magnitude of the loss of sound energy per cycle. These are (*1*) the amount of energy shared and returned out of phase (in the chemical equilibrium (8) this is proportional to $\Delta H°$) and (*2*) the time constant τ. The behavior described above corresponds to acoustic relaxation in a liquid, and for a two-state conformational equilibrium a single relaxation process occurs.

The behavior of the acoustic parameters during a single relaxation process is shown in Figure 4. At any given temperature the quantity α/f^2 will decrease with increasing frequency, f, in accordance with eq. (11)

$$\frac{\alpha}{f^2} = \frac{A}{1 + (f/f_c)^2} + B \tag{11}$$

where A is a relaxation parameter, f_c is the relaxation frequency ($= 1/2\pi\tau$), and B represents contributions to α/f^2 which are not related to the relaxation. These contributions are shear viscosity and any excess absorption due to a relaxation process with a relaxation frequency much higher than f_c. The loss per cycle or absorption per unit wavelength μ relating to the relaxation is

$$\mu = \alpha'\lambda \tag{12}$$

where α' is the excess absorption for the relaxation process. Thus,

$$\mu = (\alpha - Bf^2)\lambda$$
$$= \frac{Acf}{1 + (f/f_c)^2} \tag{13}$$

when $f = f_c (\omega\tau = 1)$, μ reaches a maximum value μ_m given by eq. (14). The

$$\mu_m = \tfrac{1}{2}Acf_c \tag{14}$$

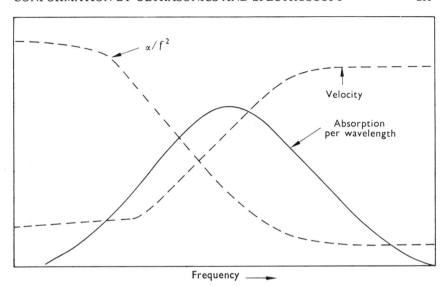

Fig. 4. The variation of the ultrasonic parameters α/f^2, μ, and c with frequency during a single relaxation process. (From E. Wyn-Jones, *Roy. Inst. Chem. Rev.* **2**, No. 1, 61 (1969), by courtesy of the Editor.)

dispersion in the sound velocity, c, is given by the expression

$$c^2 - c_0^2 = \left(\frac{2\mu_m}{\pi}\right) c_0 c_\infty \frac{(\omega\tau)^2}{1 + (\omega\tau)^2} \qquad (15)$$

where the subscripts 0 and ∞ refer to the sound velocity at low and high frequencies.

3. The Kinetics of a Two-State Equilibrium Perturbed by an Ultrasonic Wave

Consider a simple two-state unimolecular conformational equilibrium $A \rightleftharpoons B$. The initial concentration of the species A and B may be designated by a_0 and b_0. The concentrations may now be considered to be changed to new values a and b at a time t after the application of a sudden impulse. The equilibrium conditions corresponding to the new conditions (i.e., when the reaction has finally adjusted itself) may be defined as being $\bar{a}$ and $\bar{b}$.

On application of a sudden impulse the initial concentrations a_0 and b_0 will start moving towards the new equilibrium values $\bar{a}$ and $\bar{b}$. Now suppose that after a time t the actual values of a and b differ from the new equilibrium values $\bar{a}$ and $\bar{b}$ by x such that

$$x = a - \bar{a} \quad \text{and} \quad x = \bar{b} - b$$

The net forward rate at a time t can be expressed as:

$$\frac{-dx}{dt} = k_{12}a - k_{21}b \tag{16}$$

At equilibrium this is zero:

$$k_{12}\bar{a} - k_{21}\bar{b} = 0 \tag{17}$$

hence,

$$\frac{\bar{b}}{\bar{a}} = \frac{k_{12}}{k_{21}} = \text{equilibrium constant} = K$$

since $a = \bar{a} + x$, $b = \bar{b} - x$; therefore, substitution into eq. (16) gives:

$$\frac{-dx}{dt} = k_{12}(\bar{a} + x) - k_{21}(\bar{b} - x)$$

From eq. (17)

$$\frac{-dx}{dt} = (k_{12} + k_{21})x$$

Let $k = k_{12} + k_{21}$, then

$$\frac{-dx}{dt} = kx$$

and

$$\int_{x_0}^{x} \frac{-dx}{x} = \int_{t=0}^{t} k\, dt$$

where x_0 = value of x immediately after disturbance at $t = 0$; thus,

$$\frac{x}{x_0} = e^{-kt}$$

Since the relaxation time, τ, is the time required for the actual concentration to decrease to $1/e$th of its original value, $k\tau = 1$. From this, the relation (eq. (18)) between the rate constants and the relaxation time, τ, may be obtained. It is useful to express the above equations in a slightly different

$$\tau = \frac{1}{k_{12} + k_{21}} \tag{18}$$

form. Let y be the difference in actual concentrations and the initial values ($y = a - a_0$, $y = b_0 - b$) and $\bar{y}$ be the difference in equilibrium concentrations and the initial values ($\bar{y} = \bar{a} - a_0$, $\bar{y} = b_0 - \bar{b}$), then

$$\frac{-dy}{dt} = (y - \bar{y})(k_{12} + k_{21}) \tag{19}$$

thus

$$\tau \frac{dy}{dt} + y = \bar{y} \tag{20}$$

CONFORMATION BY ULTRASONICS AND SPECTROSCOPY

The above equations apply to a system receiving a sudden impulse and then undergoing a relaxation process. In acoustics we are concerned with a system undergoing a periodic change induced by the sound wave.

The pressure variation of a sound wave can be described by eq. (1) which may be expressed in the form

$$p = A'e^{i\omega t} \tag{21}$$

where A' is an amplitude factor [$=p_0 \exp(-\alpha x) \exp i\omega(-x/c)$]. If it is assumed that the pressure and temperature variation are synchronous, then the driving force for eq. (20) will have the same form as the pressure variation described by eq. (21). Thus, it follows that $\bar{y}$ defines both the pressure variation of the sound wave and also the variation of the equilibrium if it is

$$\bar{y} = Ae^{i\omega t} \tag{22}$$

assumed that it can instantaneously follow the alternation of the sound wave. However, due to the finite time of exchange, the equilibrium lags behind this driving force according to eq. (23). Since A, as defined by eq. (22), is complex,

$$y = Be^{i\omega t} \tag{23}$$

then B is also complex and

$$\frac{dy}{dt} = Be^{i\omega t} i\omega \tag{24}$$

$$= i\omega y$$

Using eq. (20) we have

$$i\omega y \tau + y = \bar{y} \tag{25}$$

From eq. (25)

$$y = \frac{\bar{y}}{1 + i\omega\tau}$$

$$= \bar{y}\frac{1}{1 + \omega^2\tau^2} - \bar{y}\frac{i\omega\tau}{1 + \omega^2\tau^2}$$

$$= G\bar{y} \tag{26}$$

where G is a complex function given by

$$G = G_r + iG_i \tag{27}$$

with

$$G_r = \frac{1}{1 + \omega^2\tau^2} \quad \text{and} \quad G_i = \frac{-\omega\tau}{1 + \omega^2\tau^2}$$

Alternatively, eq. (25) can be expressed in the form:

$$d\bar{y} = dy(1 + i\omega\tau) \tag{28}$$

The complex function G is known as the transfer function, the real (G_r) and imaginary (G_i) parts of this function can be related to the dispersion and absorption of the sound wave. It is normal to use the well known trigonometric relationships to express G in terms of the phase angle ϕ.

$$G = (\cos \phi - i \sin \phi)\rho$$
$$= \rho e^{-i\phi} \tag{29}$$

where ρ is the dynamic effective amplitude factor given by:

$$\rho = (G_r^2 + G_i^2)^{1/2}$$
$$= (1 + \omega^2\tau^2)^{-1/2}$$

with $\cos \phi = 1/(1 + \omega^2\tau^2)^{1/2}$, $\quad \sin \phi = \omega\tau/(1 + \omega^2\tau^2)^{1/2}$
then

$$\tan \phi = \frac{-G_i}{G_r} = \omega\tau$$

The driving force is given by eq. (22), and using eqs. (26) and (29) we have,

$$y = \frac{e^{-i\phi}}{(1 + \omega^2\tau^2)^{1/2}} A e^{i\omega t}$$
$$= \frac{A e^{i(\omega t - \phi)}}{(1 + \omega^2\tau^2)^{1/2}} \tag{30}$$

The physical meaning of eq. (30) is that y lags behind $\bar{y}$ with a phase angle ϕ, which increases with frequency. This is represented diagrammatically in Figure 5, the dotted line being the periodic variation of the pressure or the instantaneous equilibrium ($\bar{y}$) and the solid line the actual response of the equilibrium to the perturbation (y). Initially, at very low acoustic frequencies, the two are in phase, but as the periodicity approaches that of the relaxation time, τ, then a phase lag is introduced which reaches a maximum at $\omega\tau = 1$ or $\phi = 45°$. At very high frequencies ($\omega \gg 1/\tau$) y will not be able to follow the rapid oscillations of $\bar{y}$ and finally becomes constant.

Two general methods of measurement follow from consideration of the stationary state solutions of eq. (30) (see ref. 8).

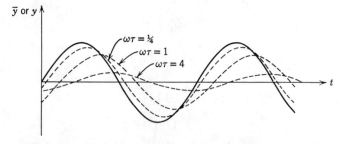

Fig. 5. Relaxational response for a period forcing function: (———) $\bar{y}$; (- - - -) y.

1. One method is concerned with the real part of the transfer function $G_r = 1/(1 + \omega^2\tau^2)$ or its square root, ρ. For $\omega = 1/\tau$, the quantities G_r and ρ amount to $\frac{1}{2}$ and $(\frac{1}{2})^{1/2}$, respectively. It will be shown that the quantities G_r and ρ can be related to the measurable quantities of dispersion of the absorption coefficient, eq. (11), and velocity of the sound wave, eq. (15), which have the form shown in Figure 4. These quantities are directly related to the real part of the ratio of the instantaneous values G_r or to the effective amplitudes ρ of concentration parameters $\bar{y}$ and y or two variables to which $\bar{y}$ and y are related. All the methods utilizing the frequency dependence of the real part of the transfer function G_r are termed "dispersion methods."

2. The other group of methods is related to the imaginary part $G_i = [-\omega\tau/(1 + \omega^2\tau^2)]$. Any phase shift between two conjugate variables, such as y and $\bar{y}$ illustrated in Figure 5, will result in a loss of energy. The loss of energy per wavelength ($\alpha\lambda$) will show a maximum, as indicated previously, at $\omega = 1/\tau$. Methods are available which can measure this loss directly and the relaxation time may be evaluated using eqs. (13) and (14) and the above condition that $\omega = 1/2\pi f_c$ at the maximum. Since the basis for all methods utilizing G_i is the measurement of energy absorption, this group can be termed "absorption methods."

4. Thermodynamics of a Two-State Equilibrium Perturbed by an Ultrasonic Wave

To define a system completely we require the knowledge of three independent thermodynamic variables (17). If X, Y, Z are three of the thermodynamic variables P, V, T, and S, and y is a variable (defined above) describing the state of the chemical reaction, then, for small perturbations about equilibrium at a frequency ω, we can form frequency-dependent thermodynamic parameters of the form $(dX/dY)_{Z,\omega}$. A parameter of this sort gives the amplitude of the sinusoidal fluctuations in the variable X as a result of the forced fluctuation in Y, while Z is held constant; the subscript ω indicates that the fluctuations are maintained at a frequency ω.

The chemical reaction, however, provides an additional degree of freedom, so the equation of state will be of the form

$$X = X(Y, Z, y)$$

from which

$$dX = \left(\frac{dX}{dY}\right)_{Z,y} dY + \left(\frac{dX}{dZ}\right)_{Y,y} dZ + \left(\frac{dX}{dy}\right)_{Y,Z} dy \qquad (31)$$

and

$$\left(\frac{dX}{dY}\right)_{Z,\omega} = \left(\frac{dX}{dY}\right)_{Z,y} + \left(\frac{dX}{dy}\right)_{Y,Z}\left(\frac{dy}{dY}\right)_{Z,\omega}$$

The first term on the right-hand side, for which y is held constant, represents the condition in which the reaction is frozen. Practically this can be achieved by raising the frequency, ω, until the reaction cannot follow the fluctuations in Y (see Fig. 5), so this term will be called the "infinite frequency component." The relation of dy to the driving force $d\bar{y}$ at constant ω has been defined previously by eq. (28) and likewise by eq. (26). As indicated previously, $\bar{y}$ defines the pressure variation of the sound wave and also the alternation of the instantaneous equilibrium, and y defines the way the actual equilibrium lags behind the driving force ($\bar{y}$) (eq. (23)). Using expansion (31) we can write relationships for the thermodynamic variables as follows:

a. Adiabatic Compressibility

$$\beta_\omega = \frac{-1}{V}\left(\frac{dV}{dP}\right)_{S,\omega} = \beta_\infty + \Delta\beta_\omega \tag{32}$$

The capital delta symbol is used to define the difference between the instantaneous and infinite values.

$$\Delta\beta_\omega = \frac{-1}{V}\left(\frac{dV}{dy}\right)_{S,P}\left(\frac{dy}{dP}\right)_{S,\omega} \tag{33}$$

b. Specific Heat

$$C_{p\omega} = T\left(\frac{dS}{dT}\right)_{P,\omega} = C_{p\infty} + \Delta C_{p\omega} \tag{34}$$

$$\Delta C_{p\omega} = T\left(\frac{dS}{dy}\right)_{P,T}\left(\frac{dy}{dT}\right)_{P,\omega} \tag{35}$$

c. Isothermal Compressibility

$$\beta_{T\omega} = \frac{-1}{V}\left(\frac{dV}{dP}\right)_{T,\omega} = \beta_{T\infty} + \Delta\beta_{T\omega} \tag{36}$$

$$\Delta\beta_{T\omega} = \frac{-1}{V}\left(\frac{dV}{dy}\right)_{P,T}\left(\frac{dy}{dP}\right)_{T,\omega} \tag{37}$$

d. Coefficient of Thermal Expansion

$$\theta_\omega = \frac{1}{V}\left(\frac{dV}{dT}\right)_{P,\omega} = \theta_\infty + \Delta\theta_\omega \tag{38}$$

$$\Delta\theta_\omega = \frac{1}{V}\left(\frac{dV}{dy}\right)_{P,T}\left(\frac{dy}{dT}\right)_{P,\omega} \tag{39}$$

Using Maxwell's relations

$$\left(\frac{dV}{dT}\right)_P = -\left(\frac{dS}{dP}\right)_T \tag{40}$$

and
$$\left(\frac{dT}{dP}\right)_S = \left(\frac{dV}{dS}\right)_P$$

the coefficient of thermal expansion becomes

$$\theta_\omega = \frac{-1}{V}\left(\frac{dS}{dP}\right)_{T,\omega} = \theta_\infty + \Delta\theta_\omega \tag{41}$$

$$\Delta\theta_\omega = \frac{1}{V}\left(\frac{dS}{dy}\right)_{P,T}\left(\frac{dy}{dP}\right)_{T,\omega} \tag{42}$$

At this stage it is convenient to define two parameters related to the equilibrium being considered. The quantities $\Delta V°$ and $\Delta H°$ are, respectively, the volume and the enthalpy change associated with equilibrium $A \rightleftharpoons B$ and defined by:

$$\Delta V° = \left(\frac{dV}{dy}\right)_{P,T} \quad \text{and} \quad \Delta H° = T\left(\frac{dS}{dy}\right)_{P,T} \tag{43}$$

It is possible to expand eq. (33) for the incremental adiabatic compressibility $\Delta\beta_\omega$ as follows:

$$\Delta\beta_\omega = \frac{-1}{V°}\left(\frac{dV}{dy}\right)_{S,P}\left(\frac{dy}{dP}\right)_{S,\omega}$$

$$= \frac{-1}{V°}\left[\left(\frac{dV}{dy}\right)_{P,T} - \left(\frac{dV}{dS}\right)_{P,y}\left(\frac{dS}{dy}\right)_{P,T}\right]\left[\left(\frac{dy}{dP}\right)_{T,\omega} + \left(\frac{dy}{dT}\right)_{P,\omega}\left(\frac{dT}{dP}\right)_{S,\omega}\right] \tag{44}$$

$$= VT\Delta C_p \left(\frac{\theta_\infty}{C_{p\infty}} - \frac{\Delta V°}{V\Delta H°}\right)\left(\frac{\theta_\omega}{C_{p\omega}} - \frac{\Delta V°}{V\Delta H°}\right) \tag{45}$$

but

$$\theta_\infty - \frac{\Delta V° C_{p\infty}}{V\Delta H°} = \theta_\omega - \Delta\theta_\omega - \frac{\Delta\theta_\omega}{\Delta C_{p\omega}}(C_{p\omega} - \Delta C_{p\omega}) \tag{46}$$

$$= \theta_\omega - \frac{\Delta V° C_{p\omega}}{V\Delta H°} \tag{47}$$

because from eqs. (33) and (43)

$$\frac{\Delta V°}{V\Delta H°} = \frac{\Delta\theta_\omega}{\Delta C_{p\omega}} \tag{48}$$

and from eqs. (33) and (43) we have

$$\frac{T\Delta V°}{\Delta H°} = \frac{\Delta\beta_T}{\Delta\theta} \tag{49}$$

therefore,

$$\frac{\Delta V°}{\Delta H°} = \left(\frac{V\Delta\beta_{T\omega}}{T\Delta C_{p\omega}}\right)^{1/2} \tag{50}$$

Employing the thermodynamic relation

$$VT\theta^2 = C_p\beta(\gamma - 1) \tag{51}$$

where γ is the ratio of the specific heats C_p/C_v at constant pressure and volume, together with eqs. (45)–(49) we find:

$$\frac{\Delta\beta_\omega}{\beta_\omega} = \frac{C_{p\omega}}{C_{p\infty}}\left\{\left(\frac{\Delta\beta_{T\omega}}{\beta_\omega}\right)^{1/2} - \left[(\gamma - 1)\frac{\Delta C_{p\omega}}{C_{p\omega}}\right]^{1/2}\right\}^2 \tag{52}$$

$$\approx \left\{\left(\frac{\Delta\beta_{T\omega}}{\beta_\omega}\right)^{1/2} - \left[(\gamma_\omega - 1)\frac{\Delta C_{p\omega}}{C_{p\omega}}\right]^{1/2}\right\}^2 \tag{53}$$

when $\Delta C_p \ll C_p$. This can be written as

$$\frac{\Delta\beta_\omega}{\beta_\omega} \approx (\gamma_\omega - 1)\left(\frac{\Delta C_{p\omega}}{C_{p\omega} - \Delta C_{p\omega}}\right)\left(1 - \frac{\Delta V^\circ}{V^\circ}\frac{C_{p\omega}}{\theta_\omega \Delta H}\right)^2 \tag{54}$$

which also reduces to

$$\frac{\Delta\beta_\omega}{\beta_\omega} = \left(\frac{\Delta\beta_{T\omega}}{\beta_\omega}\right)\left(1 - \frac{\Delta H^\circ V\theta_\omega}{\Delta V^\circ C_{p\omega}}\right)^2 \tag{55}$$

It can be seen that we have derived a relationship between the compressibilities, specific heat, molar volume, the volume change and the enthalpy difference associated with the equilibrium being perturbed, $A \rightleftharpoons B$.

To investigate the frequency dependence of this function and thus relate it to the perturbation produced by the sound wave, we will consider expressing the adiabatic compressibility in the form of a function containing ω by the use of eq. (28) for dy:

$$\left(\frac{dy}{d\bar{y}}\right)_\omega = \frac{1}{(1 + i\omega\tau)}$$

therefore,

$$\Delta\beta_\omega = \frac{-1}{V}\left(\frac{dV}{dy}\right)_{S,P}\left(\frac{dy}{d\bar{y}}\right)_\omega\left(\frac{d\bar{y}}{dP}\right)_S$$

$$= \frac{-1}{V}\left(\frac{dV}{dy}\right)_{S,P}\left(\frac{d\bar{y}}{dP}\right)_S\frac{1}{1 + i\omega\tau} \tag{56}$$

But at zero frequency $dy = d\bar{y}$, and we can define the relaxing part of the adiabatic compressibility as

$$\Delta\beta = \frac{-1}{V}\left(\frac{dV}{dy}\right)_{S,P}\left(\frac{d\bar{y}}{dP}\right)_S \tag{57}$$

where the symbol $\Delta\beta$ without subscript ω is reserved for the frequency-independent value of the property. Equation (56) thus becomes:

$$\Delta\beta_\omega = \frac{\Delta\beta}{1 + i\omega\tau} \tag{58}$$

Using eq. (58) we can reexpress eq. (32) in the form

$$\beta_\omega = \beta_\infty + \frac{\Delta\beta}{1 + i\omega\tau} \tag{59}$$

The adiabatic compressibility at zero frequency, β_0, is

$$\beta_0 = \beta_\infty + \Delta\beta \tag{60}$$

and combination with eq. (59) gives:

$$\begin{aligned}\beta_\omega &= (\beta_0 - \Delta\beta) + \frac{\Delta\beta}{1 + i\omega\tau} \\ &= \beta_0 - \frac{\Delta\beta\omega^2\tau^2}{1 + \omega^2\tau^2} - \frac{i\omega\tau}{1 + \omega^2\tau^2} \\ &= \frac{\beta_0 + \omega^2\tau^2\beta_\infty}{1 + \omega^2\tau^2}\left(1 - i\frac{\omega\tau\Delta\beta}{\beta_0 + \omega^2\tau^2\beta_\infty}\right)\end{aligned} \tag{61}$$

Equation (1) is equivalent to

$$p = p_0 \exp i\omega\left(t - \frac{x}{c'}\right)$$

where c' is the complex velocity of sound given by

$$\frac{1}{c'} = \frac{1}{c}\left(1 - \frac{i\alpha c}{\omega}\right) \tag{62}$$

with c' and c related to the compressibilities by, respectively,

$$(c')^2 = 1/\rho\beta_\omega \quad \text{and} \quad c^2 = 1/\rho\beta$$

Squaring eq. (62) and using the Laplace equations gives

$$\beta_\omega = \beta\left(1 + \frac{\alpha^2 c^2}{\omega^2} - \frac{2i\alpha c}{\omega}\right)$$

Now by definition

$$\mu = \alpha\lambda$$
$$= \frac{2\pi\alpha c}{\omega}$$

and since we are dealing with relaxation processes in which μ is small

$$\frac{\mu}{\pi} \ll 1$$

and, therefore,

$$\beta_\omega = \beta\left(1 - i\frac{\mu}{\pi}\right) \tag{63}$$

If we compare the imaginary parts of eqs. (61) and (63) it follows that

$$\frac{\mu}{\pi} = \frac{\omega \tau \Delta \beta}{\beta_0 + \omega^2 \tau^2 \beta_\infty}$$

and if $\Delta \beta \ll \beta_0$

$$\frac{\mu}{\pi} = \left(\frac{\Delta \beta}{\beta_0}\right) \frac{\omega \tau}{1 + \omega^2 \tau^2} \tag{64}$$

When $\omega \tau = 1$, μ reaches its maximum value μ_m given by

$$\mu_m = \frac{\pi \Delta \beta}{2 \beta}$$

and from eq. (54) by dropping the subscript ω

$$\frac{2 \mu_m}{\pi} = \frac{(\gamma - 1) \Delta C_p}{(C_p - \Delta C_p)} \left(1 - \frac{\Delta V^\circ}{V} \frac{C_p}{\Delta H^\circ \theta}\right)^2 \tag{65}$$

From eqs. (28), (35), and (43)

$$\Delta C_p = \Delta H^\circ \frac{d\bar{y}}{dT}$$

and $\bar{y}$ can be identified with one of the isomeric equilibrium concentrations, say $\bar{b}$ in the equilibrium $A \rightleftharpoons B$. Thus $d\bar{y}/dT = d\bar{b}/dT$ and since $\bar{b} = 1/[1 + \exp(-\Delta G^\circ/RT)]$ it follows that

$$\Delta C_p = R \left(\frac{\Delta H^\circ}{RT}\right)^2 \frac{\exp(-\Delta G^\circ/RT)}{[1 + \exp(-\Delta G^\circ/RT)]^2} \tag{66}$$

Thus we have derived expressions (65) and (66) relating the acoustic relaxation parameter μ_m to the thermodynamic equilibrium parameters ΔH°, ΔV°, and ΔS° of the equilibrium $A \rightleftharpoons B$. The kinetics of the equilibrium are related to the relaxation parameter f_c by eq. (18).

If we rearrange eq. (64), it follows that the value of α/f^2 for the relaxation process is given by

$$\frac{\alpha}{f^2} = \left(\frac{\Delta \beta}{\beta_0} \frac{\pi}{2cf_c}\right)\left(\frac{1}{1 + (f/f_c)^2}\right) = \frac{A}{1 + (f/f_c)^2} \tag{67}$$

where

$$A = \frac{\pi}{2} \frac{\Delta \beta}{\beta_0 c f_c}$$

or

$$\mu_m = \tfrac{1}{2} A c f_c$$

In practice there are classical contributions to the actual value of α/f^2 from shear and pure volume viscosities and also from any other relaxation processes occurring at frequencies much higher than f_c (18). These contributions are denoted by the classical value B (cf. eq. (5)). Thus, eq. (67) in its true form (eq. (11)) is

$$\frac{\alpha}{f^2} = \frac{A}{1 + (f/f_c)^2} + B$$

C. Experimental

1. Apparatus and Evaluation of Relaxation Parameters

Most of the ultrasonic absorption and velocity measurements in liquids which exist as equilibrium mixtures of conformational isomers have been carried out with the pulse technique which is normally used in the frequency range 5–300 MHz. Figure 6 shows a block diagram of a typical apparatus. A train of pulses is produced by the pulse generator and fed into the transmitter which is set at the desired frequency. These pulses are used to excite this oscillator, which in turn produces bursts of oscillations at the desired frequency, f. The resulting radio-frequency pulses are fed to a transducer which is acoustically coupled to a fused quartz delay line. The acoustic pulses

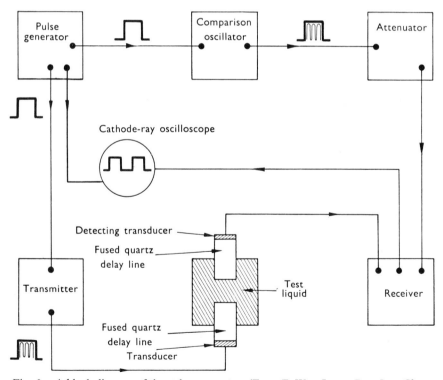

Fig. 6. A block diagram of the pulse apparatus. (From E. Wyn-Jones, *Roy. Inst. Chem. Rev.*, **2**, No. 1, 61 (1969), by courtesy of the Editor.)

produced by the transducer pass through the quartz delay line, through the liquid under test, and into a second quartz delay line, the upper end of which is attached to the detecting transducer. This second transducer retransforms the acoustic pulse into an electric pulse which is amplified and demodulated at the receiver and finally displayed on the screen of a cathode-ray oscilloscope. A second train of pulses, produced by the pulse generator, is used to excite a comparison oscillator which is set at the same frequency, f, as the transmitter. These radio-frequency pulses, after suitable delay, are fed into an attenuator and then to the receiver and finally are displayed at the side of the acoustic pulse on the oscilloscope.

The intensity and, hence, amplitude of the acoustic pulse is changed by varying the distance between the launching and detecting transducers. This change of amplitude can be measured by visually comparing the height of the acoustic pulse with that of the "comparison" pulse which is attenuated by a known amount. By plotting the attenuation against path length, the absorption coefficient α follows from eq. (1) since

$$\alpha = \frac{\ln (p_0/p)}{x}$$

The quantity $\ln (p_0/p)$ is directly proportional to the attenuator readings and x is the acoustic path in the liquid. The velocity measurements are made by using the apparatus as an acoustic interferometer. This is done by counting the number of beats between two pulses, one of which has traveled one path and the other three paths in the liquid sample over a known change in path length. Using this technique values of α/f^2 and velocities accurate to within $\pm 2\%$ can be obtained. Details of other ultrasonic absorption equipment can be found in several articles (8,19). In practice the acoustic measurements are carried out at several spot frequencies in the temperature range which spans the relaxation region, i.e., where the dispersion in α/f^2 (cf. eq. (11)) is a maximum. The results are plotted as α/f^2 versus temperature at various frequencies as shown in Figures 7 and 8. The values of A, B, and f_c in eq. (11) are then derived at five-degree temperature steps over the working temperature range. The values of α/f^2 are read from the plots (Figs. 7 and 8) and are fitted into eq. (11) using a minimization procedure (20) which ensures that the quantity

$$\frac{[(\alpha/f^2)_{\text{obs}} - (\alpha/f^2)_{\text{exp}}]^2}{(\alpha/f^2)_{\text{obs}}}$$

is a minimum at all frequencies. The values of A, B, and f_c found from this calculation are then used to check whether a plot of

$$\left[\frac{A}{(\alpha/f^2 - B)} - 1\right]^{1/2}$$

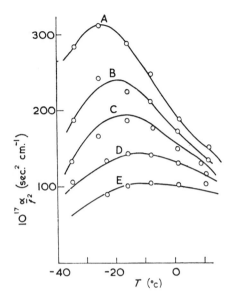

Fig. 7. Ultrasonic absorption data for 2-chloro-2-methylbutane: (A) 25.3; (B) 35.3; (C) 45.5; (D) 65.4; (E) 105 MHz/sec. (From P. J. D. Park and E. Wyn-Jones, J. Chem. Soc., A, 1969, 648, by courtesy of the Editor.)

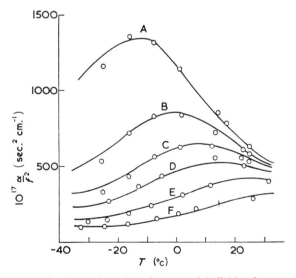

Fig. 8. Ultrasonic absorption data for *meso*-2,3-dichlorobutane: (A) 15; (B) 25; (C) 35; (D) 45; (E) 65; (F) 105 MHz/sec. (From P. J. D. Park and E. Wyn-Jones, J. Chem. Soc., A, **1969**, 648, by courtesy of the Editor.)

against f is linear, passes through the origin, and gives the same value of f_c as that found from the minimizing procedure. Figures 7 and 8 are somewhat modified in that the solid lines represent the calculated values of α/f^2 and the points represent experimental values. In the relaxation processes involving the perturbation of conformational equilibria the dispersion in velocity is less than the experimental error, and in most cases studied the sound velocities are found to vary linearly with temperature. With these sound velocities the values of μ_m are then calculated using eq. (14).

2. Determination of Energy Parameters

a. Thermodynamic. The expressions which relate the relaxation parameters and velocity to the thermodynamic equilibrium parameters are:

$$2\frac{\mu_m}{\pi} = \frac{(\gamma - 1)\Delta C_p}{(C_p - \Delta C_p)}\left(1 - \frac{\Delta V^\circ}{V}\frac{C_p}{\Delta H^\circ \theta}\right)^2 \tag{65}$$

$$\Delta C_p = R\left(\frac{\Delta H^\circ}{RT}\right)^2 \frac{\exp(-\Delta G^\circ/RT)}{[1 + \exp(-\Delta G^\circ/RT)^2]} \tag{66}$$

$$c^2 = (\gamma - 1)\frac{JC_p}{\theta^2 V \rho T} \tag{67}$$

In the above equations the unknown molecular parameters are ΔH°, ΔS°, and ΔV°, whereas the experimental data available are μ_m and $d\mu_m/dT$. Therefore, in order to solve for ΔH° and ΔS° ($\Delta G^\circ = \Delta H^\circ - T\Delta S^\circ$) the major assumption: $(\Delta V^\circ/V)(C_p/\theta) \ll \Delta H^\circ$ and also the approximation $\Delta C_p \ll C_p$ are made. Rearranging eqs. (65), (66), and (9) we can obtain (see ref. 21)

$$F = \left[\frac{2\mu_m J C_p^2}{\pi R \theta^2 c^2 \rho V T}\right]^{1/2} \tag{68}$$

$$= \frac{\Delta H^\circ}{RT}\frac{\exp(\Delta G^\circ/2RT)}{1 + \exp(\Delta G^\circ/RT)} \tag{69}$$

and therefore

$$\frac{\Delta G^\circ}{RT} = 2\ln\frac{\zeta_1 + 1 + [F_1^2 + (\zeta_1 + 1)^2]^{1/2}}{F_1} \tag{70}$$

and

$$\frac{\Delta H^\circ}{RT} = 2(F_1^2 + (\zeta_1 + 1)^2)^{1/2} \tag{71}$$

where $\zeta_1 = (T_1/F_1)(dF/dT)_1$, the subscript 1 referring to measurements at a particular temperature. In many examples the experimental value of F in eq. (68) cannot be determined experimentally because there are no data available on the "static" parameters C_p and θ. In this case a further assumption that C_p/θ is independent of the temperature is made (22) and the temperature variation of F in eq. (68) is now governed by the experimental quantity $(\mu_m/Tc^2)^{1/2}$. The theoretical variation of F with temperature can be evaluated by plotting the right-hand side of eq. (69) against $\Delta H°/RT$. A family of curves is obtained depending on the value used for $\Delta S°$, the entropy difference. In all these curves F reaches a maximum at a certain value of $\Delta H°/RT$. The variation of both $F_{(\max)}$ and $\Delta H°/RT$ corresponding to $F_{(\max)}$ is shown in Figure 9. Information about the equilibrium thermodynamic parameter $\Delta H°$ can now be obtained by comparing the experimental variation of $(\mu_m/Tc^2)^{1/2}$ with the curves shown in Figure 9. Three cases can be distinguished:

1. If $(\mu_m/Tc^2)^{1/2}$ increases with increasing temperature, then it is evident from Figure 9 that $\Delta H° > (2.3-3.2)RT$ and thus it is safe to assume that

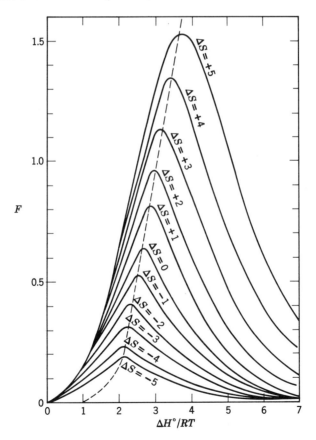

Fig. 9. Variation of F with temperature.

$\Delta G°$ is so large that the term $\exp(-\Delta G°/RT)$ is small compared to unity. In this case an approximate value of $(\Delta H°/R)$ can be obtained from the slope of the plot $\log(T\mu_m/c^2)$ against reciprocal temperature.

2. If the plot of $(\mu_m/Tc^2)^{1/2}$ against temperature reaches a maximum then it is evident that $\Delta H° = (2.3–3.2)RT$ where T is the maximum temperature. The exact value of $\Delta H°$ can be determined if $\Delta S°$ is known.

3. If μ_m/Tc^2 decreases with increasing temperature, then from Figure 9 it follows that $\Delta H° \leq (2.3–3.2)RT$ where T corresponds to the lowest temperature.

b. Kinetic. From eq. (18) and $\tau = 1/2\pi f_c$ it follows that

$$k_{21} = 2\pi f_c/(1 + K) \quad \text{and} \quad k_{12} = 2\pi f_c(1 + K^{-1}) \tag{72}$$

where K (the equilibrium constant) $= k_{12}/k_{21}$. The temperature dependence of the rate constants is governed by the Arrhenius equation:

$$k_{12} = A_{12} \exp(-E_{12}/RT) \tag{73}$$

or the Eyring absolute rate equation

$$k_{21} = \kappa(kT/h)\exp(-\Delta H^\ddagger_{12}/RT)\exp(\Delta S^\ddagger_{12}/R) \tag{74}$$

In these equations A is the Arrhenius frequency factor, E the activation energy, $\Delta H^\ddagger$ the activation enthalpy or potential barrier, $\Delta S^\ddagger$ the entropy of activation, κ the transition probability, and k and h have their usual meaning. In eq. (74) $\Delta H^\ddagger_{12}$ can be determined from the slope of the plot of $\log(k_{12}/T)$ against $1/T$. When the equilibrium constant K is not known it has been shown that the slope of the plot of $\log(f_c/T)$ against reciprocal temperature will still yield a very accurate estimate of $\Delta H^\ddagger_{12}$ (23).

In eqs. (73) and (74) the relationships between the various parameters are

$$\Delta H^\ddagger = E - RT$$

and

$$\Delta S^\ddagger = R\left[\ln\frac{hA}{\kappa kT} - 1\right]$$

An interesting point regarding the F curves drawn in Figure 9 is that very small amounts of this quantity can be measured and the limiting value of $\Delta H°$ that can be observed is ca. 7 kcal/mol. This means that an ultrasonic relaxation will be observed in two-state systems where the higher energy form is only present in quantities of the order of a fraction of a percentage.

D. Substituted Ethanes

1. Barriers

The potential energy diagram for 1,2-dichloroethane is shown in Figure 1. The two equivalent higher-energy minima in this diagram correspond to

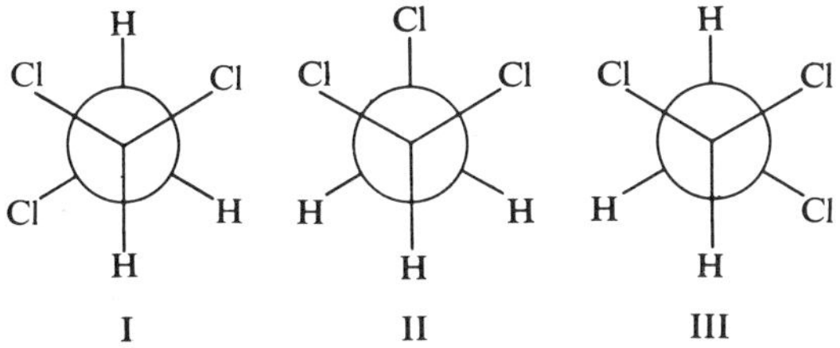

Fig. 10. Conformations of 1,1,2-trichloroethane.

the energies of the optically active *gauche* isomers, which make this state degenerate with an entropy contribution of $R \ln 2$ in excess of the more stable *anti* isomer. On the other hand, in 1,1,2-trichloroethane (see Fig. 10) the more stable isomeric state is degenerate with an entropy of $R \ln 2$ less than the stable form. It is almost traditional in this field to assume that the differences in entropy arising from contribution to vibrational and rotational terms are negligible. There is, in fact, some recent experimental data to support this assumption (24).

The potential energy barriers for the less stable to more stable isomeric transitions in several ethane derivatives which exist as two-state equilibria are listed in Table I. The second column in this table indicates the sign of the statistical entropy differences, a + sign indicating that the optically active pair consists of the higher energy isomers and a − sign showing the reverse case.

The most striking features of these energy barriers are the changes that occur in all the series in which the halogen atom is altered. In all these different series the barrier increases as the size of the halogen atom increases. This indicates the importance of steric repulsive forces, since in the eclipsed transition state the nonbonded distances between the halogen atom and its nearest nonbonded neighbor are always less than the sum of their contact distances.

The importance of electrostatic forces arising from the polar substituents can be seen if we consider 1,1,2-trichloroethane and 2-methylbutane, where the barriers are, respectively, 5.8 and 4.7 kcal/mol. If the barrier was entirely determined by steric forces, a higher value would be expected for 2-methylbutane as compared with 1,1,2-trichloroethane, since the effective volumes of the methyl groups are greater than those of the chlorine atoms.

In recent kinetic studies two mechanisms have been postulated to describe the exchange rate of the interconversion of rotational isomers in ethane derivatives. In the first, an isomer becomes thermally activated to an eclipsed state, which corresponds to the maximum in the potential energy curve and is then immediately deactivated to the stable isomer in the adjacent potential minimum. In the second mechanism the transition form which results from thermal excitation is followed by a state in which the molecule undergoes

Table I
Potential Energy Parameters for Ethane Derivatives

Molecule	$-\Delta S^\circ$ $= R \ln 2$	Barrier $\Delta H^{\ddagger}_{12}$ kcal/mol	Ref.
1,2-Dichloroethane $ClCH_2CH_2Cl$	+	3.2	25
1,2-Dibromoethane $BrCH_2CH_2Br$	+	4.2	25
n-Propyl bromide $CH_3CH_2CH_2Br$	−	3.6	26
n-Butane $CH_3CH_2CH_2CH_3$	+	3.4	27
1,1,2-Trichloroethane Cl_2HCCH_2Cl	−	5.8	26
1,1,2-Tribromoethane Br_2HCCH_2Br	−	6.4	26
2-Methylbutane $(CH_3)_2CHCH_2CH_3$	−	4.7	28
Isobutyl chloride $(CH_3)_2CHCH_2Cl$	−	3.8	29
Isobutyl bromide $(CH_3)_2CHCH_2Br$	−	4.7	30
Isobutyl iodide $(CH_3)_2CHCH_2I$	−	5.4	29
1,2-Dichloro-2-methylpropane $(CH_3)_2CClCH_2Cl$	+	4.5	29
1,2-Dibromo-2-methylpropane $(CH_3)_2CBrCH_2Br$	+	5.5	29
1,1,2,2-Tetrabromoethane $Br_2CHCHBr_2$	+	4.3	31
2,3-Dimethylbutane $(CH_3)_2CHCH(CH_3)_2$	+	2.8	28
1,2-Dibromo 1,1,2,2-tetrafluoroethane BrF_2CCF_2Br	+	6.6	32
2-Chloro-2-methylbutane $(CH_3)_2CClCH_2CH_3$	−	3.9	24
2-Bromo-2-methylbutane $(CH_3)_2CBrCH_2CH_3$	−	5.6	24
2-Iodo-2-methylbutane $(CH_3)_2CICH_2CH_3$	−	8.4	24

(*Continued*)

TABLE I (Continued)

	$-\Delta S°$ $= R \ln 2$	Barrier $\Delta H_{12}^{\ddagger}$ kcal/mol	Ref.
meso-2,3-Dichlorobutane CH$_3$CClHCClHCH$_3$	+	5.1	24
meso-2,3-Dibromobutane CH$_3$CBrHCBrHCH$_3$	−	6.4	24
2,3-Dibromo-2,3-dimethylbutane Br(CH$_3$)$_2$CC(CH$_3$)$_2$Br	+	6–7.5	33
1-Fluoro-1,1,2,2-tetrachloroethane FCl$_2$CCCl$_2$H	−	8.3	34
1,2-Dibromo-1,1-dichloroethane BrCl$_2$CCH$_2$Br	+	7.7	35
1,2-Dibromo-1,1-difluoroethane BrF$_2$CCH$_2$Br	+	5.1	35

"free internal rotation" for some time before being deactivated to any of the potential minima with equal probability. These mechanisms were proposed by Newmark and Sederholm (36,37), and have been used by Gutowsky and his collaborators (38) in dealing with nmr exchange rates in fluorinated ethanes in the liquid phase.

The first mechanism is the simplest possible and corresponds to the case where the transition probability, κ, in the Eyring rate equation (eq. (74)) is unity. From this description of internal rotation Alger, Gutowsky, and Vold (38), have considered the isomeric transition in a substituted ethane, A → B, and argue that the major contribution to the entropy of activation, $\Delta S_{AB}^{\ddagger}$, will come from the difference in vibrational terms between the eclipsed transition state and isomer A. The only difference in these vibrational terms is the torsional mode which is absent in the eclipsed state. Thus, if we denote ΔG_{osc} as the free energy associated with the torsional oscillations

$$\Delta G_{osc}^{\ddagger} = T \Delta S_{AB}^{\ddagger}$$

and substitute in the Eyring rate equation (eq. (74)) we obtain

$$k_{AB} = \left(\frac{kT}{hQ_{osc}}\right) \exp\left(\frac{-\Delta H_{AB}^{\ddagger}}{RT}\right)$$

where Q_{osc} is the partition function of the torsional oscillations of isomer A. If we assume that during the torsional oscillation the two ends of the molecule twist back and forth about the C—C bonds and also that the molecules are in torsional energy levels well below the barrier, then

$$Q_{osc} = \frac{e^{-U/2}}{(1 - e^{-U})}$$

where $U = (h/kt)(3/2\pi)(\Delta H_{AB}^{\ddagger}/2I_r)^{1/2}$, and I_r is the reduced moment of inertia of the molecule. It also follows that U is small enough such that Q_{osc} has essentially the high temperature classical value of $kT/h\nu_0$ and, therefore,

$$k_{AB} = \nu_0 \exp\left(\frac{-\Delta H_{AB}^{\ddagger}}{RT}\right) \tag{75}$$

where ν_0 is the frequency of the torsional oscillations of isomer A. Equation (75) is identified with the Arrhenius rate equation (eq. (73))

$$k_{AB} = A \exp\left(\frac{-E_{AB}}{RT}\right)$$

where A is the Arrhenius "A" factor and E the activation energy.

In the second description of internal rotation where the transition state undergoes "free" internal rotation, the molecule has a probability of one-third of returning at random to one of the three potential energy minima. This model leads to a calculated frequency factor of $\nu_0/3$ in eq. (75). In this model of delayed deactivation the experimental results must be independent of the highest barrier to rotation because, if the energy of a molecule exceeds the second highest barrier, then the molecule can rotate to an angle corresponding to any of the three potential minima without having to pass over the highest barrier.

In practice the A factor of the Arrhenius equation can be determined directly from the ultrasonic relaxation data using eq. (72) and the Arrhenius rate equation (eq. (73)). The torsional frequency of the isomer A can also be measured using far infrared or Raman spectrometry. In Table II some experimentally determined A factors and measured torsional frequencies are listed for a number of substituted ethanes. On the whole the order of magnitude of the A factor and torsional frequencies are the same, indicating that these models of internal rotation are basically correct, although several details still remain to be worked out for a full understanding of the problem.

In the simple ethane derivatives listed in Table I, the stable conformers are usually characterized by the description *anti* or *gauche*, depending on the relative position of the substituents about the C—C bond. In long-chain alkyl halides the number of possible rotational isomers increases rapidly because of the possibility of rotation about more than one C—C bond. It is therefore obvious that the simple *anti* and *gauche* notation used above will not adequately describe the rotational isomers in these longer-chain molecules. Mizushima et al. (39) introduced the notations P, S, and T to identify

TABLE II

Comparison of A Factors and Torsional Frequencies

Molecule	$A \times 10^{-12}$, sec^{-1}	$\nu_0 \times 10^{-12}$, sec^{-1}	Method
1,2-Dichloroethane	0.1–1	3.8	U/S*
1,2-Dibromoethane	1–10	2.7	U/S
1,2-Dibromo-2-methylpropane	1.5	4.0	U/S
1,2-Dichloro-2-methylpropane	0.4	3.3	U/S
meso-2,3-Dichlorobutane	4.0	3.0	U/S
meso-2,3-Dibromobutane	30	3.3	U/S
2-Chloro-2-methylbutane	0.7	3.4	U/S
2-Bromo-2-methylbutane	23	3.4	U/S
2-Iodo-2-methylbutane	1170	3.5	U/S
1-Fluoro-1,1,2,2-tetrachloroethane	2.2	2.58	U/S
	0.8	2.58	nmr
1,2-Difluoro-1,1,2,2-tetrachloroethane	0.5	2.4	nmr

*U/S = ultrasonic method.

the isomers of some primary, secondary, and tertiary halo-alkanes. These notations were used with subscripts H, C, etc. to denote the atom or group *anti* to the halogen atom. For example, the *anti* and *gauche* forms of *n*-propyl chloride (see Fig. 11) now become P_C and P_H, respectively. It has been found by Gates and co-workers (40) that this notation is satisfactory for the longer-chain primary monohalogenoalkanes as only the rotational conformation about the C_1—C_2 bond can be detected spectroscopically.

For secondary alkyl halides this nomenclature is inadequate and extensions have been proposed by Shipman and co-workers (41), Caraculacu and co-workers (42), and Gates et al. (40). The last authors carried out a thorough spectroscopic study of the 2-halobutanes and also the 2- and 3-halopentanes and -octanes. Their results showed that the conformations which could be

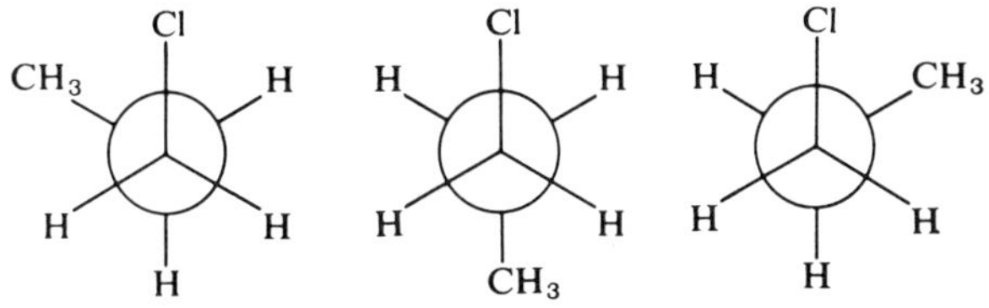

Fig. 11. Possible staggered conformations for *n*-propyl chloride.

detected spectroscopically were those formed by rotation about the C—C bonds adjacent to the carbon–halogen bond. It was concluded that rotation about any other C—C bonds in the molecule produces only second-order effects of molecular crowding, and the resulting isomeric forms could not be detected spectroscopically.

The rotational conformations in these molecules can be visualized by considering the projected drawings of the planar carbon backbone forms (fully extended forms) put forward by Gates et al. (40) and shown for 3-chloropentane in Figure 12. In this figure all the different isomers formed by rotation about the α- and α'-C—C bonds are considered. In order to describe the various isomers using the secondary S notation it is necessary to denote a hydrogen atom on the α- and α'-carbon atom by hydrogen prime (H'). The different isomers can then be described by the notation S_{AB} where A and B are the atoms or groups attached to the α- and α'-carbon atom and in a *trans* or *anti* position relative to the halogen atom.

In the first conformer illustrated in Figure 12, A and B are both H'; therefore, the isomer is the $S_{H'H'}$ form. In the 3-haloalkanes it was found that the S_{HH} form does not exist due to steric hindrance. In the 2-haloalkanes the substituents on one of the α-carbon atoms are all hydrogen atoms; therefore, only the forms $S_{CH'}$, $S_{HH'}$, and $S_{H'H'}$ exist. Thus, in a molecule such as (±)-

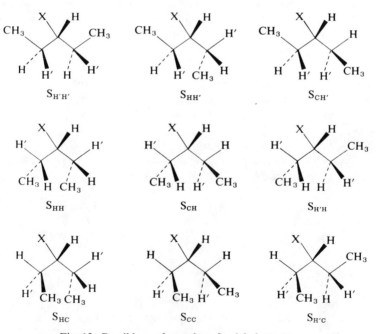

Fig. 12. Possible conformations for 3-halopentanes.

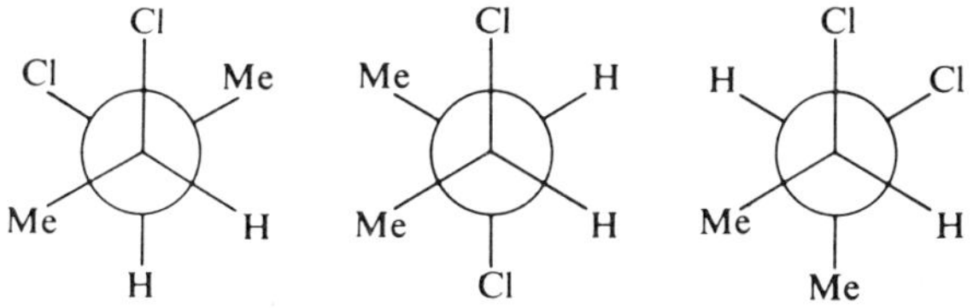

Newman projections showing three conformers of 2,3-dichlorobutane:
S_{HH}, S_{HCl}, S_{HMe}

(1)

2,3-dichlorobutane (1) the isomeric equilibrium is of the kind:

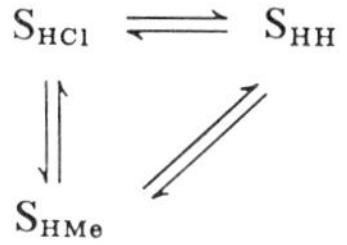

$$S_{HCl} \rightleftharpoons S_{HH}$$
$$\updownarrow \quad \nearrow$$
$$S_{HMe}$$

whereas in 3-chlorohexane the situation is more complex. Although in this molecule only three or four conformers have been detected spectroscopically (i.e., with populations in excess of ca. 5%), the ultrasonic technique is sensitive to about 0.5% population. Thus, the isomeric equilibrium scheme could be very complicated, e.g.,

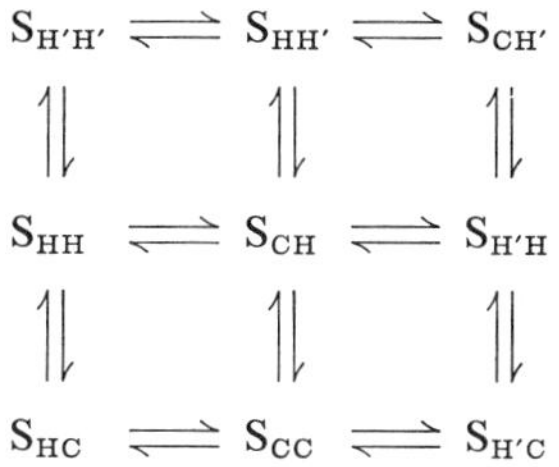

$$S_{H'H'} \rightleftharpoons S_{HH'} \rightleftharpoons S_{CH'}$$
$$\updownarrow \quad\quad \updownarrow \quad\quad \updownarrow$$
$$S_{HH} \rightleftharpoons S_{CH} \rightleftharpoons S_{H'H}$$
$$\updownarrow \quad\quad \updownarrow \quad\quad \updownarrow$$
$$S_{HC} \rightleftharpoons S_{CC} \rightleftharpoons S_{H'C}$$

In multistep conformational equilibria of the type shown above a spectrum of relaxation times and frequencies should be found, with the number of relaxation times corresponding to the number of independent steps (43). Because of the coupling between each state, the relaxation times will not be the same as those found by treating each step as an isolated two-state equilibrium.

In the molecules listed in Table III only one relaxation process was observed in the frequency region 2–300 MHz, and in all cases it is possible to interpret these data in terms of a single relaxation time f_c corresponding to values (at 25°C) in the 2–300 MHz range. Hence, either the relaxation times associated with the various equilibria are so close together that they cannot be resolved experimentally or the relaxation times are so different that all but one are well outside the frequency range of 2–300 MHz. The most reasonable conclusion is the former, viz., that there are several relaxation times sensibly equal for each of the rotational isomeric mixtures studied.

TABLE III

Molecules with the Three Isomers having Different Energies

Molecule	Barrier $\Delta H^{\ddagger}_{12}$, kcal/mol	Ref.
1,2-Dichloropropane $ClCH_2CHClCH_3$	4.7	26
1,2-Dibromopropane $BrCH_2CHBrCH_3$	4.9	26
1-Chloro-2-bromopropane $ClCH_2CHBrCH_3$	5.5	44
sec-Butyl chloride $CH_3CH_2CHClCH_3$	4.4	29
sec-Butyl bromide $CH_3CH_2CHBrCH_3$	4.8	29
sec-Butyl iodide $CH_3CH_2CHICH_3$	5.3	29
dl-2,3-Dichlorobutane $CH_3CHClCClHCH_3$	5.0	24
dl-2,3-Dibromobutane $CH_3CHBrCBrHCH_3$	3.7	24
2-Chloropentane $CH_3CHClCH_2CH_2CH_3$	4.4	44
2-Bromopentane $CH_3CHBrCH_2CH_2CH_3$	4.8	44
2-Methylpentane $(CH_3)_2CHCH_2CH_2CH_3$	3.9	28
3-Chloropentane $CH_3CH_2CHClCH_2CH_3$	3.4	44
3-Bromopentane $CH_3CH_2CHBrCH_2CH_3$	3.6	44
3-Methylpentane $CH_3CH_2CH(CH_3)CH_2CH_3$	4.1	28
1,2-Dibromopentane $BrCH_2CHBrCH_2CH_2CH_3$	3.5	44
1,4-Dibromopentane $BrCH_2CH_2CH_2CHBrCH_3$	4.7	44
2-Bromohexane $CH_3CHBrCH_2CH_2CH_2CH_3$	4.2	44
3-Bromohexane $CH_3CH_2CHBrCH_2CH_2CH_3$	3.6	44
1,2-Dibromobutane		

(*Continued*)

TABLE III (*Continued*)

Molecule	Barrier $\Delta H^{\ddagger}_{12}$, kcal/mol	Ref.
BrCH$_2$CHBrCH$_2$CH$_3$	3.4	44
n-Pentane		
CH$_3$CH$_2$CH$_2$CH$_2$CH$_3$	3.9	27
n-Hexane		
CH$_3$CH$_2$CH$_2$CH$_2$CH$_2$CH$_3$	2.6	27
1,2-Dibromo-1,1,2-trifluoroethane		
BrF$_2$CCFHBr	4.6	35

This conclusion is supported by the following arguments. A literature survey shows that the relaxation times and, hence, relaxation frequencies of a two-state isomeric equilibrium are characteristic of the type of conformational changes. This is illustrated by Table IV which lists the f_c values found for different types of conformational change. The f_c values for the compounds in Table III are in accord with expectation and in line with those previously observed for two-state systems involving internal rotation about C—C bonds in mono- and dihalogenoalkanes. Since the data for these molecules fit the behavior expected for a single relaxation process, the usual procedures led to

TABLE IV

Approximate Relaxation Frequencies for Conformational Equilibria (at 25°C)

Substance	Bond around which rotation occurs	f_c, MHz	Ref.
Hydrocarbons	C—C	5–300	27
Unsaturated aldehydes	C—C	10–100	45
Vinyl ethers	C—O	200	45
Esters	C—O	1–5	46
Amines	C—N	10–30	47
Cyclohexanes	Chair inversion	0.1–3	21
Cyclic sulfites	Chair inversion	20–40	48
Monosubstituted 1,3-dioxanes	Chair inversion	1–10	49
Monohalogenoalkanes	C—C	5–300	29
Dihalogenoalkanes	C—C	5–300	29
Trihalogenoalkanes	C—C	15–150	29

the barriers listed in Table III. To a first approximation these can be regarded as an average value for the potential barrier hindering rotation in the molecule about bonds adjacent to the carbon–halogen link.

2. Enthalpy Differences

The ultrasonic enthalpy differences for a number of liquid ethane derivatives which exist in two-state isomeric equilibria are in Table V. These are compared with the corresponding spectroscopic values. Even allowing for the expected tolerances of ± 400 cal/mol in these values, there is no agreement between the corresponding spectroscopic and acoustic enthalpies. In general it must be accepted that the enthalpy differences obtained by spectroscopic methods are the more reliable. This view is supported by the following arguments. In many simple molecules in the liquid state, values of $\Delta H°$ obtained from infrared, Raman, and dipole moment data are in excellent agreement (1). In addition, the vapor state infrared values agree with those from microwave, dipole moment, and heat capacity data (1,2). It has also been shown that infrared vapor and liquid $\Delta H°$ values can be explained in terms of polar and dipolar interactions (1); in addition there is also very good correlation between recent nmr enthalpies and spectroscopic values (51). The discrepancies revealed by Table V are a good indication that the ultrasonic enthalpy differences for ethane derivatives are in error. It is therefore desirable to investigate the relationship leading up to the determination of these enthalpies and some progress in this direction has been achieved by Crook and Wyn-Jones (52) in a study of the relaxation data in 1,1,2-trichloroethanes in different solvents.

The development of eqs. (65) and (66) requires several approximations which are based, quite correctly, on the small values of μ_m and a negligible velocity dispersion. These equations are also based on the *a priori* assumption that the observed relaxation process is entirely due to a two-state isomeric equilibrium obeying first order kinetics and occurring in ideal solution. Provided $\Delta C_p \ll C_p$ (an assumption which is reasonable since μ_m is very small) eqs. (65) and (66), when combined, become:

$$\frac{2\mu_m C_p}{(\gamma - 1)\pi} = nR\left(\frac{\Delta H°}{RT}\right)^2 \frac{\exp(-\Delta G°/RT)}{[1 + \exp(-\Delta G°/RT)]^2} \left(1 - \frac{\Delta V°}{V}\frac{C_p}{\theta \Delta H°}\right)^2 \quad (76)$$

where n is the mole fraction of solute.

A useful way to check the above assumptions is to investigate the concentration dependence of the quantity $2\mu_m C_p/(\gamma - 1)\pi$. This concentration dependence is predicted by eq. (76) by use of the above assumption in their derivation. In Figure 13 typical plots of $2\mu_m C_p/(\gamma - 1)\pi$ versus mole fraction, n, are drawn and show, for dilute solutions, that the linearity predicted by

TABLE V

Enthalpy Differences between Rotational Isomers

Molecule	$-\Delta H°$ (ultrasonic), kcal/mole	Ref.	$-\Delta H°$ (infrared), kcal/mol	Ref.
n-Propyl bromide	1.3	26	0, 0.44, 0.5	50
1,1,2-Trichloroethane	2.1	26	0.1, 0.22, 0.29	50
1,1,2-Tribromoethane	1.6	26	0.5	50
2-Methylbutane	0.9	28	0	50
Isobutyl chloride	1.31	29	0.36	29
Isobutyl bromide	0.70	30	0.26	29
1,2-Dichloro-2-methyl-propane	2.01	29	0	29
1,2-Dibromo-2-methyl-propane	0.97	29	0.74	29
1,1,2,2-Tetrabromoethane	0.9	31	0.9, 0.75	50
2,3-Dimethylbutane	1.0	28	0.1	50
1,2-Dibromo-1,1,2,2-tetra-fluoroethane	1.6	32	0.93	50

eq. (76) is obeyed. In addition, the linearity of these plots at the weaker concentrations also shows that the equilibrium parameters $\Delta H°$, $\Delta S°$, and $\Delta V°$ are independent of concentration.

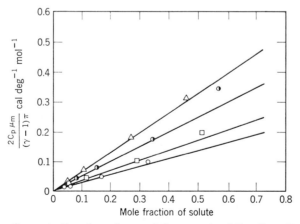

Fig. 13. Concentration dependence of the quantity $2C_p\mu_m/(\gamma - 1)\pi$: (△) nitrobenzene; (◐) n-heptane; (□) mesityl oxide; (○) methyl cyanide; (●) pure liquid. (From K. R. Crook and E. Wyn-Jones, *J. Chem. Phys.*, **50**, 3448 (1969), by courtesy of the Editor.)

TABLE VI

Molecular Equilibrium Parameters

Solvent	Slope of lines, Fig. 13, cal/mol-deg.	$-\Delta H^\circ$, cal/mol	$\Delta V^\circ/V$	$(\Delta V^\circ/V) \times (C_p/\theta)$, cal/mol
p-Xylene	0.51	306	−0.03	− 410
Mesityl oxide	0.34	225	−0.01	− 350
n-Heptane	0.51	419	−0.01	− 320
Ethyl acetate	0.57	107	−0.01	− 120
Methyl cyanide	0.29	146	−0.03	− 370
Trichloroethylene	0.33	625	−0.01	− 59
Nitrobenzene	0.68	24	−0.02	− 760
Pure liquid	0.61	421	−0.06	− 1300

From the slopes of the plots in Figure 13 the quantity

$$R\left(\frac{\Delta H^\circ}{RT}\right)^2 \frac{\exp(-\Delta G^\circ/RT)}{[1+\exp(-\Delta G^\circ/RT)]^2} \left(1 - \frac{\Delta V^\circ}{V}\frac{C_p}{\theta \Delta H^\circ}\right)^2$$

was derived. Using ΔH° values found from nmr studies and with $\Delta S^\circ = R \ln 2$ values of $\Delta V^\circ/V$ were derived. These values were then used to check whether the assumption

$$(\Delta V^\circ/V)(C_p/\theta) \ll \Delta H^\circ$$

used in the manipulation of eqs. (9), (65), and (66) to yield ultrasonic ΔH° values, is valid. The values listed in Table VI show that this assumption cannot be used for this molecule and probably accounts, to a large extent, for the discrepancies found in Table V.

E. Tertiary Amines

Ultrasonic relaxation studies have been carried out in a number of tertiary amines. The first of these studies was carried out by Heasell and Lamb (53), who assigned the relaxation process to a perturbation of a conformational equilibrium between two of the isomers shown in Figure 14. The assignment of this relaxation process was supported by the fact that no relaxation was observed in trimethylamine, where all the conformers formed by internal rotation about the C—N bond have the same energy. In addition, Litovitz and Carnevale (54) confirmed the intramolecular nature of the relaxation process when they found that the relaxation frequency was independent of pressure up to 3000 atm. The latter data also show that there is a

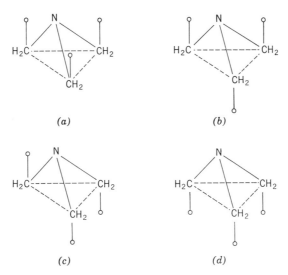

Fig. 14. Possible conformations for trialkylamines.

zero volume change in going from either stable state to the transition state; this in turn is a good indication that the assumption $(\Delta V^\circ/V)(C_p/\theta) \ll \Delta H^\circ$ can be used in connection with the derivation of thermodynamic parameters in the internal rotation in tertiary amines. The energies associated with internal rotation in several tertiary amines are in Table VII. The choice of which pair of conformers is responsible for the relaxation process is not straightforward since physicochemical evidence of the structure of these molecules is sparse. The available information is summarized below:

From a study of molecular models Heasell and Lamb (53) found that there is a great deal of steric hindrance in conformer d; this steric hindrance appears to decrease as the number of alkyl groups that are folded up towards the nitrogen lone pair electrons increases. Thus, from a steric point of view the most stable conformers are a and b. In addition, these forms will be stabilized further by the electrostatic attraction between the polarizable alkyl groups that are folded back and the lone pair electrons on the nitrogen. Furthermore, there is evidence from spectroscopic (55) and thermochemical (56) experiments that trimethylamine is a weaker base than trimethylamine and also primary and secondary amines. In order to explain these data it was proposed that at least one of the alkyl groups is folded back so that it projects towards the lone pair of the nitrogen atom.

On the basis of this evidence Heasell and Lamb and also Krebs and Lamb (31) concluded that the most likely conformers taking part in the relaxation process are a and b. However, from the ultrasonic data alone it

TABLE VII

Potential Energy Parameters for Amines

Molecule	$-\Delta H°$, kcal/mol	$-\Delta S°$, eu	$\Delta H^{\ddagger}_{12}$, kcal/mol	Ref.
Triethylamine	3.4	4.7	6.8	53
	5.9	13	10.8	58
Tripropylamine	1.5	1.1	4.5	47
Tributylamine	0.9	−0.9	4.3	31
Tripentylamine	1.2	−0.9	4.8	47
Trihexylamine	1.1	1.3	4.3	47
Triisopentylamine	1.1	1.1	3.0	47
Triallylamine	1.5	2.1	5.9	47
Trimethallylamine	1.3	1.9	4.3	47
3-N-Diethylaminopropiononitrile	1.2	0.7	4.5	47
3-N-Diethylaminopropylamine	1.3	1.8	6.3	47
4-N-Diethylaminobutane-2-one	1.4	2.0	6.5	47
N,N-Diethylcyclohexylamine	1.4	1.7	3.7	47
3-N-Diethylaminoethylamine	1.6	1.9	5.9	47
1-Diethylamino-2-chloropropane	1.1	0.4	5.8	47
3-N-Dipropylaminopropiononitrile	1.2	0.7	5.9	47

cannot be deduced which is the most stable isomer. These authors studied the molecular models of a and b and concluded that the entropy of a is greater than that of b since there is more freedom of movement in the alkyl groups of a. In the experimental results the sign of the entropy difference changes from positive to negative in going from triethyl- to trihexylamine. This indicates that the stable isomer changes from a to b as the sign of the entropy difference changes. In the higher tri-n-alkylamines parachor calculations (57) have revealed that the more stable form is the conformer b, in agreement

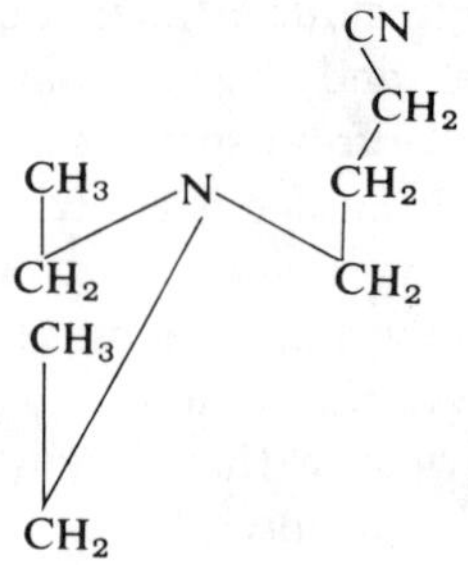

$$\begin{array}{c} CH_3 \\ | \\ CH_2 \\ | \\ CH_3 \end{array} N \begin{array}{c} CN \\ | \\ CH_2 \\ | \\ CH_2 \\ | \\ CH_2 \\ | \\ CH_2 \end{array}$$

(2)

with the ultrasonic data. In asymmetric substituted amines such as 3-*N*-diethylaminopropiononitrile (2), the number of rotational conformers about the C—N bonds increases considerably compared with the tri-*n*-alkylamines. Since the ultrasonic method does not give any information about the structure of the conformers taking part in a relaxation process and in view of the lack of other data on internal rotation in these molecules, it is not appropriate to make any further comment about the observed barriers in these molecules.

In order to explain the results on two studies carried out on triethylamine covering the temperature range $-60°$ to $40°C$ and frequency range 10–250 MHz sec^{-1}, Padmanabhan and Heasell (58) proposed that the entropy difference between the isomers and the entropy of activation were temperature dependent. These temperature-dependent entropy terms were introduced to account for the nonlinearity in the thermodynamic and kinetic plots. These authors concluded that the entropy of the higher energy isomer, b of Figure 14, decreases with decreasing temperature, presumably as the rotations of the methyl groups become more restricted. In all the other work on amines, however, the kinetic plots were linear, even when expanded, and therefore indicate that the energy and entropy of activation are independent of temperature.

The main factors which hinder the rotation about the C—N bonds in these molecules are: nonbonded steric interactions between the bulky groups; electrostatic interaction which mainly exists between the lone pair electrons on the nitrogen atom and the polarizable alkyl substituents; and possibly repulsive forces between adjacent bonds. The extent of the various contribution from the effects cannot be resolved from Table VII. It is worth noting, however, that the barrier heights obtained here are of the same order of magnitude as those found for rotation about C—C bonds in substituted ethanes. The values found for rotation about the C—N bonds in acid amides from nmr spectroscopy (7a) are much higher, being in the range 15–30 kcal/mol. In these compounds the σ-bonds associated with the N atom are nearly coplanar and so the situation is favorable for the delocalization of π-electrons. This would increase the bond order of the CN link from that of unity found in the tertiary amines to a value (59) near to 1.6. This increase in bond order would tend to make rotation more difficult in the amides and, hence, lead to a higher value for the barrier height.

F. Esters

Historically esters were among the first molecules that were studied using the ultrasonic technique. The early measurements were made by Biquard (60) in 1936 and several papers have subsequently appeared on the subject. The ultrasonic relaxation in esters has been attributed to a perturbation of the equilibrium between the planar isomers **3** and **4** formed by internal rotation about the C—O bond adjacent to the carbonyl link. The

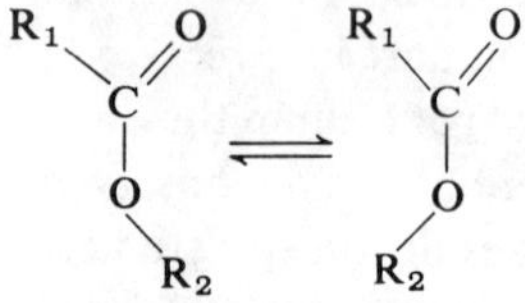

$$\underset{\underset{(3)}{cis}}{\overset{R_1}{\underset{O}{\overset{\diagdown}{\underset{R_2}{C}}}}\overset{O}{\underset{}{\diagup\diagdown}}} \quad \rightleftharpoons \quad \underset{\underset{(4)}{trans}}{\overset{R_1}{\underset{O}{\overset{\diagdown}{\underset{R_2}{C}}}}\overset{O}{\underset{}{\diagup\diagdown}}}$$

ultrasonic results for those molecules where a complete ultrasonic treatment has been carried out are in Table VIII. Although Tabuchi (61) appears to have made a complete study on ethyl formate, Piercy and Subrahmanyam (62,63) have pointed out that since only velocity dispersion was measured, the results are not considered accurate enough for quantitative treatment.

In connection with the derivation of the equilibrium thermodynamic parameters $\Delta H°$ and $\Delta S°$, it is worth pointing out that Slie and Litovitz (65) found $(\Delta V°/V) = 0$ for ethyl acetate by carrying out ultrasonic measurements over the temperature range 40–80°C and pressure range 1—1000 atm.

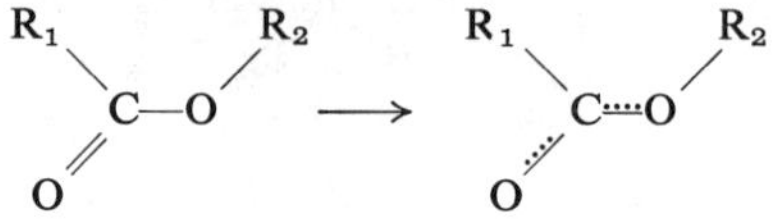

$$\underset{}{\overset{R_1}{\diagdown}}\underset{O}{\overset{}{\diagup}}C\text{—}O\overset{R_2}{\diagup} \quad \longrightarrow \quad \underset{}{\overset{R_1}{\diagdown}}\underset{O}{\overset{}{\diagup}}C\cdots O\overset{R_2}{\diagup}$$

(5)

Piercy and Subrahmanyam (62,63) have suggested that the barrier in esters is mainly due to the partial double bond character of the C—O bond adjacent to the carbonyl link. This arises because of the delocalization of the π-electrons in the system (5). In an attempt to determine the cause of the energy difference between the isomers, Bailey and North (64) studied esters of the type shown above and arrived at the following conclusions:

The enthalpy difference $\Delta H°$ was found to increase steadily with the size of $R_1 + R_2$ from less than 1 kcal/mol for the formates up to 8 kcal/mol for the C_4 groups. This was accompanied by a decrease in $\Delta H_{21}^{\ddagger}$ from almost 8 kcal/mol in the formates to less than 1 kcal/mol in the C_4—C_5 esters. In this series, however, $\Delta H_{12}^{\ddagger}$ was almost constant. These results mean that increasing the size of the groups increases the energy of the upper state, at the same time leaving the energy of the lower and transition states almost unaltered. Thus, altering R_1 and R_2 changes the energy of the *trans* state, whereas the *cis* and transition states are unaffected. Furthermore, the effect is roughly dependent on the size of R_1 *and* R_2 rather than on either R_1 or R_2

TABLE VIII
Potential Energy Parameters for Esters R_1COOR_2

Ester	No. of C atoms $R_1 + R_2$	$\Delta H°$, kcal/mol	$\Delta S°$, cal/deg-mol	$\Delta H^{\ddagger}_{21}$, kcal/mol	$\Delta S^{\ddagger}_{21}$, cal/deg-mol	Ref.
Methyl formate	1	0.4	1.0	7.8	−1.8	64
	1	2.3	—	7.8	−8.7	63
Ethyl formate	2	0.5	2.0	6.1	−9.6	64
	2	2.3	—	8.0	−8.8	63
Methyl acetate	2	4.2	0	5.9	−2.3	64
n-Propyl formate	3	3.7	1.6	6.7	−6.8	64
Ethyl acetate	3	4.5	−1.0	4.2	−8.6	64
	3	2.9	—	5.7	—	65
	3	1.6	—	5.7	—	66
Methyl propionate	3	5.1	1.6	4.9	−10.2	64
Ethyl propionate	4	5.8	1.4	1.2	−18.5	64
Isopropyl formate	3	3.7	—	5.8	4.3	62

alone. These results mean that steric effects are important in determining enthalpy differences, but not effects such as hydrogen bonding between R_2 (alcohol) and C=O or dielectric stabilization (the *trans* state has the higher dipole moment).

TABLE IX

Two-State Enthalpy Parameters for Diesters
ROOC(CH$_2$)$_n$COOR (in kcal/mol) (67)

Ester	n	$\Delta H°$	$\Delta H^{\ddagger}_{21}$	$\Delta H^{\ddagger}_{12}$
Dimethyl Oxalate	0	3.8	4.0	7.8
Malonate	1	4.3	2.3	6.6
Succinate	2	5.2	1.5	6.7
Glutarate	3	5.3	1.1	6.4
Adipate	4	1.7	0.9	2.6
Diethyl Oxalate	0	0.5	6.0	6.5
Malonate	1	4.9	2.3	7.2
Succinate	2	5.1	1.5	6.6
Adipate	4	1.9	1.1	3.0

Bailey and North (67) have also studied a series of diesters; their results are summarized in Table IX. The experimental data were analyzed for a single relaxation associated with a two-state equilibrium. These results show that the enthalpy difference in both dimethyl and diethyl malonates and succinates are very close to those found for the corresponding simple esters—methyl acetate and methyl propionate, respectively. These findings were interpreted as follows:

When $n = 0$, the enthalpy difference depends on the interaction between the two ester groups and for $n = 1, 2,$ and 3, $\Delta H°$ depends on the interaction between one of the ester groups and the intervening (CH$_2$)$_n$ chain. In the case of the adipates ($n = 4$), it is possible to bring the O and C=O groups into a position where dipole–dipole interactions tend to stabilize a "cyclic conformation" (**6**) in which the internal rotation of the ester group is considerably hindered.

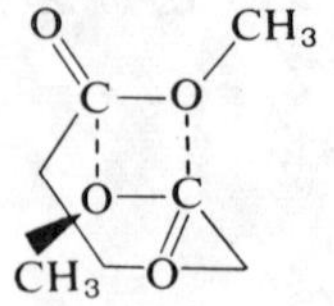

(6)

The only explanation for the ultrasonic relaxation in methyl formate is a perturbation of the equilibrium between the *trans* and *cis* isomers shown above. In ethyl formate, on the other hand, there are four possible conformations due to internal rotation about the C—C and C—O bonds (7–10).

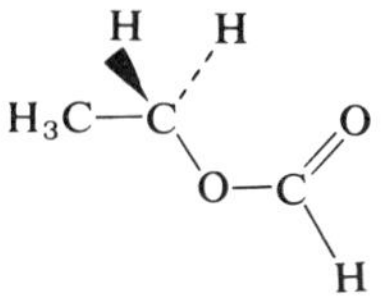

           (7)   (8) (and mirror image)

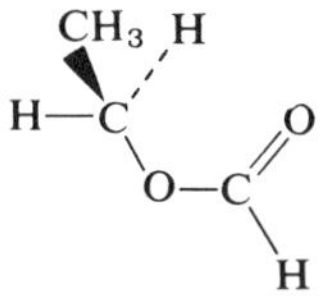

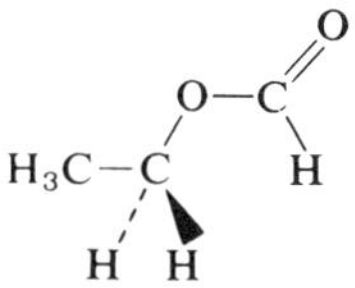

           (9)   (10) (and mirror image)

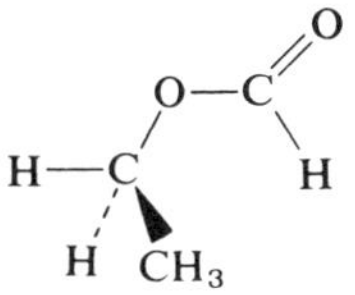

However, the evidence presented in Section IID shows that the ultrasonic relaxation frequency is characteristic of the particular conformational change concerned and, in view of the work discussed above, there is no doubt that ultrasonic relaxation in esters is caused by internal rotation about the C—O bond. In microwave experiments (68,69) carried out in the vapor state, only the *cis* isomer was detected in methyl formate, whereas in ethyl formate isomers **7** and **8** formed by internal rotation about the C—C bond were detected. The potential barrier for the isomeric transition **7→8** was found to be 1.2 kcal/mol; in ultrasonic terms this corresponds to a relaxation frequency higher than 1 GHz and well outside the experimental range used in the above work.

G. Acyclic and Heterocyclic Ring Inversions

Ultrasonic relaxation in cyclohexane derivatives (**11**) (21,70,71) and also some heterocyclic molecules (72–76), such as substituted 1,3-dioxanes (**12**), cyclic sulfites (**13**), and a cyclic sulfate (**14**), has been attributed to a perturbation of the equilibrium between the axial and equatorial chair conformations. The energy parameters for the molecules that have been studied to date by ultrasonics are collected in Table X. In the cyclohexanes, 1,3-dioxanes, and 4-methyltrimethylene sulfate the more stable conformer is the equatorial isomer, whereas in the cyclic sulfites the chair conformer in which the S=O bond is axial is the more stable.

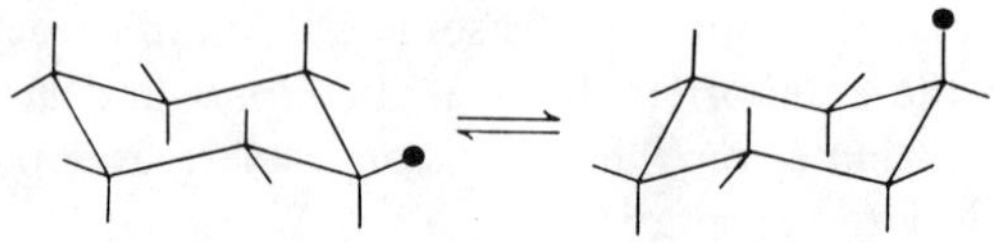

Equatorial ⇌ Axial

cyclohexane derivatives
(11)

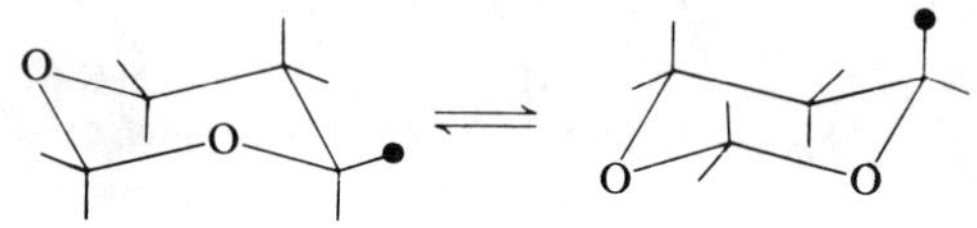

Equatorial ⇌ Axial

dioxanes
(12)

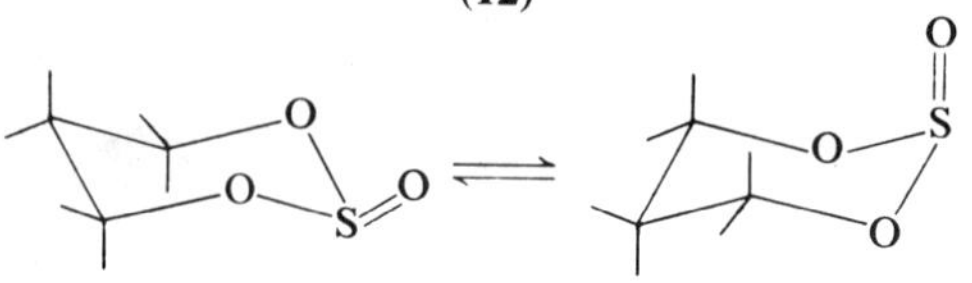

Equatorial ⇌ Axial

cyclic sulfites
(13)

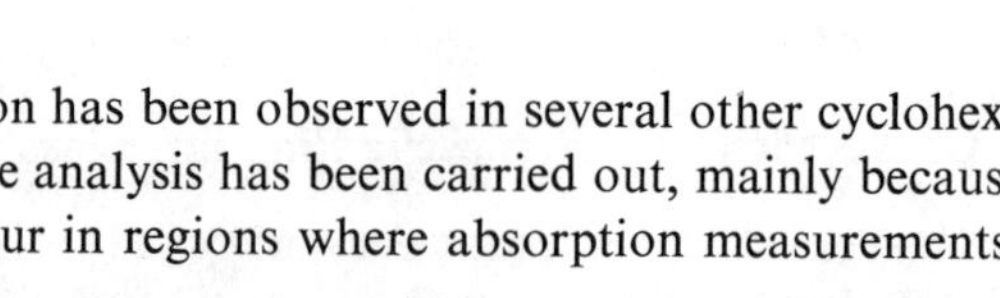

4-methyltrimethylene sulfate
(14)

A relaxation has been observed in several other cyclohexane derivatives, but no complete analysis has been carried out, mainly because the relaxation frequencies occur in regions where absorption measurements are difficult to effect (70,71).

The enthalpy differences in Table X are subject to the limitations imposed by the assumption concerning the volume change. In methylcyclohexane Piercy (21) found $\Delta V°/V \approx 0.4\%$, which means that this assumption is reasonable. The enthalpy difference in 4-methyl-1,3-dioxane (73) compares well with a value of $\Delta G° = 3$ kcal/mol found from equilibration studies (77), and in the 4-phenyl derivative an even larger free-energy difference has been predicted in view of the apparent ring deformation when the phenyl group is in an axial position (78). Indeed the nmr spectrum of 4-phenyl-1,3-dioxane

TABLE X

Energy Parameters for Ring Inversion in Cyclic Systems

Molecule	$-\Delta H°$, kcal/mol	$-\Delta S°$, eu	$\Delta H^{\ddagger}_{12}$, kcal/mol	$\Delta S^{\ddagger}_{12}$ eu	Ref.
Methylcyclohexane	2.9	3.2	10.3	—	82
	3.6	4.0	10.8	—	21
Chlorocyclohexane	—		12±2	—	21
Bromocyclohexane	—		12±2	—	21
4-Methyl-1,3-dioxane	2.5	—	9.2	5.0	
4-Phenyl-1,3-dioxane	5.3	—	4.5	−9.0	
4-Methyltrimethylene sulfate	1.6	−0.5	4.6	−6.3	74
Trimethylene sulfite	1.3	−2.5	5.5	−4.0	76
4-Methyltrimethylene sulfite	1.2	—	4.8	—	76
5,5-Dimethyltrimethylene sulfite	1.5	−1.5	4.3	−7.0	76
5,5-Diethyltrimethylene sulfite	1.3	—	4.2	—	76
5-Methyl-5-ethyltrimethylene sulfite	1.4	—	4.3	—	76
4,6-Dimethyltrimethylene sulfite	1.2	—	5.8	—	76
4,5,6-Trimethyltrimethylene sulfite	1.2	—	3.6	—	76

was analyzed entirely in terms of the equatorial isomer (79). The high $\Delta G°$ values of trimethylene and neopentyl sulfites as well as those of 4-methyltrimethylene sulfate are consistent with recent nmr, dipole moment, and spectroscopic data, which indicate a conformational equilibrium heavily weighted towards the axial form in the sulfite and the equatorial isomer in the sulfate (73,76).

In the current theories of ring inversion in six-membered cyclic molecules, it is generally accepted that the transition state is a half-chair conformation with four adjacent ring atoms in a plane (80). In cyclohexane there is only one possible half-chair form which has this cyclohexene-like structure, whereas in 1,3-dioxane there are three possible half-chair forms, all of which differ in energy. Because of the asymmetry in the molecules listed in Table X there are six nonequivalent half-chair conformations as shown for trimethylene sulfite (**15–20**). Thus, in these molecules there are up to six possible paths for the ring inversion to occur and none of these necessarily corresponds to the values of $\Delta H^{\ddagger}_{12}$ listed in Table X. For example, in some molecules, such

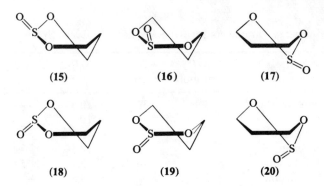

(15) (16) (17)
(18) (19) (20)

as 1,3-dioxane, the spread of activation energies between the possible transition states is only a small percentage of the total barrier (81). In these molecules one could thus expect all transition states to be involved in the ring inversion. On the other hand, in other molecules one of the above half-chair forms may have a much lower energy than the others. This, in turn, means that during ring inversion this form becomes the favored transition state for ring inversion. In molecules of this type it is possible that the entropy of activation will serve as a guide to the actual number of transition states through which the molecule passes during ring inversion. For example, if the isomerization proceeds via all six possible half-chair forms one would, on purely statistical grounds, expect the entropy of the transition state to be more positive by $R \ln 6$ than that of the chair forms. In trimethylene and neopentyl sulfites, 4-methyltrimethylene sulfate, and 4-phenyl-1,3-dioxane, the negative entropy of activation suggests that the ring inversion may proceed via one or two favored half-chair states.

According to Hendrickson (80), the energy of a particular conformation in six-membered cyclic molecules arises from three factors: torsional energy resulting from changing the dihedral angles between bonds to neighboring carbon atoms from the 60° of a staggered conformation; bond angle strain due to distortion of valency angles away from the tetrahedral angle; and repulsion between nonbonded atoms or groups. Hendrickson (80) and Harris and Spragg (81) have shown that for cyclic and heterocyclic molecules the major contribution to the ring inversion barrier comes from torsional strain, which accounts for at least 60% of the total barrier. In cyclic molecules this factor can be estimated qualitatively by reference to the torsional barrier in propane and dimethyl ether (for 1,3-dioxanes), which are, respectively, 3.20 and 2.72 kcal/mol. There is some correlation between these values and the barriers quoted in Table X.

In the frequency range 5–300 MHz ultrasonic experiments failed to detect any relaxation processes arising from ring inversions in ethylene sulfite, tetramethylene sulfite, ethylene carbonate, and propylene carbonate. In 4-chloromethyl-1,3-dioxolan-2-one a relaxation process was attributed to a perturbation of the equilibrium between the isomers **21** and **22**, formed by

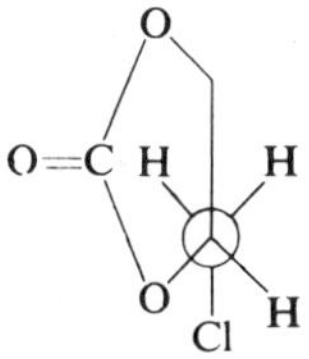

(21) 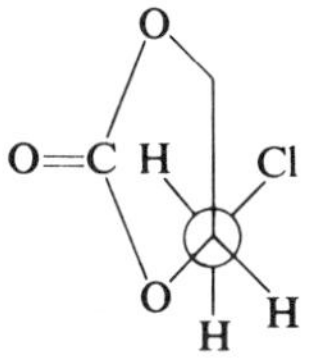   (22) 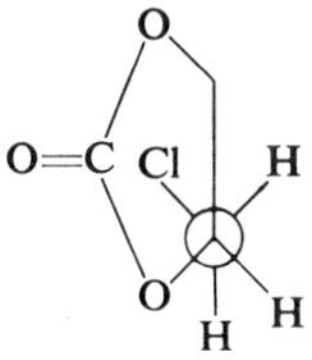   (23)

internal rotation about the C—C bond (74). A potential barrier of 5.1 kcal/mol was found which compares well with those found for internal rotation of the chloromethyl group in some ethane derivatives in Table I.

H. α,β-Unsaturated Aldehydes and Ketones

Internal rotation in α,β-unsaturated aldehydes and ketones was first observed by de Groot and Lamb through ultrasonic measurements (83,84). The relaxation process they observed in these molecules was attributed to a perturbation of the equilibrium between two planar isomers of the molecules formed by internal rotation about the C—C single bond. For example, in molecules such as **24**, a resonance structure of the type **25** has often been

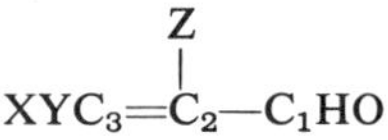

$$XYC_3=C_2-C_1HO$$
(24)

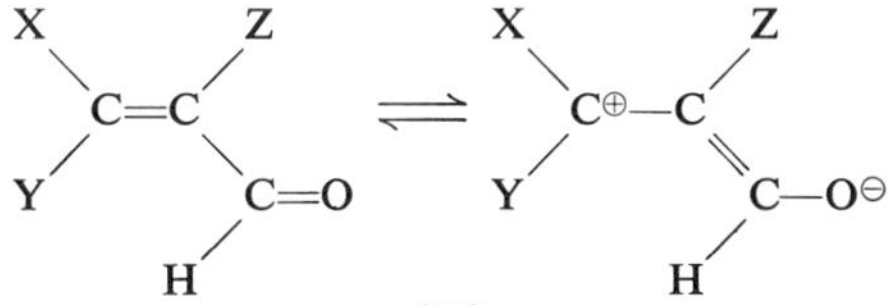

(25)

invoked to explain their chemical properties. This quantum-mechanical resonance favors the planar form of the molecule and, therefore, will stabilize the two rotational isomers **26** and **27** formed by internal rotation about the

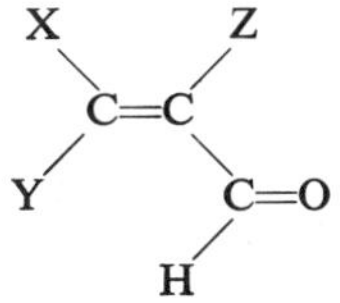

s-trans   s-cis
(26)   (27)

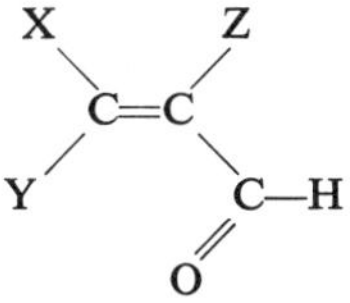

C_1—C_2 bond. The potential energy parameters associated with internal rotation in these types of molecules and also in closely related molecules are given in Table XI.

TABLE XI

Energy Parameters for Unsaturated Aldehydes and Ketones

Molecule	$\Delta H^{\ddagger}_{12}$, kcal/mol	$-\Delta H°$ kcal/mol	Ref.
Acrolein	4.96	2.06	83
Crotonaldehyde	5.51	1.93	83
Cinnamaldehyde	5.62	1.5	83
Methacrolein	5.31	3.07	83
Furacrolein	5.10	1.2	83
α-Methylcinnamaldehyde	5.15	1.3	84
α,β-Dimethylacrolein	5.55	1.3	84
2-Ethyl-3-propylacrolein	5.17	1.3	84
Citral	4.46	—	84
3(2-Furyl)acrylophenone	6.09	1.3	84
2-Furylideneacetone	7.09	1.4	84
α-n-Hexylcinnamaldehyde	4.72	1.6	84
Furan-2-aldehyde	11.0	1.3	20
Thiophene-2-aldehyde	10.3	1.3	20

In the aldehydes, dipole moment studies (85) have indicated that the s-trans isomer is the more stable, and thus the enthalpy differences listed in Table XI are the differences in energy between the s-cis and s-trans forms. The influence of the various interactions which affect the potential energy parameters has been aptly illustrated in this work. For example, in acrolein (X = Y = Z = H), crotonaldehyde (X = CH_3, Y = Z = H), and cinnamaldehyde (X = C_6H_5, Y = Z = H) the barrier increases, respectively, from 4.96 to 5.51 to 5.62 kcal/mol. This has been attributed to the substitution of electron-donating groups at the C-3 atom which strengthen the conjugation. The electron-donating power of the phenyl group is greater than that of methyl, which, in turn, is greater than that of hydrogen, this being reflected in the energy barriers. Furthermore, in the related molecules **28**, where the

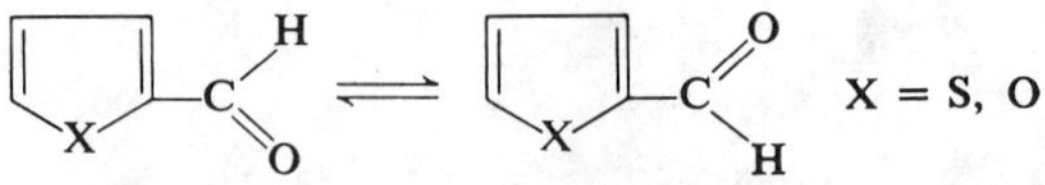

(28a) ⇌ (28b)   X = S, O

electron mobility and, hence, conjugation are even stronger, the barriers are of the order of 10–11 kcal/mol. On the other hand, the enthalpy differences between the *s-cis* and *s-trans* forms decrease as the conjugation is increased. This can be explained as follows. As the conjugation increases, the electrostatic forces also increase; this in turn stabilizes the *s-cis* isomer, thus decreasing the enthalpy difference. In methacrolein (**29**) the enthalpy difference

$$\begin{array}{c} H \\ \diagdown \\ C=C \\ \diagup \\ H \end{array} \begin{array}{c} CH_3 \\ \diagup \\ \diagdown \\ C=O \\ \diagup \\ H \end{array}$$

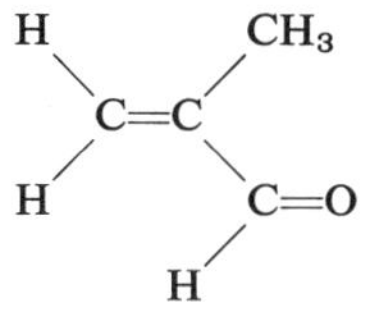

(**29**)

is very large ($\Delta H° = 3.1$ kcal/mol) compared to acrolein; this can be explained in terms of an electrostatic interaction in the *s-trans* form between the polarizable methyl group on C-2 and the oxygen atom, thus making this isomer even more stable than the corresponding isomer in acrolein. The influence of steric hindrance in these molecules is well illustrated in mesityl

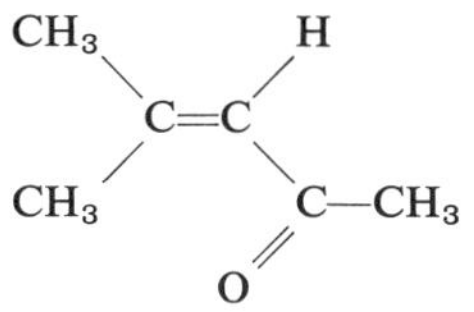

$$\begin{array}{c} CH_3 \\ \diagdown \\ C=C \\ \diagup \\ CH_3 \end{array} \begin{array}{c} H \\ \diagup \\ \diagdown \\ C-CH_3 \\ \diagup \\ O \end{array}$$

(**30**)

oxide (**30**), where the *s-trans* form is completely blocked and thus no ultrasonic relaxation was observed in this molecule.

I. Other Systems

Ultrasonic relaxation arising from the perturbation of a conformational equilibrium has also been observed in other systems. Although the results have not yet been fully analyzed, it is of interest to point out the type of isomerism involved. These include internal rotation in propionaldehyde (**31**) (83), methyl vinyl ether (**32**) (83), *o*-chloroanisole (**33**) (84), trimethyl phosphate (**34**) (84), 2,2-dimethyl-1,3-dioxane (**35**) (86), and possibly *N*-methylacetamide (**36**) (86).

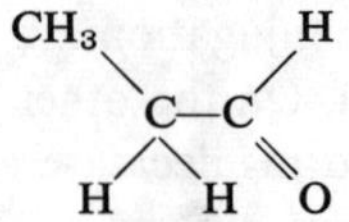

(31)

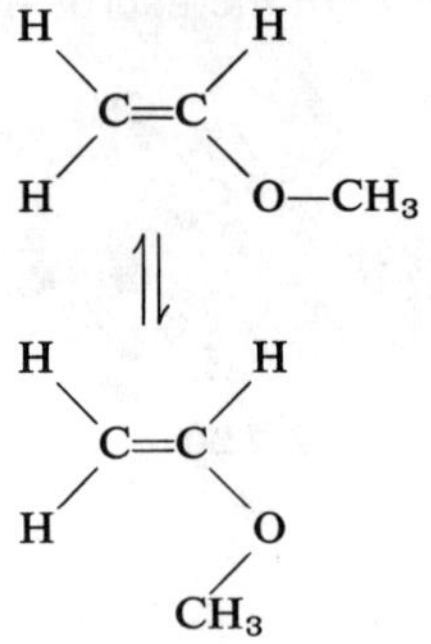

(32)

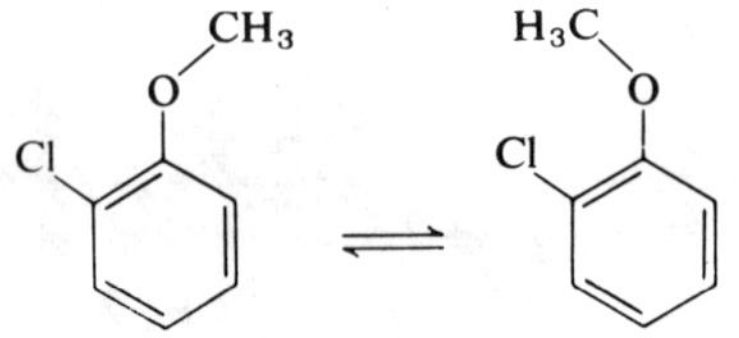

(33)

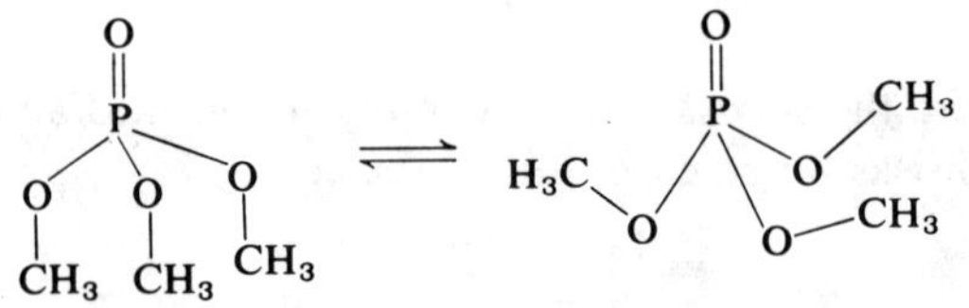

(34)

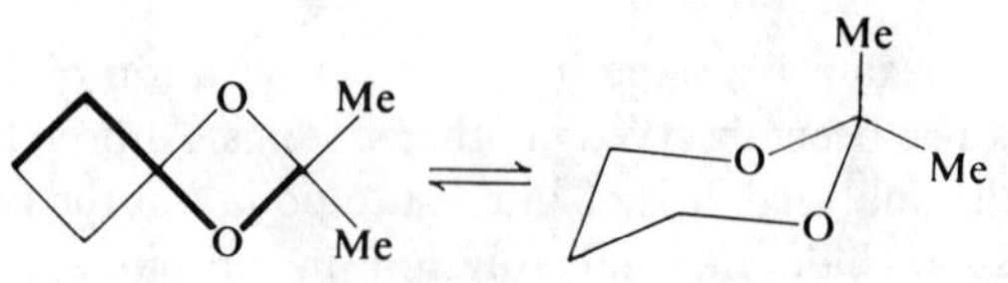

(35)

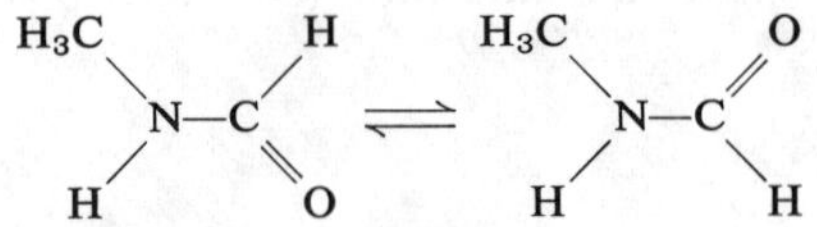

(36)

III. VIBRATIONAL SPECTROSCOPY

A. Introduction

The use of vibrational spectroscopy as a tool to study conformational problems in small molecules has been described in excellent articles by Mizushima (1) and Sheppard (2). As described previously, the conformational equilibria that exist in 1,2-dichloroethane (Fig. 1) or furan-2-aldehyde (Fig. 2) can be represented by a two-state unimolecular process of the type:

$$A \rightleftharpoons B \qquad (77)$$

and provided the potential barrier is of the order 3–28* kcal/mol, the above equilibrium may be considered to be "dynamic." The magnitude of the rate of isomeric exchange in the above equilibrium will be a function of the energy barrier hindering rotation. To a good approximation the temperature dependence of the forward and reverse rates of isomerization can be described by the Arrhenius rate equation (eq. (73)):

$$k = A \exp\left(\frac{-E}{RT}\right)$$

For the unimolecular isomerization (eq. (77)) it is safe to assume that the Arrhenius A factor will lie in the range 10^{11}–10^{14} sec^{-1}, and if the energy barrier to rotation is in the range 3–28* kcal/mol, the extreme limits for the rate constant k are 7×10^{11} to 9×10^{-4} sec^{-1}; these correspond to actual lifetimes of 1.5×10^{-12} to 1×10^{3} sec. Since the time constant for the interaction of electromagnetic radiation with a molecular dipole is ca. 10^{13} sec^{-1} it follows that the lifetimes of the isomeric species A and B will be long enough so that their vibrational spectra can be measured. This means that the vibrational frequencies of the separate isomers can be obtained using infrared and Raman spectroscopy.

Normally in the vapor or liquid states the molecules exist as an equilibrium mixture of isomers and thus the recorded spectra in these phases will contain the vibrational frequencies of all the isomers, provided their populations are in excess of ca. 4%.† In practice it is found that many of the vibrational modes of the individual isomers occur in similar environments and their frequencies will be coincident; in addition, in many isomers the vibrational

---

*The figure 28 kcal/mol is arbitrary in the sense that the exchange rates in a conformational equilibrium with a potential barrier of this magnitude will be very slow indeed (of the order of hours), and this barrier is thus well beyond the limit where separate isomers, such as the *cis* and *trans* isomers of 1,2-dichloroethylene, become isolable at room temperature.

† We estimate that at least 4% or so of the higher-energy isomer must be present in order that this isomer may be detected spectroscopically with some confidence.

modes will have frequencies which are accidentally degenerate. In the two-state equilibrium, eq. (77)

$$\frac{c_B}{c_A} = \exp\left(\frac{-\Delta G°}{RT}\right)$$

where the c's are the isomeric populations and $\Delta G°$ is the Gibbs free energy. According to the above equation the population of isomer B will increase as the temperature of the liquid is lowered. At the melting point, however, there will be a discontinuity because of the change of phase. At this temperature, corresponding to the liquid → crystalline phase transition, all the molecules in the liquid with conformation A will be converted into the more stable form B. This is the normal behavior of the rotational isomers of ethane derivatives (1,2) and, in most cases, only the stable isomer exists in the solid crystalline phase because of the normal thermodynamic distribution of energy and also because of the stabilizing influences of the intermolecular forces of the crystal packing. The above process can be followed spectroscopically by observing absorption bands belonging to each of the stable isomers, and thus the vibrational spectrum will be considerably simplified in the solid crystalline phase.

In some molecules symmetry considerations also facilitate vibrational assignment. For example, the *anti* isomer of 1,2-dichloroethane has a center of symmetry and thus the rule of mutual exclusion will hold for the infrared and Raman frequencies of this isomer (1,2). In more complicated and less symmetric molecules the frequency of the carbon–halogen stretching mode has been used as a successful criterion for isomeric structure and stability (40–42). The frequency of this mode is sensitive to the conformation of the molecule. The influence of solvent on the vibrational frequencies can also be used as a guide for vibrational assignment (87).

B. Enthalpy Differences

1. Theory

Most of the enthalpy differences reported in the literature are those between the rotational isomers of ethane derivatives and the following theory will be based on the internal rotation in a molecule such as 1,2-dichloroethane, where the isomeric equilibrium is a two-state process such as that described by eq. (77). The equilibrium constant K is given by

$$\frac{c_B}{c_A} = K = \exp\left(\frac{-\Delta G°}{RT}\right) \tag{78}$$

where c_B and c_A refer to the relative populations of the *anti* and *gauche* isomers, respectively; actually, $c_A = c'_A + c''_A$ where $c'_A (= c''_A)$ is the population of any one of the individual optical *gauche* isomers, $\Delta G°$ is the Gibbs free energy difference related to the enthalpy ($\Delta H°$) and the entropy ($\Delta S°$) differences by the relation

$$\Delta G° = \Delta H° - T\Delta S°$$

In many publications, eq. (78) is treated by a statistical mechanics approach leading to

$$\frac{c_B}{c_A} = \frac{\pi_{fB}}{2\pi_{fA}} \exp\left(\frac{-\Delta E°}{RT}\right)$$

where $\Delta E°$ is the internal energy difference between the isomers and π_{fB} and π_{fA} are the usual products of the partition functions for the *anti* and *gauche* isomers. The factor 2 is introduced to account for the statistical weight of the *gauche* isomer. To a first approximation $\Delta E° \approx \Delta H°$, for $\Delta H° = \Delta E° + P\Delta V°$, where P is pressure and $\Delta V°$, the volume change between the isomers, is normally considered to be negligible. It follows that

$$\Delta S° = R \ln\left(\frac{\pi_{fB}}{\pi_{fA}}\right) - R \ln 2$$

The translational partition functions for the two isomers are equal in magnitude and it is also usual to assume that the partition functions for the overall rotation and vibration are approximately the same, making $\Delta S° = -R \ln 2$. There is some recent experimental evidence to justify this approximation (24).

In the infrared or Raman spectra of a conformational mixture, many of the vibrational bands of the separate isomers will have different frequencies and, since the intensity of a vibrational band is proportional to the population it follows that by choosing two corresponding bands one belonging to each isomer,

$$I_A = \alpha_A c_A l \quad \text{and} \quad I_B = \alpha_B c_B l \tag{79}$$

where the I's are the band areas proportional to band intensities of isomers A and B in eq. (77), the α's are the integrated absorption coefficients, and l is the cell length. The expression for the equilibrium constant K can be written as

$$K = \frac{c_B}{c_A} = \frac{I_B \alpha_A}{I_A \alpha_B}$$

and using the van't Hoff isochore, it follows that

$$\ln\left(\frac{I_B}{I_A}\right) = \frac{-\Delta H°}{RT} + \ln\frac{\alpha_B}{\alpha_A} + \frac{\Delta S°}{R} \tag{80}$$

By measuring intensities at different temperatures, $\Delta H°$ can be derived from the slope of a plot of $\ln(I_B/I_A)$ against reciprocal temperature. This treatment, which is widely used, assumes that $\Delta H°$, $\Delta S°$, and the quantity $\ln(\alpha_B/\alpha_A)$ are independent of temperature. In most of the work carried out in this field, optical densities have been used instead of integrated intensities (1,2).

If c is the total concentration, then $c = c_A + c_B$ at all temperatures, and from eq. (79) (see ref. 88)

$$(I_B/\alpha_B)l + (I_A/\alpha_A)l = c$$

Thus,

$$I_B = \frac{-\alpha_B}{\alpha_A} I_A + \alpha_B lc$$

Assuming the α's are independent of temperature, if I_B is plotted against I_A at different temperatures, the quantity $-\alpha_B/\alpha_A$ will be the slope of the resulting straight line. The entropy difference, $\Delta S°$, can then be found from eq. (80). This procedure can only be applied when the integrated intensities of the absorption bands are known. A distinct method to determine $\Delta S°$ has been used by Mizushima (1) in which the optical densities (or intensities) of three bands, one belonging to each of the isomers and the third being an absorption band whose frequency is common to both isomers, are compared.

2. Results

The enthalpy difference for a number of molecules determined from infrared and Raman spectroscopy are listed in Table XII. This table is to be considered as a supplement to data that have already been tabulated elsewhere (2,7). In order to provide an appreciation of some of the more important aspects of these enthalpy differences, the following section contains a rationalized summary.

a. Enthalpy Difference and Change of Phase. In order to explain the difference between the vapor and liquid phase enthalpy differences for 1,2-dichloroethane and 1,2-dibromoethane, Morino and Mizushima applied the Onsager model of the molecular electric field. These authors considered purely dipolar effects and arrived at the relation (1)

$$\Delta H_v° - \Delta H_l° = xh$$

where $\Delta H°$ is the enthalpy difference, the subscripts v and l referring to vapor and liquid phase, respectively; $x = (\epsilon - 1)/(2\epsilon + 1)$ where ϵ is the dielectric constant of the liquid and $h = (\mu_A^2/a_A^3) - (\mu_B^2/a_B^3)$, μ_A and μ_B being the dipole moments and a_A and a_B being the molecular radii of isomers A and B, respectively. Although the above equation appeared to explain the data for relatively simple molecules, modifications have been attempted by Wada (102) and Volkenstein and Brevdo (103). Recently, Abraham and his collaborators have reviewed the solvent problem (104,105) in connection with

TABLE XII

Enthalpy Differences Determined from Infrared Measurements

Molecule[a]	Enthalpy difference, kcal/mol ($-\Delta H°$)		Ref.
	Gas phase	Liquid phase	
1,2-Dichloroethane	1.15 ± 0.15	0.0	2
	1.24 ± 0.05	0.0 ± 0.06	89
	—	0.4	90
	1.27 ± 0.1	0.92 (n-hexane)	5
	—	0.51 (CCl_4)	5
	—	0.36 (CS_2)	5
	—	−0.26 (methanol)	5
1,2-Dibromoethane	1.54 ± 0.25	0.72 ± 0.07	2
	1.78 ± 0.05	0.77 ± 0.06	89
	1.65 ± 0.10	0.96 (CCl_4)	5
	—	0.88 (CS_2)	5
	—	0.48 (methanol)	5
2-Chlorobutane			
$S_{H'H'} - S_{HH'}$	—	0.650 ± 0.30	91
$S_{H'C} - S_{H'H'}$	—	0.500 ± 0.20	91
$S_{H'H'} - S_{H'C}$	—	0.100 ± 0.10	91
2-Chloropentane			
$S_{CH'} - S_{H'H'}$	0.780 ± 0.03	—	92
2-Bromopentane			
$S_{CH'} - S_{H'H'}$	0.940 ± 0.07	—	92
3-Chloropentane			
$S_{CH'} - S_{H'H'}$	0.560 ± 0.03	—	92
$S_{H'C'} - S_{CH'}$	—	0.130 ± 0.06	42
$S_{H'H} - S_{CC}$	—	0.910 ± 0.17	42
$S_{H'C} - S_{CC}$	—	0.260 ± 0.02	42
$S_{CH'} - S_{CC}$	—	0.130 ± 0.04	42
3-Bromopentane			
$S_{CH'} - S_{H'H'}$	0.870 ± 0.05	—	92
2-Bromohexane			
$S_{CH'} - S_{H'H'}$	0.920 ± 0.03	0.440 ± 0.10	92
3-Bromohexane			
$S_{CH'} - S_{H'H'}$	0.740 ± 0.08	0.360 ± 0.150	92
Isobutyl chloride	0.230 ± 0.02	0.370 ± 0.15	93
Isobutyl bromide	0.300 ± 0.03	0.260 ± 0.120	93
1-Cyano-2-chloroethane	—	−0.380 ± 0.03	94
1-Cyano-2-bromoethane	—	−0.540 ± 0.03	94
Ethane-1,2-dithiol	0.640	—	95

(*continued*)

TABLE XII (continued)

Molecule[a]	Enthalpy difference, kcal/mol ($-\Delta H°$)		Ref.
	Gas phase	Liquid phase	
1,2-Dimethylthioethane	1.100	0.100	96
2-Chloroethyl mercaptan	0.520	0.160	97
2-Bromoethyl mercaptan	0.950	0.620	97
Isopropyl mercaptan	1.42	—	98
1,1-Dichloro-3-fluorobutadiene	—	0.844 ± 0.100	99
1,1-Dibromo-3-fluorobutadiene	—	0.706 ± 0.100	99
2-Chloro-2-methylbutane	1.1	0.36	100
2-Bromo-2-methylbutane	1.4	0.38	100
2-Iodo-2-methylbutane	—	0.69	100
meso-2,3-Dichlorobutane	1.30 ± 0.30	0.64 ± 0.1	101
(+)-2,3-Dichlorobutane			
$S_{HCl} - S_{HH}$	0.045	−0.07 + 0.16	101
$S_{HH} - S_{HCH3}$	1.52 ± 0.27	−0.39 ± 0.09	101
	1.58 ± 0.10	−0.41 ± 0.15	101
$S_{HCl} - S_{HCH3}$	1.64 ± 0.12	−0.42 ± 0.06	101
	1.83 ± 0.30	−0.42 ± 0.21	101
meso-2,3-Dibromobutane	1.43 ± 0.22	−0.43 ± 0.07	101
	1.52 ± 0.30	−0.5 ± 0.05	101
(±)-2,3-Dibromobutane			
$S_{HH} - S_{HCH3}$	1.56 ± 0.13	−0.43 ± 0.13	101
$S_{HBr} - S_{HCH3}$	1.8 ± 0.35	0.45 ± 0.04	101
	1.66 ± 0.33	0.36 ± 0.06	101
$S_{HBr} - S_{HH}$	0.18 ± 0.36	0.84 ± 0.06	101
	0.08 ± 0.20	0.75 ± 0.16	101

[a] The notations S_{XY} are described in ref. 40 and on pp. 234–237 of this chapter (cf. Fig. 12).

some nmr work and have determined the effects of both dipolar and quadrupolar electric fields in the medium. Using the Onsager theory, they arrived at the equation:

$$\Delta H_v° - \Delta H_l° = \frac{kx}{1 - lx} + \frac{3hx}{5 - x} \qquad (81)$$

where $l = 2(\eta_d^2 - 1)/(\eta_d^2 + 2)$, η_d being the refractive index and $h = (q_B^2 - q_A^2)/a^5$, the q's being the quadrupole moments of the isomers. This theory has been found to work satisfactorily in dealing with enthalpy differences in various solvents measured by nmr spectroscopy, and in many cases good correspondence was found between $\Delta H_v°$ derived from the nmr work and that obtained experimentally using infrared methods.

b. Interactions Affecting Enthalpy Differences. The effect of steric repulsive forces as well as electrostatic (dipole–dipole) repulsive forces on enthalpy differences is illustrated by Mizushima's findings regarding conformational populations in 1,2-dichloro- and 1,2-dibromo-ethane and n-butane in the vapor phase. The fact that $-\Delta H°$ for the dibromo compound was greater than for the dichloro compound was attributed to nonbonded steric repulsive forces arising from the bulkier bromine atoms. Since nonbonded steric repulsive forces are proportional to the van der Waal's radii of the substituents, the steric forces in 1,2-dibromoethane and n-butane should be very close. However, the enthalpy difference in 1,2-dibromoethane is in fact greater than that in n-butane; this difference has been attributed to the dipole–dipole repulsive forces in the polar compound. From the data (Table XII) the steric and electrostatic contributions to the conformational enthalpy difference in the dihaloethanes were estimated.

An electrostatic attractive force between the alkyl group and the electronegative halogen atom has been invoked (106) to explain the enthalpy differences in the n-propyl and several other monoalkyl halides. This interaction arises in the *gauche* isomer (Fig. 11) between the nonbonded alkyl group and the halogen atom and is opposite to the steric repulsive force. The overall effect is a decrease in energy in the *gauche* form and thus a smaller enthalpy difference in comparison with the 1,2-dihaloethanes or with n-butane.

In the 2-haloethanols (**37**) (107,108) the *gauche* isomer was found to be

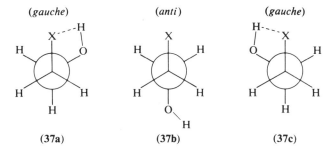

more stable (presumably) because of an internal hydrogen bond, the interaction being strongest in 2-fluoroethanol where the *gauche* form predominates (108a) by 2.07 kcal/mol (108b).

In many halogenated ethanes, especially those containing fluorine atoms, the trends in the enthalpy differences have been difficult to understand. One interesting point is brought out by the electron diffraction work of Iwasaki (109), who found that fluorine substitution at a carbon atom causes the other carbon–halogen bonds to be shortened. Kagarise (110), who has studied the enthalpy differences in several halogenated molecules, has also pointed out that the steric contributions in fluorinated ethanes would be expected to be

TABLE XIII

Sheppard's Rule

Molecule	$-\Delta H°$, kcal/mol	Reference molecule	$-\Delta H°$ of reference molecule, kcal/mol
1,2-Dichloro-2-methyl-propane	1.3	$ClCH_2$—CH_2Cl	1.2
1,2-Dibromo-2-methyl-propane	1.7	$BrCH_2CH_2Br$	1.6
Isobutyl chloride	0.23	$CH_3CH_2CH_2Cl$	0
Isobutyl bromide	0.30	$CH_3CH_2CH_2Br$	0
1,2-Dichloropropane	1.0	$ClCH_2CH_2Cl$	1.2
2,3-Dichloro-2,3-dimethylbutane	1.3	$ClCH_2CH_2Cl$ $(CH_3)_2CHCH(CH_3)_2$	1.2, 0 ; 1.2
2,3-Dibromo-2,3-dimethylbutane	1.6	$BrCH_2CH_2Br$ $(CH_3)_2CHCH(CH_3)_2$	1.6, 0 ; 1.6
1,2,2-Trichloropropane	2.0	$ClCH_2CCl_2CH_3$	3.0
1,1,2-Trichloroethane	1.5	$ClCH_2CHCl_2$	3.0
2-Chloropentane $S_{CH'} - S_{H'H'}$	0.8	$CH_3CH_2CH_2CH_2CH_3$	0.5–0.7
2-Bromopentane $S_{CH'} - S_{H'H'}$	0.9	$CH_3CH_2CH_2CH_2CH_3$	0.5–0.7
3-Chloropentane $S_{CH'} - S_{H'H'}$	0.6	$CH_3CH_2CH_2CH_2CH_3$	0.5–0.7
3-Bromopentane $S_{CH'} - S_{H'H'}$	0.9	$CH_3CH_2CH_2CH_2CH_3$	0.5–0.7
2-Bromohexane $S_{CH'} - S_{H'H'}$	0.9	$CH_3CH_2CH_2CH_2CH_2CH_3$	0.8
3-Bromohexane $S_{CH} - S_{H'H'}$	0.7	$CH_3CH_2CH_2CH_2CH_2CH_3$	0.8

small since the disparity in van der Waal's radii between fluorine and other halogens is less than between hydrogen and halogen.

c. Sheppard's Rules. Sheppard (2) has proposed rules to account for energy differences in the vapor phase. The rules refer to the net interaction energy of groups not directly bonded and are applied as follows: (*a*) Interactions between nonbonded groups of the same kind (e.g., methyl/methyl, halogen/halogen) contribute to the overall enthalpy difference an amount equal to that found for the same arrangement of groups in the substituted ethanes with only one methyl or one halogen substituent on each carbon;

interactions involving hydrogen are neglected. (b) Interactions between nonbonded methyl/halogen groups make a small (assumed negligible) contribution to the enthalpy difference (cf. the n-propyl halides).

The enthalpy differences calculated using these rules together with the appropriate reference molecule are listed in Table XIII; the agreement with those obtained experimentally from spectroscopic studies is good, which emphasizes that nonbonded interactions between similar groups do not change greatly from molecule to molecule and that interactions involving hydrogen and those between a methyl and a halogen atom are small.

C. Potential Barriers

1. Introduction

In molecules such as CH_3CH_2Cl or CCl_3CH_2Cl the potential energy diagram, as the CCl_3 group rotates with respect to the $—CH_2Cl$ groups, is shown in Figure 15. The energy is described by a threefold symmetric function similar to that in the well known example of ethane. On the other hand, in 1,2-dichloroethane and furan-2-aldehyde the presence of rotational isomers introduces nonequivalent maxima and minima in the potential energy diagram as shown in Figures 1 and 2. The shapes of the curves in Figures 1, 2, and 15 can be reproduced by a periodic function of the type:

$$V(\phi) = \sum_{n=1}^{n} \left(\frac{V_n}{2}\right)(1 - \cos n\phi) \qquad (82)$$

where $V(\phi)$ represents the potential energy corresponding to an azimuthal angle ϕ, the number n is an integer ($= 1, 2, 3, \ldots$), and V_n defines the periodicity of the individual potential energy terms in the Fourier series (eq. (82)). For example, the threefold potential energy diagram for CH_3CH_2Cl (Fig. 15) can be reproduced by taking the V_3 term alone in eq. (82). Thus,

$$V(\phi) = \frac{V_3}{2}(1 - \cos 3\phi)$$

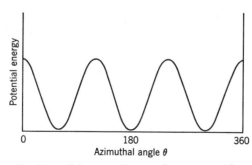

Fig. 15. Potential energy diagram for a symmetric rotor.

On the other hand, the threefold potential energy diagram for 1,2-dichloroethane (Fig. 1) is somewhat modified because of the enthalpy difference between the *gauche* and *anti* forms. This shape can be reproduced by using a potential function of the kind

$$V(\phi) = \frac{V_1}{2}(1 - \cos \phi) + \frac{V_2}{2}(1 - \cos 2\phi) + \frac{V_3}{2}(1 - \cos 3\phi)$$

with the V_3 term being much larger than V_1 and V_2 in order to retain the threefold symmetry. In a similar way Figure 2 can be reproduced by the same equation, but with $V_2 \gg V_3, V_1$ in order to obtain the twofold potential. The pertinent motion involved during internal rotation in these molecules is torsion about the C—C bond and for the *anti* isomer ($\phi = 0$) in 1,2-dichloroethane the torsional energy levels are shown by the horizontal lines in the diagram (Fig. 1). The $1 \leftarrow 0$ torsional transition is one of the fundamental vibrational modes in the molecule and these frequencies usually occur in the far-infrared region 40–250 cm^{-1}. If the torsional frequency of an isomer is 100 cm^{-1} and the barrier to internal rotation is ≈ 5.8 kcal/mol, then, if we consider simple harmonic motion, it follows that the molecule must be excited to over its 20th torsional energy level (1 cm^{-1} = 2.9 cal/mol) before the barrier to internal rotation is surmounted. It is therefore safe to assume that the molecules with potential barriers of this order will have torsional energy levels well below the top of the barrier. If we neglect tunneling effects and assume that the torsional motion is simple harmonic, the torsional frequency is given by:

$$\nu = \frac{1}{2\pi c}\left(\frac{f}{I_r}\right)^{1/2} \tag{83}$$

where f is the force constant for torsional motion and I_r is the reduced moment of inertia of the molecule. By definition

$$f = \frac{d^2 V(\phi)}{d\phi^2} \tag{84}$$

where $V(\phi)$ is given by eq. (82). The potential energy curve for 1,2-dichloroethane in Figure 1 can be reproduced by the first three terms of eq. (82) (see ref. 1); thus

$$V(\phi) = \frac{V_1}{2}(1 - \cos \phi) + \frac{V_2}{2}(1 - \cos 2\phi) + \frac{V_3}{2}(1 - \cos 3\phi) \tag{85}$$

If we assume that in the *anti* isomer ($\phi = 0$) of this molecule the amplitude of the torsional oscillations are small, eq. (85) reduces to (111,112)

$$V(\phi) = \left[V_1 + 4V_2 + 9V_3\right]\phi^2 \tag{86}$$

and by use of eqs. (83), (84), and (86)

$$\nu = \frac{1}{2\pi c}[(V_1 + 4V_2 + 9V_3)/2I_r]^{1/2} \tag{87}$$

This equation relates the torsional frequency of the *anti* isomer to the potential energy parameters V_1, V_2, and V_3. The torsional frequency ν can be determined using spectroscopic techniques, and the reduced moment of inertia I_r is calculated using methods described in the literature (113).

Normally in the calculation of I_r ideal values have to be assumed for bond angles and bond lengths. In many publications the quantity

$$(V_1 + 4V_2 + 9V_3)$$

is referred to as V^*, the total potential energy term. In other cases, such as substituted ethanes, the term V_2 is assumed to be very small and is neglected (111). Thus, for 1,2-dichloroethane the enthalpy difference $\Delta H°$ is given by $\Delta H° = 3V_1/4$ and the potential barrier $\Delta H_{ag}^{\ddagger}$ for the *anti* → *gauche* isomerization is:

$$\Delta H_{ag}^{\ddagger} = \frac{V_3}{2}(1 - \cos 3\phi) + \frac{V_1}{2}(1 - \cos \phi) \tag{88}$$

where $\phi = 60°$. In the case of symmetric rotors with potential energy curves similar to Figure 15, the above method for determining the torsional barrier can be used by setting the terms $V_1 = V_2 = 0$. However, for these molecules more sophisticated methods of solution are available; these methods have been discussed by Lin and Swalen (114).

2. Experimental Methods

Recent developments in far-infrared spectroscopy, particularly Michelson interferometry (115) and laser Raman spectroscopy (116), have made the spectral region 10–250 cm^{-1} readily accessible and there are now commercial instruments available that utilize these developments. This means that the torsional frequencies in molecules can be readily measured. In addition, the torsional frequencies of several symmetric molecules have been measured using neutron scattering techniques (117). In some molecules this technique has an advantage over conventional spectroscopy in that the optical selection rules do not apply. For example, the torsional frequency of ethane is not active in infrared and Raman spectroscopy, but has been measured with neutron scattering spectroscopy.

3. Results

The torsional barrier for several classes of molecules are in Table XIV.* These values are subject to the limitations imposed by the assumptions

---

*This table is a supplement to that published in a recent review article (7b).

leading to the simple treatment described above. In brief, these assumptions are:

1. the torsional oscillations are simple harmonic;
2. the amplitude of the torsional oscillations is small; and
3. the torsional mode is a pure mode.

In addition, the reduced moments of inertia in several cases have been calculated by assuming ideal bond lengths and angles, even though it has been found that changes of 0.1 Å in bond lengths or 2°or in bond angles can affect the torsional barrier by as much as 500 cal/mol. Moreover, many of the values quoted in Table XIV are subject to assumptions concerning some of

TABLE XIV

Potential Barriers to Internal Rotation from Infrared Study

Molecule	Potential terms kcal/mol		Phase	Ref.
	V_1	V_3		
1,2-Dichloroethane (*anti*)	—	5.3	g	118
(*gauche*)	—	5.1	g	118
1,1,2-Trichloroethane (*sym*)	—	11.5	g	118
(*asym*)	—	7.2	g	118
1,1,2,2-Tetrachloroethane (*anti*)	—	10.2	g	118
(*gauche*)	—	12.0	g	118
1-Fluoro-1,1,2,2-tetrachloroethane (*anti*)	0.53	10.65	l	119
1,2-Difluoro-1,1,2,2-tetrachloroethane (*anti*)	0.22	10.40	l	119
1,1,2-Trichloro-trifluoroethane (*anti*)	3.08	8.25	l	119
1,1-Difluoro-1,2,2-trichloroethane (*anti*)	2.40	8.63	l	119
1,2-Dichlorotetrafluoroethane	0.67	5.27	g	120
1,2-Dibromotetrafluoroethane	0.92	8.50	l	121
1,2-Dibromotrifluoroethane	—	6.73	l	119
1,2-Dibromo-1,1-difluoroethane	1.36	3.81	l	119
1,2-Dibromo-1,1-dichloroethane	—	6.25	l	119
n-Propyl chloride (*anti*)	—	2.78	g	122
(*gauche*)	—	2.96	g	122
n-Propyl bromide (*anti*)	—	2.38	g	122
(*gauche*)	—	2.65	g	122
n-Propyl iodide (*anti*)	—	2.47	g	122
(*gauche*)	—	2.77	g	122
n-Propyl fluoride	—	3.36	g	122

(*continued*)

TABLE XIV (continued)

Molecule	Potential terms, kcal/mol $V*$	Phase	Ref.
2,3-Dichloropropene	11.6	g	123
	13.2	l	123
2-Bromo-3-chloropropene	15.8	l	123
2,3-Dibromopropene	17.1	l	123
Acrylyl fluoride	18.6	l	124
Acrylyl chloride	17.0	l	124
Anisole	6.02	l	125
p-Fluoroanisole	6.10	l	125
p-Chloroanisole	5.80	l	125
p-Bromoanisole	6.20	l	125
m-Fluoroanisole	7.02	l	125
m-Chloroanisole	6.91	l	125
m-Bromoanisole	7.59	l	125
o-Fluoroanisole	6.25	l	125
o-Chloroanisole	8.90	l	125
o-Bromoanisole	9.12	l	125

the V_n terms. For example, in threefold barriers it is usual practice to neglect the V_2 term, whereas in twofold barriers (Fig. 2) the V_3 term is usually neglected.

In view of the assumptions inherent in the determination of the torsional barriers in Table XIV no detailed discussion of these values will be given in this section. It is worth pointing out, however, that the barriers for the ethane derivatives compare well with both ultrasonic and nmr values, and those for the aldehydes can be compared with the ultrasonic values quoted in Table XI.

D. The Use of Infrared Spectroscopy as a Probe to Determine Potential Barriers (126–128)

The melting points of most liquids which exist as equilibrium mixtures of different conformers are very low (-30 to $-150°C$) and in order to record the solid infrared spectra, special low temperature cells using coolants such as Dry Ice or liquid air are used. When using these cells the liquid sample is often supercooled and a glassy-type solid is formed which normally exists as an equilibrium mixture of the different conformers. In order to convert the glassy solid to the crystalline solid, the sample is gently heated until the temperature corresponding to the glassy solid → crystalline solid transition is reached. At this temperature all the less stable conformers are converted to the most stable form. At temperatures around this transition point the above process can be followed spectroscopically by observing the disappearance of an absorption band of the less stable isomer. By measuring the rate of disappearance of this band at different temperatures and using the Arrhenius rate equation (eq. (73)), the potential barrier to internal rotation can be measured. The values found for 1,2-dichloroethane (1.7 kcal/mol), bromo- and chloro-cyclohexane (15 kcal/mol), and 3-chloropentane ($S_{H'H'} - S_{H'H} = 3.0$ kcal/mol) may be compared with those in Tables I, X, and III, respectively. However, Piercy (25) has pointed out that in the *gauche* → *anti* conversion in 1,2-dichloroethane at 120°K the Arrhenius A factor is $\sim 10^5$ sec^{-1} and not the value $\sim 10^{12}$ sec^{-1} expected for a unimolecular process (see Table II). Because of this apparent discrepancy Piercy has suggested that the activation energy measured with this method must be connected with some rate-limiting process other than the barrier to internal rotation, such as diffusion or crystal growth.

ADDENDUM IN PROOF

Energy barriers have now been determined for the isomerization (35) involving the equilibrium between a twist boat and chair form in simple 1,3-dioxanes (129). The barriers for the twist boat → chair isomerization in 2,2-dimethyl-1,3-dioxane and 1,3-dioxane-2-spirocyclopentane are 5.0 and 4.6 kcal/mol, respectively. These appear to be the first examples of direct kinetic measurements giving conformational energies of the flexible form in simple 1,3-dioxanes.

Recent studies (130) into the kinetics of coupled processes, such as those involved in the interconversion of the three isomers of 1,2-dichloro- propane (cf. Table III), have shown that the only way to explain the observation of a single ultrasonic absorption in these molecules is to assume that the Arrhenius "A" factors for one stage is much greater than for the other two stages; this also explains why many of the barriers quoted in Table II are very similar in magnitude to those in Table I.

Finally, in connection with the solution of eqs. (82)–(88), using spectroscopic data to yield potential energy barriers in ethane type molecules, attention is drawn to two papers recently published on this work (131, 132). A simple solution for the potential energy terms V_1, V_2, and V_3 for these molecules can be obtained from the following equations:

$$\nu_A = (1/2\pi c)[(V_1 + 4V_2 + 9V_3)/2I_{rA}]^{1/2}$$

$$\nu_B = (1/2\pi c)[(-\tfrac{1}{2}(V_1 + 4V_2) + 9V_3)/2I_{rB}]^{1/2}$$

$$\Delta H° = \tfrac{3}{4}(V_1 + V_2)$$

where ν_A, ν_B, and $\Delta H°$ are experimentally determined parameters.

References

1. S. Mizushima, *Structure of Molecules and Internal Rotation*, Academic Press, New York, 1954.
2. N. Sheppard, *Advan. Spectry.*, **1**, 288 (1959).
3. E. L. Eliel, N. L. Allinger, S. J. Angyal, and G. A. Morrison, *Conformational Analysis*, Interscience, New York, 1965.
4. M. Hanack, *Conformation Theory*, Academic Press, New York, 1965.
5. M. V. Volkenshtein, *High Polymers*, Vol. 17, *Configurational Statistics of Polymeric Chains*, Interscience, New York, 1963.
6. J. E. Anderson, *Quart. Rev.*, **19**, 426 (1965).
7. See, for example, (a) G. Binsch, in *Topics in Stereochemistry*, Vol. 3, E. L. Eliel and N. L. Allinger, Eds., Interscience, New York, 1968; (b) J. P. Lowe, in *Progress in Physical Organic Chemistry*, Vol. 6, A. Streitwieser and R. W. Taft, Eds., Interscience, New York, 1968, p. 1; (c) R. A. Pethrick and E. Wyn-Jones, *Quart. Rev.*, **23**, 301 (1969); (d) R. J. Abraham and K. Parry, *J. Chem. Soc.*, B, **1970**, 539.
8. M. Eigen and L. De Maeyer, in *Technique of Organic Chemistry*, Vol. 8, Part 2. S. L. Freiss, E. S. Lewis, and A. Weissberger, Interscience, New York, 1963, p. 895.
9. M. Eigen, *Discussions Faraday Soc.*, **17**, 194 (1954).
10. J. F. Hueter and R. H. Bolt, *Sonics*, Wiley, New York, 1960.
11. K. F. Herzfeld and T. A. Litovitz, *Absorption and Dispersion of Ultrasonic Waves* Academic Press, New York, 1959.
12. C. Kirchoff, *Poggendorf's Ann. Phys.*, **134**, 177 (1868).
13. C. Truesdell, *J. Rational Mech. Anal.*, **2**, 643 (1955).
14. J. K. Kincaid and H. Eyring, *J. Chem. Phys.*, **6**, 620 (1938).
15. C. Kittel, *J. Chem. Phys.*, **14**, 614 (1946).
16. T. A. Litovitz, *J. Acoust. Soc. Amer.*, **31**, 681 (1959).
17. J. H. Andrae and J. Lamb, *Proc. Phys. Soc.*, **1956**, 814.
18. J. Lamb, in *Physical Acoustics*, W. P. Mason, Ed., Academic Press, New York, Vol. II, Part A, 1956.
19. H. J. McSkimin, in *Physical Acoustics*, Vol. I, Part A, Warren P. Mason, Ed., Academic Press, New York, 1956, Chap. 4.
20. R. A. Pethrick and E. Wyn-Jones, *J. Chem. Soc.*, A, **1969**, 713.
21. J. E. Piercy, *J. Acoust. Soc. Amer.*, **33**, 198 (1961).
22. J. Lamb, *Z. Elektrochem.*, **64**, 135 (1960).
23. E. Wyn-Jones and W. J. Orville-Thomas, *Chem. Soc. Spec. Publ.*, **20**, 209 (1966).
24. P. J. D. Park and E. Wyn-Jones, *J. Chem. Soc.*, A, **1969**, 646 and references quoted therein.
25. J. E. Piercy, *J. Chem. Phys.*, **43**, 4066 (1966).
26. R. A. Padmanabhan, *J. Sci. Ind. Res. (India)*, **19B**, 336, (1960).

27. J. E. Piercy and M. G. Seshagiri Rao, *J. Chem. Phys.*, **46**, 3951 (1967).
28. J. H. Chen and A. A. Petrauskas, *J. Chem. Phys.*, **30**, 304, (1959).
29. E. Wyn-Jones and W. J. Orville-Thomas, *Trans. Faraday Soc.*, **64**, 2907 (1968).
30. A. E. Clark and T. A. Litovitz, *J. Acoust. Soc. Amer.*, **32**, 1221 (1960).
31. K. Krebs and J. Lamb, *Proc. Roy. Soc. (London)*, **A244**, 558 (1958).
32. K. R. Crook, P. J. D. Park, and E. Wyn-Jones, *J. Chem. Soc., A*, **1969**, 2910.
33. P. J. P. Park and E. Wyn-Jones, *J. Chem. Soc., A*, **1968**, 2064.
34. R. A. Pethrick and E. Wyn-Jones, *J. Chem. Phys.*, **49**, 5349 (1968).
35. R. A. Pethrick and E. Wyn-Jones, unpublished results.
36. D. S. Thompson, R. A. Newmark, and C. H. Sederholm, *J. Chem. Phys.*, **37**, 411 (1962).
37. R. A. Newmark and C. H. Sederholm, *J. Chem. Phys.*, **43**, 602 (1965).
38. T. D. Alger, H. S. Gutowsky, and R. L. Vold, *J. Chem. Phys.*, **47**, 3130 (1967).
39. S. Mizushima, T. Shimanouchi, K. Nakamura, M. Hayashi, and S. Tsuchiya, *J. Chem. Phys.*, **26**, 970 (1957).
40. P. N. Gates, E. F. Mooney, and H. A. Willis, *Spectrochim. Acta*, **23A**, 2043 (1967).
41. J. J. Shipman, V. L. Folt, and S. Krimm, *Spectrochim. Acta*, **18**, 1603 (1962).
42. A. Caraculacu, J. Štokr, and B. Schneider, *Collection Czech. Chem. Commun.*, **29**, 2783 (1964).
43. M. Eigen and K. Tamm, *Z. Elektrochem.*, **66**, 93, 107 (1962).
44. T. H. Thomas, E. Wyn-Jones, and W. J. Orville-Thomas, *Trans. Faraday Soc.*, **65**, 974 (1969).
45. M. S. de Groot and J. Lamb, *Proc. Roy. Soc. (London)*, **A242**, 36 (1957).
46. S. V. Subrahmanyam and J. E. Piercy, *J. Acoust. Soc. Amer.*, **37**, 340 (1965).
47. E. J. Williams, T. H. Thomas, E. Wyn-Jones, and W. J. Orville-Thomas, *J. Mol. Struct.*, **2**, 307 (1968).
48. R. A. Pethrick, E. Wyn-Jones, P. C. Hamblin, and R. F. M. White, *J. Mol. Struct.*, **1**, 332 (1967-8).
49. P. C. Hamblin, R. F. M. White, and E. Wyn-Jones, *Chem. Commun.*, **1968**, 1058.
50. Values quoted in ref. 2.
51. R. J. Abraham and M. A. Cooper, *J. Chem. Soc., B*, **1967**, 202.
52. K. R. Crook and E. Wyn-Jones, *J. Chem. Phys.*, **50**, 3446 (1969).
53. E. L. Heasell and J. Lamb, *Proc. Roy. Soc. (London)*, **A237**, 233 (1956).
54. T. A. Litovitz and E. A. Carnevale, *J. Acoust. Soc. Amer.*, **30**, 134 (1958).
55. M. Tamres, S. Searles, E. M. Leighly, and D. W. Mohrman, *J. Amer. Chem. Soc.*, **76**, 3983 (1954).
56. H. C. Brown, *J. Chem. Soc.*, **1956**, 1248.
57. B. A. Arbuzov and L. M. Guzhavina, *Dokl. Akad. Nauk SSSR*, **61**, 63 (1948).
58. R. A. Padmanabhan and E. L. Heasell, *Proc. Phys. Soc.*, **76**, 321 (1960).
59. F. E. Prichard and W. J. Orville-Thomas, *J. Mol. Spectry.*, **6**, 572 (1962).
60. P. Biquard, *Ann. Physik*, **6**, 195 (1936).
61. D. Tabuchi, *J. Chem. Phys.*, **28**, 1014 (1958).
62. J. E. Piercy and S. V. Subrahmanyam, *J. Chem. Phys.*, **42**, 1475 (1965).
63. S. V. Subrahmanyam and J. E. Piercy, *J. Acoust. Soc. Amer.*, **37**, 340 (1965).
64. J. Bailey and A. M. North, *Trans. Faraday Soc.*, **64**, 1499 (1968).
65. W. M. Slie and T. A. Litovitz, *J. Chem. Phys.*, **39** 1538 (1963).
66. J. M. M. Pinkerton, *Ultrasonic Conference Brussels*, Mededel Koninkl Vlaam Acad Wetenshap Belg. K.1. Welenshap 195, p. 17.
67. J. Bailey, S. Walker, and A. M. North, *J. Mol. Struct.*, **6**, in press (1970).
68. R. F. Curl, Jr., *J. Chem. Phys.*, **30**, 1529 (1959).

69. J. M. Riveros and E. Bright Wilson, Jr., *J. Chem. Phys.* **46**, 4605 (1967).
70. J. Karpovich, *J. Chem. Phys.*, **22**, 1767 (1954).
71. J. Lamb and J. Sherwood, *Trans. Faraday Soc.*, **51**, 1674 (1955).
72. R. A. Pethrick, E. Wyn-Jones, P. C. Hamblin, and R. F. M. White, *J. Mol. Struct.*, **1**, 333 (1967–68).
73. P. C. Hamblin, R. F. M. White, and E. Wyn-Jones, *Chem. Commun.*, **1968**, 1058; *J. Mol. Struct.*, **4**, 275 (1969).
74. R. A. Pethrick, E. Wyn-Jones, P. C. Hamblin, and R. F. M. White, *J. Chem. Soc.*, A, **1969**, 1852, 2552.
75. P. C. Hamblin, R. F. M. White, G. Eccleston, and E. Wyn-Jones, *Can. J. Chem.*, **47**, 2731 (1969).
76. G. Eccleston, R. A. Pethrick, E. Wyn-Jones, P. C. Hamblin, and R. F. M. White, *Trans. Faraday Soc.*, **66**, 310 (1970).
77. E. L. Eliel and Sr. M. C. Knoeber, *J. Amer. Chem. Soc.*, **88**, 5347 (1966).
78. J. E. Anderson, F. G. Riddell, and M. J. T. Robinson, *Tetrahedron Letters*, **1967**, 2017.
79. K. C. Ramey and J. Messick, *Tetrahedron Letters*, **1965**, 4423.
80. J. B. Hendrickson, *J. Amer. Chem. Soc.*, **83**, 4537 (1961).
81. R. K. Harris and R. A. Spragg, *J. Chem. Soc.*, B, **1968**, 684.
82. J. E. Piercy and S. V. Subrahmanyam, *J. Chem. Phys.*, **42**, 1475 (1965).
83. M. S. de Groot and J. Lamb, *Proc. Roy. Soc. (London)*, A**242**, 36 (1957).
84. R. A. Pethrick and E. Wyn-Jones, unpublished data.
85. J. B. Bentley, K. B. Everard, R. J. B. Marsden, and L. E. Sutton, *J. Chem. Soc.*, **1949**, 2957.
86. D. Grimshaw, G. Eccleston, and E. Wyn-Jones, unpublished data.
87. H. E. Hallam and T. C. Ray, *J. Mol. Spectry.*, **17**, 69 (1964).
88. K. O. Hartman, G. L. Carlson, R. E. Witkowski, and W. G. Fateley, *Spectrochima Acta*, **24A**, 157 (1968).
89. Yu. A. Pentin and V. M. Tatevskii, *Dokl. Akad. Nauk SSSR*, **108**, 290 (1956).
90. N. Sheppard and J. J. Turner, *Proc. Roy. Soc. (London)*, A**252**, 506 (1959).
91. L. P. Melikhova, Yu. A. Pentin, and O. D. Ul'yanov, *Zh. Struct. Khim.*, **4**, 535 (1963).
92. T. H. Thomas, E. Wyn-Jones, and W. J. Orville-Thomas, *Trans. Faraday Soc.*, **65**, 974 (1969).
93. E. Wyn-Jones and W. J. Orville-Thomas, *Trans. Faraday Soc.*, **64**, 2907 (1968).
94. E. Wyn-Jones and W. J. Orville-Thomas, *J. Chem. Soc.*, A, **1966**, 101.
95. H. Hayashi, Y. Shiro, T. Oshima, and K. Murata, *Bull. Chem. Soc. Japan*, **38**, 1734 (1965).
96. H. Hayashi, Y. Shiro, T. Oshima, and K. Murata, *Bull. Chem. Soc. Japan*, **39**, 118 (1966).
97. M. Murakami, Y. Shiro, T. Oshima, and K. Murata, *Bull. Chem. Soc. Japan*, **38**, 1740 (1965).
98. G. A. Crowder and D. W. Scott, *J. Mol. Spectry.*, **16**, 122 (1965).
99. K. O. Hartman, G. L. Carlson, R. E. Witkowski, and W. G. Fateley, *Spectrochim. Acta*, **24A**, 157 (1968).
100. P. J. D. Park and E. Wyn-Jones, *J. Chem. Soc.*, A, **1968**, 2944.
101. P. J. D. Park and E. Wyn-Jones, *J. Chem. Soc.*, A, **1969**, 422.
102. A. Wada, *J. Chem. Phys.*, **22**, 198 (1954).
103. M. V. Volkenshtein and V. I. Brevdo, *Zh. Fiz. Khim.*, **28**, 1313 (1954).
104. R. J. Abraham, L. Cavalli, and K. G. R. Pachler, *Mol. Phys.*, **11**, 471 (1966).
105. R. J. Abraham, K. G. Pachler, and P. L. Wessels, *Z. Phys. Chem.*, **58**, 257 (1968).

106. G. J. Szasz, *J. Chem. Phys.*, **23**, 2449 (1955).
107. S. Mizushima, T. Shimanouchi, T. Miyazawa, K. Abe, and M. Yasumi, *J. Chem. Phys.*, **19**, 1477 (1951).
108. (a) E. Wyn-Jones and W. J. Orville-Thomas, *J. Mol. Struct.*, **1**, 29 (1967–68); (b) P. J. Krueger and H. D. Mettee, *Can. J. Chem.*, **42**, 326 (1964).
109. M. Iwasaki, *Bull. Chem. Soc. Japan*, **32**, 205 (1959).
110. R. E. Kagarise, *J. Chem. Phys.*, **26**, 380 (1957).
111. F. B. Brown, A. D. H. Clague, N. D. Heitkamp, D. F. Koster, and A. Danti, *J. Mol. Spectry.*, **24**, 163 (1967).
112. F. A. Miller, W. J. Fateley, and R. E. Witkowski, *Spectrochim. Acta*, **23A**, 891 (1967).
113. K. S. Pitzer and W. D. Gwinn, *J. Chem. Phys.*, **10**, 428 (1942).
114. C. C. Lin and J. D. Swalen, *Rev. Mod. Phys.*, **31**, 841 (1959).
115. H. A. Gebbie and R. Q. Twiss, *Rept. Prog. Phys.*, **24**, 217 (1966).
116. R. F. Schaufele and T. Shimanouchi, *J. Chem. Phys.*, **47**, 3605 (1967).
117. J. W. White, in *Excitions*, A. B. Zahler, Ed., Cambridge Univ. Press, Mass., 1968.
118. G. Allen, P. N. Brier, and G. Lane, *Trans. Faraday Soc.*, **63**, 824 (1967).
119. R. A. Pethrick, P. J. D. Park, and E. Wyn-Jones, unpublished data.
120. F. B. Brown, A. D. Clague, N. D. Heitkamp, D. F. Koster, and A. Danti, *J. Mol. Spectry.*, **24**, 963 (1967).
121. K. R. Crook, P. J. D. Park, and E. Wyn-Jones, *J. Chem. Soc.*, *A*, **1969**, 2910.
122. K. Radcliffe and J. L. Wood, *Trans. Faraday Soc.*, **62**, 1678 (1966).
123. G. A. Crowder, *J. Mol. Spectry.*, **23**, 1 (1967).
124. K. Radcliffe and J. L. Wood, *Trans. Faraday Soc.*, **62**, 2038 (1966).
125. N. L. Owen and R. E. Hester, *Spectrochim. Acta*, **25A**, 343 (1969).
126. L. P. Melikhova, O. D. Ul'yanova, and Yu. A. Pentin, *Zh. Fiz. Khim.*, **36**, 1814 (1962).
127. V. A. Pozdyshev, Yu. A. Pentin, and V. M. Tatevskii, *Dokl. Acad. Nauk SSSR*, **114**, 583 (1957).
128. A. J. Woodward and N. Jonathan *J. Chem. Educ.*, **46**, 756 (1969).
129. G. Eccleston and E. Wyn-Jones, *Chem. Comm.*, **1969**, 1511, and unpublished data.
130. J. Rassing and E. Wyn-Jones, unpublished data.
131. A. Cunlife, *J. Mol. Struct.*, **6**, 9 (1970).
132. G. L. Carlson, W. G. Fateley, and J. Kiraishi, *J. Mol. Struct.*, **6**, 101 (1970).

The Stereochemistry of the Quaternization of Piperidines

JAMES McKENNA

Chemistry Department, The University of Sheffield, Sheffield, England

I.	Introduction	275
II.	The Reaction	276
III.	Energy Schematics	277
	A. Curtin-Hammett Analysis	278
	B. Relationship between Base-Conformer Equilibrium Constants and Quaternization Product Ratios	278
	C. Cross-Products	279
	D. Competitive Protonations	280
IV.	Stereoselectivity of Quaternizations	281
V.	Stereospecificity of Quaternizations	284
	A. Kinetics-Based Methods	285
	1. Variation in Product Ratios Associated with Introduction of the Two Exocyclic Alkyl Groups in One or the Reverse Order	285
	2. Cyclization of Appropriate α,ω-Dihalides with Secondary Amines NHMeAlk and Alkali	288
	3. The Hammond Postulate Method	288
	4. Quantitative Kinetic Analysis of Quaternization Rate Constants	290
	B. Thermodynamics-Based Method	291
	C. Spectroscopic and Related Procedures	292
	1. Nuclear Magnetic Resonance Methods	292
	2. Examination of Infrared Spectra	296
	3. X-Ray Crystallographic Analysis	297
	D. Lactonization Method	298
	E. Effect of Functional Groups on the Steric Course of Quaternization of Piperidines	299
	F. Summary of Conclusions of Stereospecificities of Quaternizations of Piperidines	300
VI.	Degradation of Diastereomeric Quaternary Salts	302
VII.	Related Work	305
	References	307

I. INTRODUCTION

When a tertiary piperidine (**1**) is treated with an alkylating group R′X, where R′ is different from the group R already present, two diastereomeric quaternary products (**2** and **3**) are possible as shown in Chart 1 and indeed a mixture is usually obtained. The stereoselectivity (measured by the product ratio) and stereospecificity (the preferred direction of quaternization) of this reaction for a wide range of piperidines have engaged the attention of several groups of workers during recent years, and some results have also been reported for the different rates of reaction of diastereomeric quaternary salts in substitutions and eliminations. A review of the field is now very timely, as the quaternizations provide a very highly illustrative example of the reactivity of conformationally mobile systems.

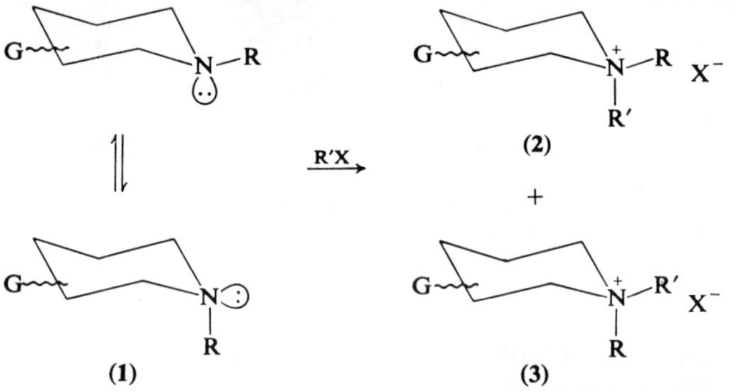

Chart 1. Diastereomeric quaternary salts from tertiary piperidines. G = biasing or reference group or fused ring.

In this review we first discuss the mechanism of the Menshutkin (quaternization) reaction and then deal in some detail with the energy schematics of the quaternization of tertiary piperidines of the type represented by formula **1** (Chart 1), which are conformationally mobile at nitrogen. Methods for the determination of the stereoselectivities and stereospecificities of the quaternizations are then examined. The first is a simple matter, but a discussion of the various procedures which have been used for the much more difficult stereospecificity problem takes up a substantial section of the review. These procedures allow configurations to be allocated to individual diastereomers of the general type **2** and **3** (Chart 1). Differences in the rate constants for displacement and elimination reactions of such diastereomers are then discussed, and finally there is a brief summary of work related to the main subject matter of the review.

II. THE REACTION

The Menshutkin reaction is one of the most thoroughly studied in organic chemistry, so that the stereochemical aspects of the tertiary piperidine quaternizations may be examined without problems arising from unexpected

QUATERNIZATION OF PIPERIDINES 277

mechanistic ambiguities. All the quaternizations discussed in this review are taken to be bimolecular nucleophilic displacements, and even in the event of this not being completely true for some of the isopropylations considered, or benzylations, the qualitative interpretations offered for the steric course of the quaternizations are of a type which would not be seriously affected if the active electrophile were sometimes in part $[R']^+$ rather than solely $R'X$.

III. ENERGY SCHEMATICS

The construction of a "free-energy surface" for a reaction or of a schematic diagram related thereto is not an appealing notion for theoretical kineticists, but with the approximation that free-energy differences (ΔG) may be equated to the corresponding potential-energy differences (ΔV), we may set out the physical-organic chemists' usual style of schematic free-energy diagram for the competitive quaternizations (Fig. 1). For comparison, formation of diastereomeric proton salts from the base conformers is also represented. For simplicity in our qualitative discussions here and elsewhere in this review the fraction of reaction of or via boat or other nonchair conformations or less-stable chair conformations of monocyclic piperidines is

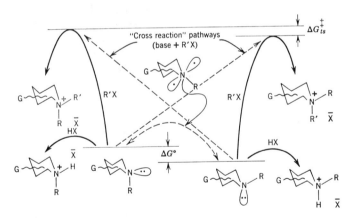

Fig. 1. Energy schematics of competitive quaternizations (and protonations). G = biasing or reference group or fused ring.

$\Delta G_{ts}^{\ddagger} \approx \Delta V_{ts}^{\ddagger} = 2.3\, RT \log_{10} P,$ . . . . (a)

where P is product ratio;

$k_{\text{obs}} = k_e N_e + k_a N_a$ . . . . (b, i)

$P = k_e N_e / k_a N_a$ or $k_a N_a / k_e N_e$ . . . . (b, ii)

where k's are rate constants, N's fractions of conformers interconverted by inversion at nitrogen, and subscripts a and e refer to axial and equatorial conformers.

not considered, although these additional reaction pathways would have to be taken into account in a detailed quantitative analysis (if such were possible). While the schematics as illustrated are very simple, there have been several unfortunate, but illustrative, errors arising directly from failures to pay adequate attention to the qualitative energetics of this reaction system.

A. Curtin-Hammett Analysis

Since quaternization rates are much slower than rates of conformational inversion at nitrogen, the Curtin-Hammett principle (1) holds and the product ratio for a quaternization is determined directly by the free-energy difference $\Delta G_{ts}^{\ddagger}$ between the competitive transition states (2) (eq. (a), Fig. 1). This was not appreciated in early work (3) on quaternization of tropines, where the product ratio was believed instead to reflect, in effect, $\Delta G°$, the free-energy difference between the reactant conformers, so that if, for example, the left-hand product in Figure 1 was thought to predominate in a particular quaternization, it was incorrectly deduced that the left-hand base conformer *therefore* also predominated.

If we could calculate the free energies (or, to a first approximation, the potential energies) of the competitive transition states in Figure 1, we would at once know (eq. (a)) the stereoselectivity and stereospecificity of the particular quaternization, but such calculations are unfortunately beyond the present scope of even classical or semiempirical procedures. Even a satisfactory qualitative interpretation of the preferred steric outcome of a quaternization determined by other procedures is difficult. However, as we shall see later, while *absolute product ratios* are difficult to interpret convincingly, the *variations in product ratios* associated with inversion of the order of introduction of different alkyl groups are much easier to discuss and have formed the basis of a very useful procedure for determination of quaternization stereoselectivities.

B. Relationship between Base-Conformer Equilibrium Constants and Quaternization Product Ratios

While the above Curtin-Hammett analysis (eq. (a) of Fig. 1) is nowadays presumably universally accepted, there appears to be an understandable tendency among workers interested in the related important question of conformational equilibria of secondary and tertiary piperidines to seek to link quaternization product ratios with conformer ratios associated with inversion at nitrogen in the reactant tertiary piperidines. In place of the early naive approach, the algebra given in eqs. (b) of Figure 1 is used (4). These equations are probably (see below) formally correct, and indeed are together essentially equivalent to eq. (a). However, since the base/conformer ratios and the reaction rate constants for individual conformers are presently both very difficult to measure or calculate, eqs. (b) have not so far proved superior to the simple Curtin-Hammett expression for prediction or interpretation.

C. Cross-Products

While it is obvious that the quaternary salts **2** and **3** in Chart 1 may be formed by nucleophilic attack on the alkylating agent via an axial and an equatorial lone electron pair, respectively, the production of cross-products by the collision processes indicated in Chart 2 appears to have been overlooked by other interested authors. These cross-reactions (5) are represented in Figure 1 by the broken straight lines with arrows pointing one way. The exact preferred positioning of these lines on the figure is a matter of some speculation, but axial bonding, for example, to the alkylating agent R'X can obviously start before the initial axially positioned exocyclic group R has

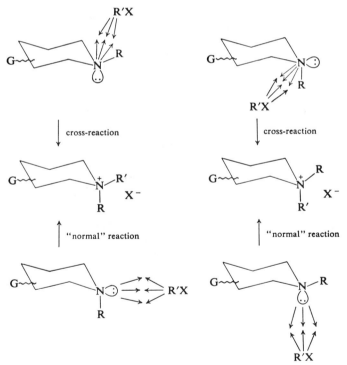

Chart 2.   "Normal" and cross-reactions in quaternizations. G = biasing or reference group or fused ring. The arrows in the formulas represent movement towards collisions between molecules, not (except incidentally) movement of electron pairs.

gone over all the way to its equilibrium position in the potential well representing the R-equatorial conformer. In Figure 1 we represent the cross-reaction pathways as being in the initial stage energetically similar to the pathways for conformational inversion for the unreacting bases, although there will of course be differences even at that stage because of interaction of the force fields of the colliding molecules. At the other extreme, we represent the cross-reaction pathways as utilizing the same reaction saddle points as the "normal" pathways, as seems very likely. Otherwise eq. (b, i) would be invalid, since two additional terms would clearly then have to be introduced into the expression for the overall quaternization rate constant. Provided that only two transition states are indeed involved for the competitive quaternizations, as shown, the concentrations of these are not affected by the additional pathways, so that the base reacts *as if* all the N—R equatorial conformer were axially quaternized and all the N—R axial conformer were equatorially quaternized; eq. (b, i) therefore holds good. However, this equation corresponds to a make-believe rather than an actual physical situation, and it will be evident from consideration of Figure 1 that each quaternization product is derived equally from each base conformer, differences in free energy of activation for reactions leading to either transition state being exactly offset by differences in reactant conformer concentrations. Hence, statements made (6) in the current literature of the type "the quaternary product is derived preferentially from the more stable base-conformer" are incorrect and may be aptly rewritten as "the quaternary product is derived preferentially by axial alkylation" (or an equivalent statement). We may note that the same make-believe aspect underlies the use of an equation analogous to eq. (b, i) of Figure 1 in the conformational analysis of cyclohexane derivatives by the Winstein-Eliel method (7); however, since in this case the products also undergo rapid conformational interconversion, no infelicitous comments similar to that noted above have been made.

The reason for the neglect of consideration of cross-reaction pathways in such examples basically appears to be the overstatic nature of transition-state theory, which is incidentally also the chief reason for the low regard for this theory by such diverse groups as modern theoretical kineticists and old-fashioned organic chemists, both of whom are in their own way well attuned to the essentially dynamic nature of chemical reactions.

D. Competitive Protonations

Before finishing our discussion of Figure 1, we may note one final point of at least strong indirect interest in our considerations of the competitive quaternizations. In the related case of competitive *protonations* of tertiary piperidines, or deuteronations of analogous appropriately biased secondary piperidines, the free-energy barrier to conformational inversion in the base is, as shown, much higher than that for proton (or deuteron) transfers. If the reverse reaction cannot take place, the simple expectation in this case of "non-Curtin-Hammett" energetics is that the product ratio should equal the reactant/conformer ratio (cf. the initial incorrect interpretation (3) of the steric course of piperidine quaternizations). However, in determinations of equilibrium constants for such conformer mixtures, needed, for example, in applications of eqs. (b) of Figure 1, the fast time scale involved cannot be forgotten, nor the fact that two liquids cannot presently be essentially completely mixed in less than about a few milliseconds. Hence, although deprotonation of the base cations will be very slow *after* mixing a liquid base with an excess of a strong liquid acid, *during* the mixing the lower acidity in local volume elements may favor fast reversible proton transfer (5,8). Thus, the final product ratio may be determined partly by thermodynamic and partly by kinetic control. It is possible that methods for avoiding this complication, apparently overlooked in a recent investigation (9), may be developed by the elimination of at least one of the two liquids in the experimental technique adopted (10).

IV. STEREOSELECTIVITY OF QUATERNIZATIONS

Product ratios may in principle be determined by various separation procedures, such as fractional crystallization or appropriate types of chromatography (which are also used if pure samples of one or both diastereomers are required for kinetic work, X-ray analysis, or similar investigations). Generally, the physical separations are rather difficult, one notable exception being the very ready separation of N-methyl-N-alkylcamphidinium salts (for the parent base structure see Table I) by fractional crystallization (11).

A much easier method of determination of product ratios of quaternary salts is by nmr spectroscopy, concentrations of the diastereomers being directly proportional to the areas of suitable signals, e.g., $\overset{+}{N}$—CH$_3$, $\overset{+}{N}$—$\underline{CH_2}$—Ph. Average ratios taken from a few spectra of the mixed salts without special quantitative effort may be quoted (12) as, for example, ~3:1 or 7–10:1. Statistical treatments of measurements on numerous spectra of several different samples of pure mixed salts typically give standard deviations for the ratio of ~5% if a regular 60-MHz instrument is used; if we take just two standard deviations (2σ) as a reasonable indication (~95% confidence limits) of random error, we may write such a very precisely determined quaternization product ratio as, for example, 3.2 ± 0.3:1. To achieve an accuracy corresponding to this degree of precision, it may be necessary to compare the spectra with those of salts containing deuterium atoms instead

TABLE I

Stereoselectivity of Quaternizations in Acetone[a]
Alk = Et, Pr[n], Pr[i], CH$_2$Ph

Base System		Stereoselectivity	
Name	Formula	>NAlk[b] + MeI >NMe + CD$_3$I	>NMe + Alk-I[b]
4-Phenylpiperidine	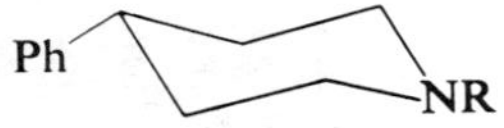 Ph–◯–NR	2–7:1[c]	~1:1[d,e]
2-Methylpiperidine	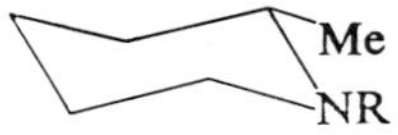 Me, ◯–NR	2–3:1 >12:1 (◯NPr[i]) ~1:1[e] (CD$_3$I)	~1:1[e] >12:1 (Pr[i]I)
trans-Decahydroquinoline	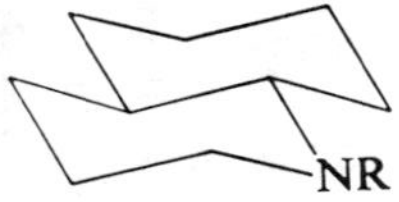 ◯◯–NR	2–3:1 >12:1 (◯NPr[i])	~1:1[e] >12:1 (Pr[i]I)
Camphidine	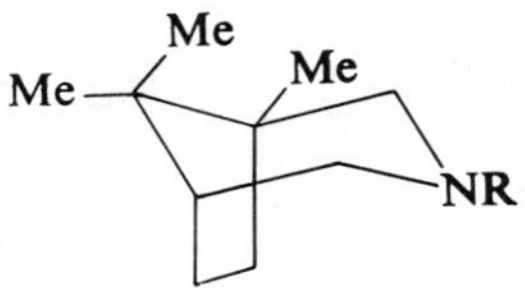 Me, Me, Me–◯◯–NR	>12:1	1:2–3 ~1:1[e] (Pr[i]I and PhCH$_2$I[f])

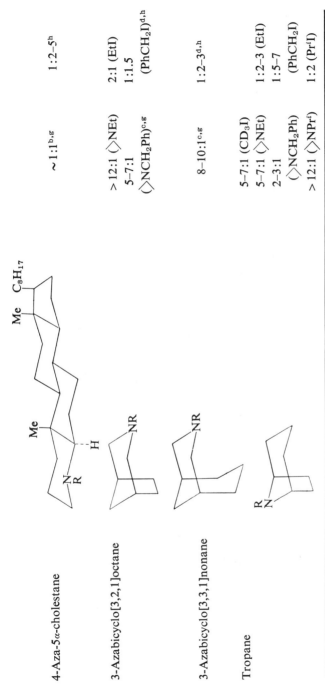

4-Aza-5α-cholestane	~1:1[b,g]	1:2–5[h]
3-Azabicyclo[3,2,1]octane	>12:1 (◯NEt)	2:1 (EtI)
	5–7:1	1:1.5
	(◯NCH₂Ph)[c,g]	(PhCH₂I)[d,h]
3-Azabicyclo[3,3,1]nonane	8–10:1[c,g]	1:2–3[d,h]
Tropane	5–7:1 (CD₃I)	
	5–7:1 (◯NEt)	1:2–3 (EtI)
	2–3:1	1:5–7
	(◯NCH₂Ph)	(PhCH₂I)
	>12:1 (◯NPr^i)	1:2 (Pr^{i}I)

[a] Refs. 12, 14, and 15. Refluxing solvent except for some very fast quaternizations, where the reactions were run at room temperature. Reactions of N-methyl bases with CD₃I were run in ether at room temperature, except in the camphidine and 4-aza-5α-cholestane series, where refluxing acetone was used.
[b] All the alkyl groups captioned in the heading were studied except for those specifically noted in individual reactions.
[c] No ⟩NPr^n.
[d] No Pr^{n}I.
[e] ~1:1 means a range from 1.5:1 to 1:1.5.
[f] The low value may be due to partial equilibration.
[g] No ⟩NPr^i.
[h] No Pr^{i}I.

of the protons being estimated, to allow for base-line ripple or small peaks which form part of other multiplets (12).

Unfortunately, however, organic and physical-organic chemists frequently either ignore error treatments altogether or use them without adequate analysis or understanding and grossly overestimate the precision of their results. One suspects that many product ratios for quaternizations quoted to two significant figures may in fact deserve only one, and many quoted to three may deserve but two. Some improvement in the standards of precision noted above may be obtained with more elaborate modern instruments, but quotation of product ratios from nmr signal areas to three significant figures seems rarely, if ever, either justified or necessary. However, moderately high precision in measurement of the quaternization product ratios is indeed required for some purposes (as is described in the sequel) giving relevance to the foregoing considerations.

In the earlier literature, particularly before the general use of nmr spectroscopy, product ratios were themselves usually overestimated, and many reactions, for example, in the tropane field, which were once thought (3,13) to be highly stereoselective are in fact much less so (12). It will be evident in the sequel (see, e.g., Table I) that while there are examples in the competitive quaternizations of quite high stereoselectivities ($>12:1$), more typical figures range from $\sim 1:1$ (zero stereoselectivity) to $\sim 2-5:1$.

The major determinants (12) in quaternization product ratios from the tertiary piperidines are the base system involved, the exocyclic alkyl groups, and the order in which the latter are introduced (e.g., $>$NMe + Alk-I or $>$N-Alk + MeI). Variations in solvent and leaving group are generally of much lesser importance (2b), as illustrated later. There have been very few detailed investigations of the temperature coefficients of the product ratios, but only small effects are expected (or observed) (2b).

V. STEREOSPECIFICITY OF QUATERNIZATIONS

This is the major problem requiring solution and, therefore, the one which has most engaged the attention of interested authors. The numerous methods used for determination of configuration of diastereomeric pairs of quaternary salts may be subdivided into those based on (*a*) kinetic considerations, (*b*) thermodynamic considerations, (*c*) spectroscopic and related (X-ray) procedures, and (*d*) a chemical procedure (possibility of lactonization of one but not the other diastereomer). We cannot discuss all the methods in the same detail, and in the following account appropriate weight has been given to theoretical and practical reliability, importance in the previous development of the field, potential for future research, and general applicability by organic stereochemists; X-ray crystallographic analysis, for example, is hardly so applicable at present.

A. Kinetics-Based Methods

Three of these are based on consideration of product ratios (stereoselectivities) and variations therein, and the fourth is a more elaborate method of kinetic analysis.

1. Variation in Product Ratios Associated with Introduction of the Two Exocyclic Alkyl Groups in One or the Reverse Order

The argument here (2a,12,14–16) is the simple one that if there are marked differences in stereoselectivity between the reactions $>$N-Alk + MeI and $>$NMe + Alk-I (Alk being an alkyl group different from and, therefore, larger than CH_3 or CD_3), the former reaction will give more axial (or less equatorial) quaternization than the latter, which will give more equatorial (or less axial). The quaternizations must all, of course, be run under reasonably comparable conditions. In practice, the alkyl groups used in the method have been mostly primary, isopropyl often introducing some anomalies and tertiary butyl being expected to be too strongly biased a group for exhibition of a satisfactory differential effect.

The basis of this argument is that in the relevant steric environments alkyl groups other than methyl have a greater *effective* size than methyl; the differentiation, partly entropic and partly potential energetic, is relatively small for methyl versus primary alkyl groups, but it is expected to be magnified for partly attached groups in competitive transition states because of the bipyramidal geometry around the alkyl carbon at which bimolecular nucleophilic displacement is taking place (or the analogous near-planar arrangement of the three C^+ bonds in an attacking carbonium ion, should such a species be involved).

The application of the method is illustrated in Table I; in the discussion of the results (12,14–16) we first ignore ratios of salts containing the atypical group isopropyl and the results for trideuteriomethylation of N-methyl bases and we consider the comparison of product ratios from the reactions $>$N-Alk + MeI and $>$NMe + Alk-I (Alk = Et, Pr^n, and $PhCH_2$). The figures, which are quoted to a precision adequate for the application of the method (cf. our earlier comments), show that in the case of the first four systems methylation is predominantly axial, and alkylations have low stereoselectivities in the case of the first three, but the fourth (camphidine) undergoes markedly preferred axial alkylation, although with lower stereoselectivity than is the case for the inverse methylation reactions. With the same pair of exocyclic groups, the predominant isomer formed on methylation is the minor isomer in the alkylation of the N-methyl base (and vice-versa), a characteristic but not universal result (note the inverse ratios are mostly quoted for inverted quaternization sequences). Slight variations around the 1:1 ratio for alkylations of the N-methyl bases in the first three systems do not seem profitable to discuss, except to note that sometimes in these reactions there is encountered a weak preference for equatorial quaternization.

In the 4-aza-5α-cholestane system (14) we have, by contrast, near-zero stereoselectivities in the methylations and, from the increase in stereoselectivity, preferred equatorial ethylation and benzylation of the N-methyl base. The next system, 3-azabicyclo[3,2,1]octane (15), is interesting in that although three of the four reactions noted in Table I correspond to preferred axial quaternization, ethylation of the N-methyl base shows some preference for equatorial stereospecificity. The figures for 3-azabicyclo[3,3,1]nonane could clearly be taken as indicating preferred axial quaternization, in line with the commonest deduction for the preceding cases. However analysis (16) of the nmr spectra of the N-methyl base methiodide **4** clearly indicates that in this, and, by ready extrapolation, in other quaternary salts of the same base system, the piperidine ring has a boat conformation. (By contrast, the major component **5** of an equilibrium mixture of the N-methyl hydrochlorides evidently has a piperidine-chair.) Hence *boat-axial* (i.e., equatorial, from a configurational standpoint) stereospecificity was deduced (16) for quaternizations in this system. This deduction was subsequently confirmed by independent evidence (17) discussed below.

The tropane figures are in part unsatisfactory in that for salts with ethyl and benzyl substituents no clear prediction of stereospecificity is evidently possible. However, the results for the methyl-isopropyl salts strongly suggested a preference for axial quaternizations (which would therefore also be expected for reactions leading to the other salts). Application of other methods (discussed below) for determination of configuration of diastereomeric quaternary salts to derivatives of tropane or simple tropines (e.g., **6** and **7**) also give unsatisfactory or conflicting results, and the preferred direction of quaternization of these bases is at present an open question (12) undergoing detailed reexamination (18–20).

While isopropyl sometimes falls in line with the other alkyl groups in the pattern of stereoselectivities exhibited, there are evident anomalies (12) in relation to N-methyl-N-isopropyl salts in the 2-methylpiperidine and *trans*-decahydroquinoline series, where one salt is formed predominantly in each

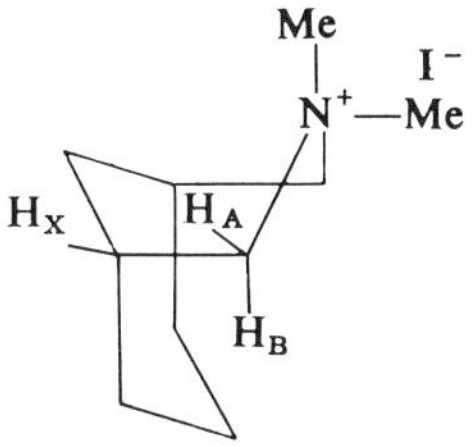

(4)

$J_{AB} = 13.5; J_{AX} = 10.3;$
$J_{BX} = 2.1$ Hz (in C_5H_5N/D_2O)

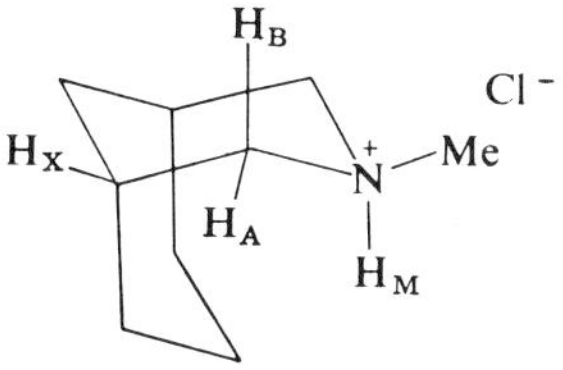

(5)

$J_{AB} = 12.8; J_{AX} = 0-2;$
$J_{BX} = 3.2; J_{AM} = 3.5;$
$J_{BM} = 10.4$ Hz (in CF_3CO_2H)

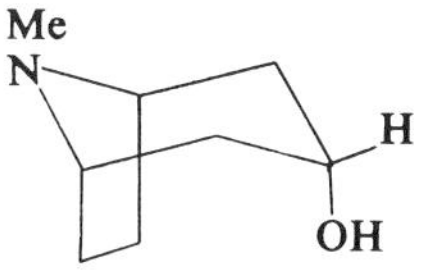

(6)

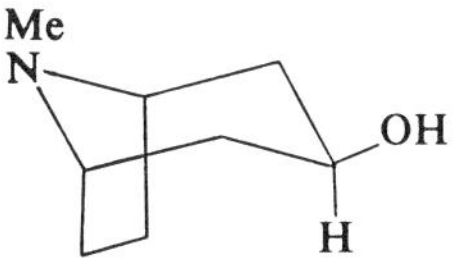

(7)

$Ph(CH_2CH_2Br)_2$ + MeNHEt + K_2CO_3 $\xrightarrow[\text{MeOH}]{\text{in hot}}$
(8)

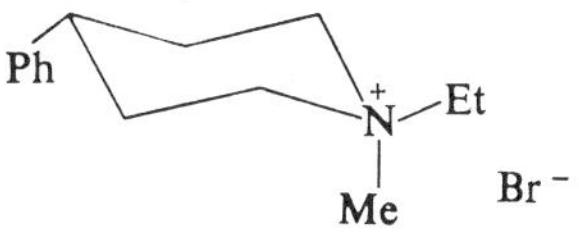

(9) + (10) (ratio 1.5:1)

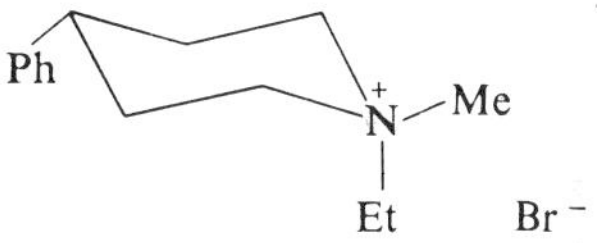

system irrespective of the order of introduction of the two exocyclic groups. This result has been explained (12) in terms of the unusually high equatorial preference for an isopropyl substituent when there is an adjacent equatorial C-alkyl substituent.

It is interesting that the reaction >N-Me + CD_3I usually resembles >N-Alk + MeI rather than >NMe + Alk-I in its stereoselectivity pattern (and hence in deduced stereospecificity). This is in accordance with the point of view given above that differences in effective bulk of methyl and primary alkyl groups are magnified for the partly attached groups in transition states, and one might expect, in general, that stereoselectivities would vary more with variation of the quaternizing alkyl groups than of the N-alkyl groups originally present.

2. Cyclization of Appropriate α,ω-Dihalides with Secondary Amines NHMeAlk and Alkali

It was proposed (21) that the major component of the mixture of diastereomeric quaternary salts produced by kinetic control in such a cyclization (e.g., **8 → 9 + 10**) should be the thermodynamically more stable (usually with *N*-methyl axial and *N*-alkyl equatorial), so an additional method for assignment of configuration to such salts might be possible. The mixtures so examined had previously also been obtained by quaternizations in the 2-methyl- and 4-phenylpiperidine series. The theoretical basis of this proposed method is not as satisfactory as one might wish, and, in any case, in practice the product ratios observed were usually too near unity for assured deductions of configuration. In all the cases examined (21), however, the salts previously assigned the thermodynamically more stable configuration by other methods did indeed predominate in the reaction mixtures from the cyclizations.

3. The Hammond Postulate Method

This method (6,22) is based on examination of the relatively small variations in quaternization product ratios associated with variation of leaving group or solvent or in introduction of polar groups into the nucleophile (i.e., the piperidine). The initial argument was that a better nucleophile, leaving group, or ionizing solvent means a more "reactant-like" transition state and thus a longer $N \cdots R'$ partial bond (where R'X is the quaternizing agent) and increased axial or reduced equatorial quaternization, and, vice-versa, one may expect less axial or more equatorial quaternization for slower reactions with poorer nucleophiles, leaving groups, or ionizing solvents. It was later suggested (23) that product-ratio variations in examples where there were differences in quaternization rates associated with differences in steric compression between transition states could also be interpreted on the basis of the Hammond postulate.

This suggested method is very profitable to discuss, but difficult to recommend at least in its present form. The major theoretical objection is that the Hammond postulate in its original form (24) was really never intended for such detailed applications and indeed appears quite unsuitable for them. Development of a more useful general theory to broaden (in effect) the scope of the original postulate is very difficult, but several attempts have been made recently (25,26). Some of the quaternization results (not the steric-variation examples) may be at least partly reinterpreted on the basis of the Swain-Thornton postulate (25) which states that a better nucleophile lengthens an adjacent partially formed bond. As is the case with other reactions of organic compounds, it is possible to *visualize* a "reaction coordinate" in the quaternizations only if we are prepared to reduce the problem to three dimensions by considering a structureless nucleophile N attacking a reagent R'X, where R' is also structureless, via a linear transition state N····R'····X, so that the only three variables are the N····R' and the R'····X bond lengths and the potential energy of the system. Emphasis (25) has been placed on the necessity to consider to what extent the initial transition state may be moved *along* the initial reaction coordinate and to what extent at *right angles* to it by introduction of substituent variations. All these considerations are certainly relevant if the theoretical basis of the proposed method for determination of quaternization stereospecificities is to be satisfactorily upgraded and the method suitably restated. The steric-variation examples (23) need particularly close scrutiny, since reduction in the rate of a nucleophilic displacement by increased steric compression around the displacement center may increase or decrease the length of the bond between nucleophile and displacement center and may move the transition state in *either* direction along the reaction coordinate, or indeed not move it along this coordinate at all, but only at right angles (cf. the reaction of radiolabeled iodide with methyl and *t*-butyl iodides, where we have a "half-way" transition state in each case).

Application of the method makes demands for precision in the nmr spectroscopy used to determine product ratios far beyond those required in method *1* described above, where only quite large variations are considered. As an example, we quote in Table II product ratios (6) for alkylations of

TABLE II

Stereoselectivities of Methylations of 4-Substituted 1-Ethylpiperidines at 30° (6)

4-Substituent	Methylating agent	Solvent	Product ratio, ±0.2
Phenyl	MeI	MeCN	2.5:1
Phenyl	MeOSO$_2$Ph	MeCN	3.7:1
Cyclohexyl	MeI	MeCN	3.4:1
Cyclohexyl	MeOSO$_2$Ph	MeCN	4.3:1
Phenyl	MeI	C$_6$H$_6$	2.3:1
Phenyl	MeOSO$_2$Ph	C$_6$H$_6$	3.4:1
Cyclohexyl	MeI	C$_6$H$_6$	3.3:1
Cyclohexyl	MeOSO$_2$Ph	C$_6$H$_6$	3.7:1

some 4-substituted piperidines from which the deduction of preferred axial quaternizations is made from the premises underlying the method stated above; a similar deduction was made (19) for pseudotropine (7). No detailed statistical treatment of random errors was given in the initial publication, but the 2σ limits quoted in the table appear to be in accordance both with the information (about average deviation) given there and with general experience. One significant figure has therefore been dropped from the originally quoted ratios.

The method as proposed fails when applied to a comparison of product ratios from three N-methylpiperidines on quaternization on the one hand with ethyl iodide and on the other with the very reactive triethyloxonium fluoroborate, which experimentally gave markedly more *equatorial* ethylation (27). It is clear that a more detailed theoretical analysis of the method and redefinition of its scope would be very much in order.

4. *Quantitative Kinetic Analysis of Quaternization Rate Constants*

This suggested method (4) utilizes the two pairs of simultaneous equations (eqs. (b) of Fig. 1) with an input of (hopefully) known conformer fractions N_a and N_e in the reactant piperidine and the quaternization rate constant k_{obs}. The unknowns are k_a and k_e, and application of the method should lead to two sets of possible values for these conformer rate constants, from which a choice has to be made from considerations of internal consistency (when the method is applied to a number of related quaternizations) and of values of rate constants for model (rigid) piperidines, which can react only as single conformers. Such models, of course, cannot be satisfactory for the more flexible bases of the conformationally mobile systems under study, in which equilibrium conformer proportions (leading to values for N_a and N_e) have also in the past at least been very difficult to measure with assurance; the recommended values for key examples like N-methylpiperidine vary markedly almost from year to year. With these background disadvantages the method would be (or at least in the past would have been), at best, very difficult to apply for the elucidation of the stereospecificities of piperidine quaternizations; these would be immediately deducible if and when values for k_a and k_e became known. Application of the method to the quaternization of 1-ethyl-4-phenylpiperidine with methyl iodide led (4) to the deduction of preferred equatorial reaction (and, by extrapolation, to a similar deduction for a range of quaternizations for which the opposite stereochemical conclusion had been reached by application of other methods described in this review). However, the initial deduction was shown (6,12,28) to be invalid because of an incorrectly determined product ratio for the methylation.

Instead of being used for the initially intended purpose, this type of kinetic analysis may be (12) of considerable interest in the future for establishment of various types of useful cross correlations following determination of quaternization stereospecificities by more suitable methods and the attainment of general agreement among interested workers regarding conformer equilibrium constants associated with nitrogen inversion in tertiary piperidines.

B. Thermodynamics-Based Method

The formal analysis given above for kinetically controlled product ratios in the quaternizations requires that equilibrium between the diastereomeric products cannot readily be established under the preparative conditions employed; this would be the normal expectation, and it has been shown to be correct with the possible exception of a few quaternizations leading to salts containing N-benzyl groups (2a,12,29). Under much more forcing conditions than those used preparatively, some reversals of the quaternization have been noted, and a method has been proposed (2a) for determination of configurations of epimeric N-methyl-N-benzyl salts based on establishment of the equilibrium **11** ⇌ **12** in Chart 3; the latter salt, with equatorial benzyl, would usually be expected to predominate. In the 3-azabicyclo[3,3,1]nonane system, the salt with *boat-equatorial* benzyl should be the more stable (12).

This method has been applied (2a,12,14,15) to the appropriate salts of all the base systems shown in Table I, and the results support the configurational assignments made on other grounds, except in the case of tropane, where, for reasons not understood, no isomerization could be effected. The usual experimental procedure involves heating a chloroform solution of the quaternary salt at a temperature carefully chosen for each individual system, the aim being to bring about the isomerization with a minimum of attrition through unspecified decomposition. There is always some such attrition, however, and sometimes it is uncomfortably fast, so that the question arises as to how near the true equilibrium constant is the value measured for the N-methyl-N-benzyl epimers. There is also at least the theoretical possibility of formation of some N,N-dimethyl and N,N-dibenzyl salts as well. Another experimental drawback has been the difficulty of preparation of pure samples of individual isomers; the method is less satisfactory when applied to mixtures, as the observed overall change in composition is then less. The most clear-cut application (2a) has been to camphidinium salts, where the pure isomers are readily prepared by fractional crystallization. The equilibrium (or near-equilibrium) constant for the epimeric 4-phenyl salts is, interestingly, approximately the same ($\sim 2:1$) as that calculated (6) (at room remperature) for the conformational equilibrium **13** ⇌ **14**, although by a method (averaged nmr chemical shift) open to criticism (30).

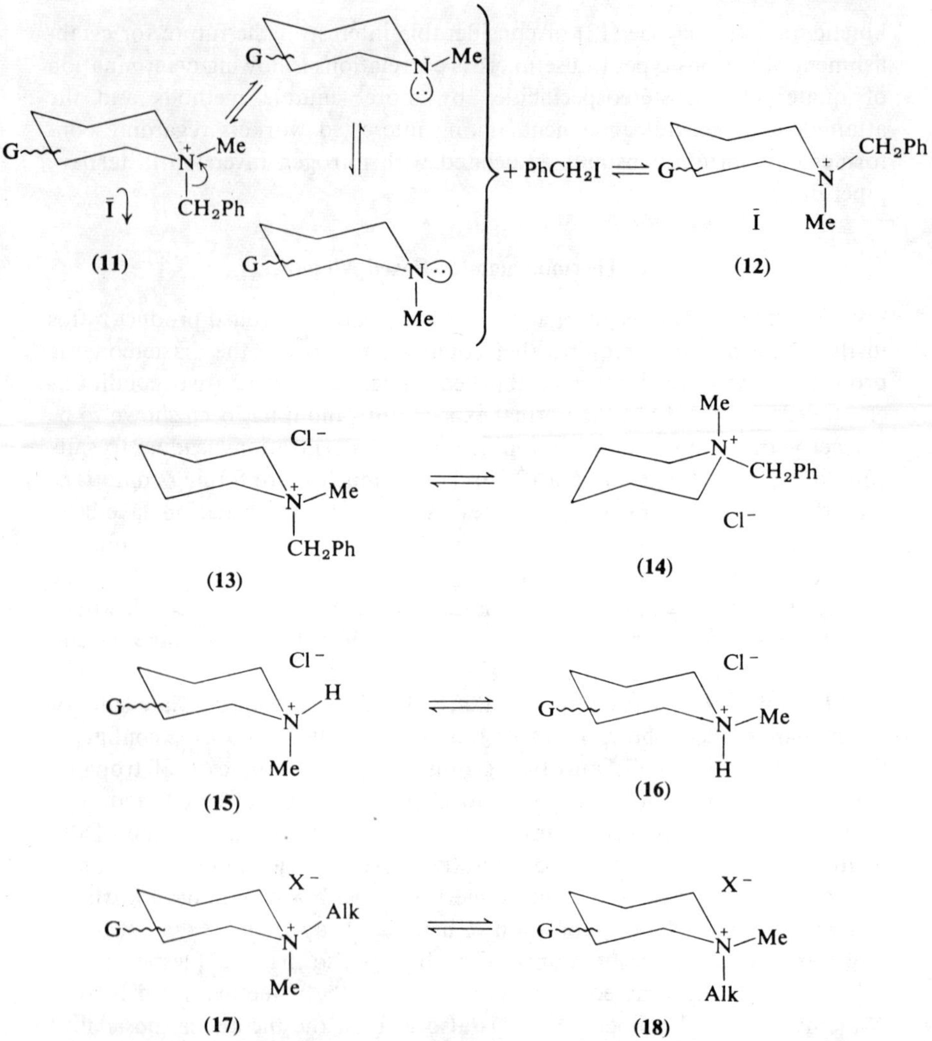

Chart 3. Illustration of determination of configuration of quaternary salts by a thermodynamic and an empirical nmr method. G = biasing or reference group or fused ring.

C. Spectroscopic and Related Procedures

1. Nuclear Magnetic Resonance Methods

These fall into two general types: (*a*) earlier empirical correlations in chemical shifts for salts (proton and quaternary) of expected similar structure, configuration, and conformation and (*b*) recent, more analytical methods. As yet there are relatively few examples of the second type, while the first is now mainly of historical interest.

The chief type of empirical correlation used has been between chemical shifts of N-methyl groups in N-methyl hydrochlorides and in N-methyl-N-alkyl quaternary salts of the same base systems. If two signals of different intensity are seen in the spectrum of a solution of the hydrochloride at equilibrium (**15** ⇌ **16**), the more intense is evidently associated with the thermodynamically more stable form, normally with N-methyl equatorial (**16**). On the assumption (12,31) that the relative positions of N-methyl signals in pairs of diastereomeric N-methyl-N-alkyl quaternary salts (**17, 18** derived from the same base system) are the same, a method for configurational assignment of these salts becomes available.

Although this empirical approach has been useful in the past and usually gives the correct configurational assignments for a pair of appropriate quaternary salts if used with circumspection, it cannot really be recommended since one is never sure in any particular case if it will lead to an incorrect conclusion. One would obviously be cautious if the possibility seemed to exist of the hydrochlorides and the quaternary salts under comparison having markedly different conformations or if epimeric salts had N-methyl signals so close together (e.g., $\delta < 0.1$ ppm) that solvent and concentration dependence of chemical shifts might be critical. Apart from these points, there is always the basic theoretical possibility that the summed bond anisotropies might give reversed axial–equatorial N-methyl signal orders for the two types of salt, even in cases where it is felt conditions might be quite favorable for the comparison.

An even simpler empirical generalization which has been used occasionally for assignment of configuration to appropriate diastereomeric quaternary salts is that the signals for axial N-methyl groups are usually at higher field than those for equatorial N-methyl groups in the epimers (20). There are exceptions to this generalization also, however (12,31,32), so again it should be employed only as a pointer to probable configuration and not for definite assignments in any particular system (4). Still another approach lies in comparisons of signal positions for, say, N-methyl groups between two or more series of quaternary salts from bases of related structure (33). This might seem to be better than comparisons of quaternary salts with proton salts, but the same sort of precautions are nevertheless necessary, and even with these there remains, as before, the possibility of incorrect assignments. A somewhat incautious comparison (33) of nmr spectra of certain quaternary 3-azabicyclo[3,3,1]nonanes and 3-azabicyclo[3,2,1]octanes (for parent structures see Table I) led to an incorrect deduction of preferred equatorial quaternization in the latter system; the former salts have piperidine-boat conformations (12,16) and the latter chair conformations (12,15).

Applied to tropane and simple tropines, the first of the empirical methods discussed above suggests (31) preferred equatorial and the second preferred axial (20a) quaternization.

Contributions to chemical shifts from individual σ-bonds are difficult to assess accurately, but if there are strongly anisotropic groups in quaternary salts of fairly easily definable preferred conformations, it is sometimes possible to reach assured configurational assignments for diastereomeric quaternary salts by ignoring the relatively weak magnetic effects of the σ-bonds and considering only the major effects associated with the strongly anisotropic groups. For example, the preferred conformation for 1-benzyl-2-methylpiperidine benziodide is readily shown (28,34) qualitatively to be that represented in Chart 4. Considering only the magnetic effects of the two benzene rings, one calculates by the Johnson-Bovey (J-B) or point-dipole (P.D.) method the indicated contributions to the chemical shifts of the four benzyl-methylene protons. The experimental observations are in good qualitative agreement with the calculations, and a clear differentiation between axial and equatorial benzyl groups is thus possible; the geminal methylene protons of the latter are those showing the much larger degree of chemical-shift magnetic nonequivalence. Quaternization of the dideuteriobenzyl base (containing $>$N—CD$_2$Ph) with benzyl iodide may thus readily be shown to take a preferred axial course. Exactly the same analysis (28,34) has been

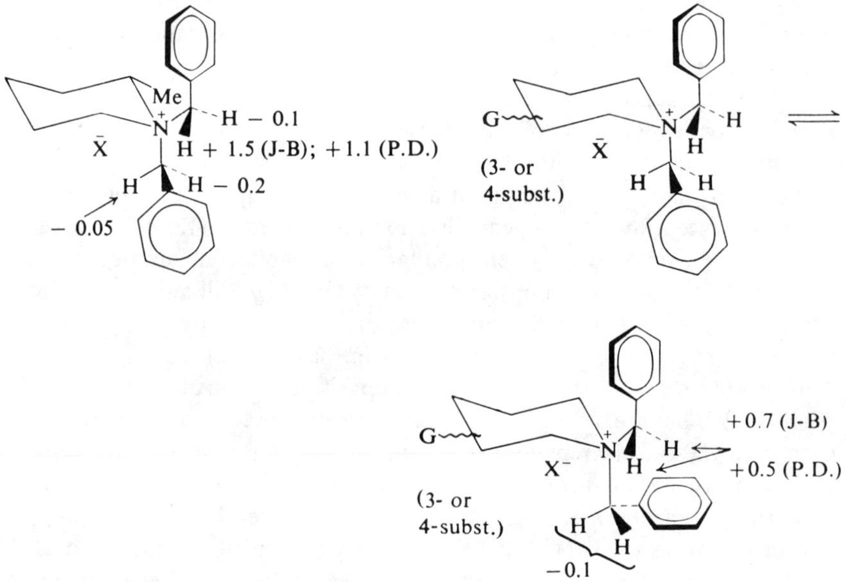

Increments in chemical shifts (ppm) due to benzene ring anisotropies as calculated by Johnson-Bovey method or point-dipole approximation.

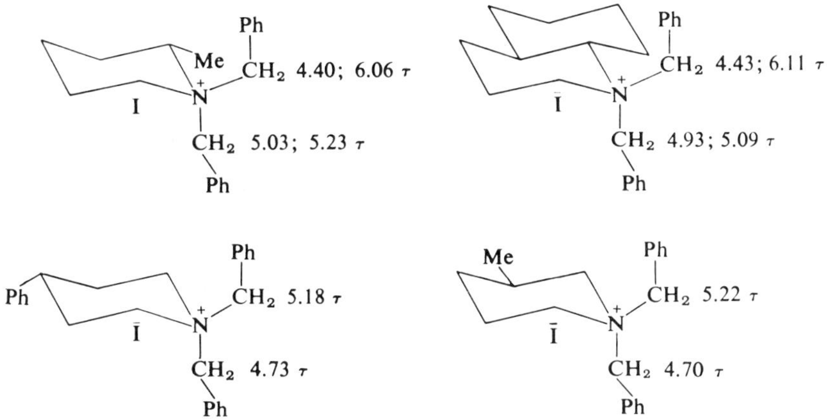

Structure 1 (top left)
Me, CH₂ 4.40; 6.06 τ
CH₂ 5.03; 5.23 τ
Ph, Ph, I⁻

Structure 2 (top right)
Ph, CH₂ 4.43; 6.11 τ
CH₂ 4.93; 5.09 τ
Ph, I⁻

Structure 3 (middle left)
Ph, Ph, CH₂ 5.18 τ
CH₂ 4.73 τ
Ph, I⁻

Structure 4 (middle right)
Me, Ph, CH₂ 5.22 τ
CH₂ 4.70 τ
Ph, I⁻

Chemical shifts (CDCl₃ solutions) observed for benzyl-methylene protons.

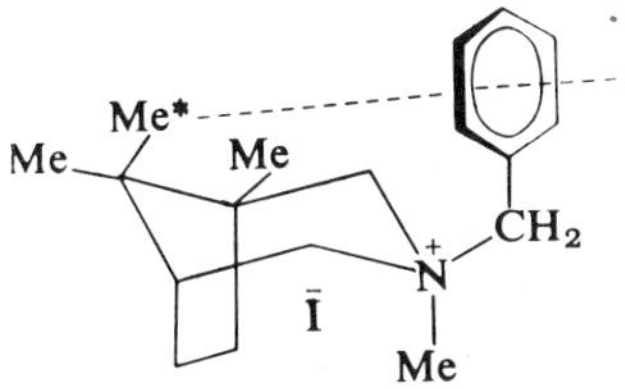

Camphidinium structure
Me*, Me, Me, CH₂, Me, I⁻

Protons of one C—Me group (asterisked) at 9.33 τ. Protons of other C—Me groups at 9.09 τ (CDCl₃ solution). In other camphidinium salts, C—Me protons at 8.85–9.10 τ. Calculated (J-B) shielding effect by benzene ring 1.5–2.5/3 = 0.5–0.8 ppm.

Chart 4. Assignment of configuration by conformational analysis and comparisons of observed and calculated chemical-shift increments.

applied in the *trans*-decahydroquinoline system, and preferred axial benzylation again demonstrated. In the 3-methyl- and 4-phenylpiperidine systems (Chart 4), *gem*-methylene protons of the 1-benzyl benziodides appear as singlets, of which that for the axial methylene group is unambiguously expected (28, 34) to be at substantially lower field than that for the equatorial. This prediction again permits a clear demonstration of preferred axial benzylation of the *N*-dideuteriobenzyl bases. The probable preferred conformations of tropanium salts prevent application of the method (34).

Configurations have been assigned (28, 34) to epimeric *N*-methyl-*N*-benzylcamphidinium salts on the basis of the expected upfield shift in one configuration for the protons of one of the three C-methyl groups (asterisked) in the conformation shown (Chart 4); two other rotameric orientations of the *N*-benzly groups in this configuration are also expected to be substantially populated, reducing the shielding effect to about one-third. In the diastereomeric salt (not shown) a slight deshielding effect by the benzene ring on the bridge methylene protons is expected and observed. The results support the conclusion reached by the application of other methods: preferred axial methylation of the *N*-benzyl base and preferred axial benzylation of the *N*-methyl base (cf. Table I).

Such analyses are evidently much superior to the earlier more empirical methods and, although clearly limited in scope, are nevertheless capable of some extension to other quaternary salts and other base systems.

2. *Examination of Infrared Spectra*

Several groups of workers (11,12,35) have noted certain regularities in the ir spectra of ranges of diastereomeric pairs of quaternary salts, and an empirical rule was suggested (11,12) that of the epimers $\text{>}\overset{+}{\text{N}}\text{MeAlk}\ \bar{\text{I}}$ the salt with equatorial *N*-methyl showed a diagnostic band in the region 885–905 cm^{-1} (KBr disks), while the epimer with *N*-methyl axial had this band in the region 840–885 cm^{-1}. Often several bands are in practice observed in the regions noted, but in such cases there is a characteristic increase in intensity of diagnostic band in the higher frequency region and/or an analogous reduction in the lower frequency region in the spectrum of the salt with *eq*-NMe as compared with *ax*-NMe (see Fig. 2).

The vibrational modes responsible for the observed characteristic bands have not been identified, so obviously this rule (if the word is indeed deserved) is just as useful as it is found to be in practice for making correlations and for (often) providing modest support for configurational conclusions reached by application of methods with a sounder theoretical basis. With

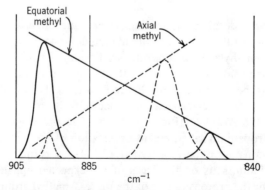

Fig. 2. Idealized representation of diagnostic bands in infrared spectra of epimeric *N*-methyl-N-alkyl quaternary salts. There are usually several additional nondiagnostic bands in the same region.

some tropanium salts configurations corresponding to preferred equatorial quaternization are predicted, while with others the rule is found to be inapplicable, as it is (11) with quaternary salts in the 2-methylpiperidine and *trans*-decahydroquinoline systems. With most quaternary salts derived from the other base systems of Table I, however, the configurations predicted (11,12,14) are in accord with those deduced by the other procedures.

3. X-Ray Crystallographic Analysis

In principle, this method is applicable to any quaternary salt which will give technically satisfactory crystals (by no means do all); bromide or iodide anion provides a satisfactory "heavy atom." In practice, for the ranges of salts falling within the scope of this review, the only fully unambiguous X-ray analysis appears to be that for 1-benzyl-4-phenylpiperidine methobromide (Fig. 3), which provides a final proof (36) for the preferred axial methylation of 1-benzyl-4-phenylpiperidine. Detailed refinement of the work on *N*-ethylcamphidine methiodide has been prevented by crystallographic disorder in the sample examined (12), while the analysis (18) of "*N*-ethyl-nortropine methobromide" appears to be open to the criticisms of inadequate sample characterization and possible inhomogeneity and is therefore presently under reinvestigation (20b).

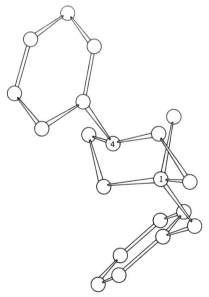

Fig. 3. Major cation from reaction of 1-benzyl-4-phenylpiperidine with methyl iodide in acetone or methanol. Examined as bromide; projection onto (001). Reproduced, with permission, from *Chemical Communications*, 1969, 339.

D. Lactonization Method

This is the most classical approach to determination of configuration of appropriate quaternary salts in that one uses a sequence of chemical reactions to get the requisite answer. As an example (17), the reactions leading to the conclusion of preferred equatorial (probably boat axial) quaternization of the hydroxy base (**19**) with methyl bromoacetate are shown in Chart 5. The method has also been applied (23,27) to the hydroxy bases **22** (R = Bu^t or

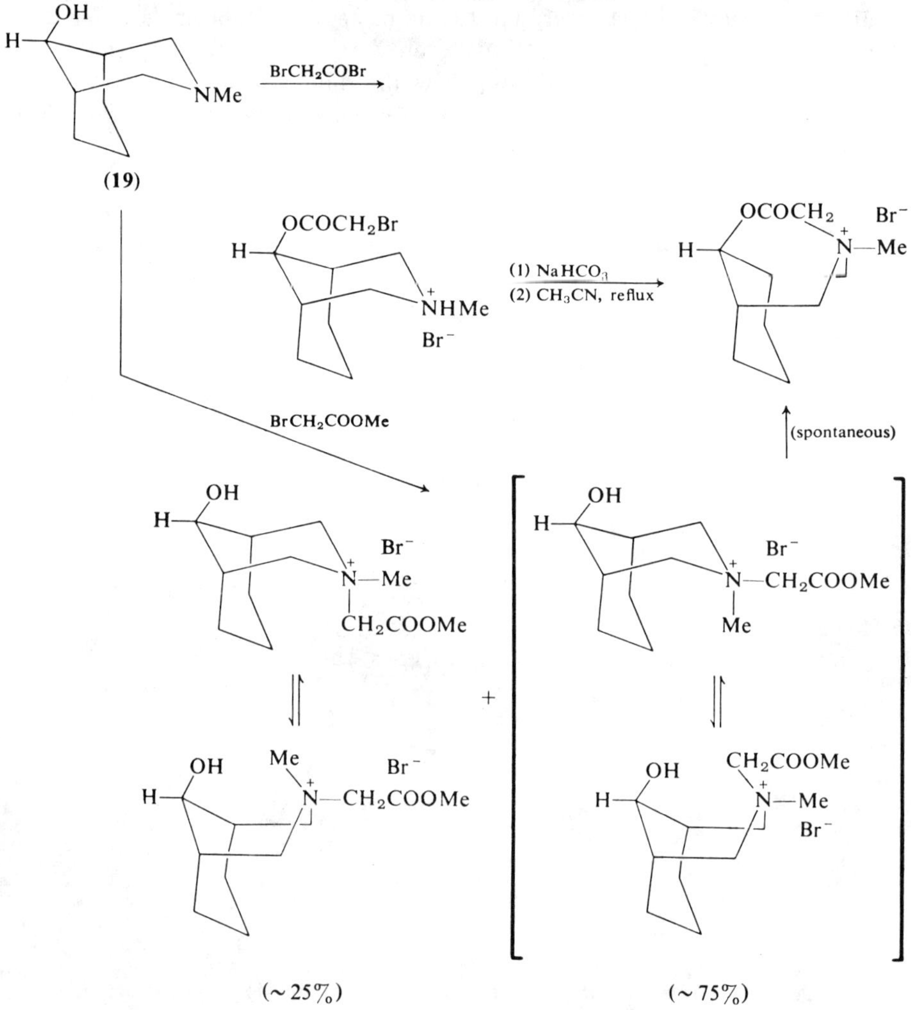

Chart 5. The lactonization method.

QUATERNIZATION OF PIPERIDINES

Ph); in these cases no marked stereoselectivity was observed, but in such a case the preparation of quaternary salts of known configuration aids the application of empirical nmr methods in the deduction of configurations of related salts. Quaternary salts from the bases **19, 20,** and **21** and methyl bromoacetate have been interrelated (17) by oxidation of the derivatives of **19** and **20** to those of **21** with ruthenium tetroxide.

Applied to pseudotropine (**7**), the lactonization method initially (3) indicated a preference for equatorial quaternization, but this conclusion was reversed (19,20a,20b) on reinvestigation of the experimental evidence. However, this evidence is unfortunately still ambiguous (20b,20c), particularly

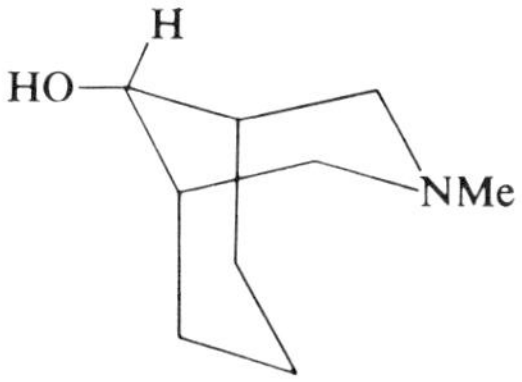

(20) 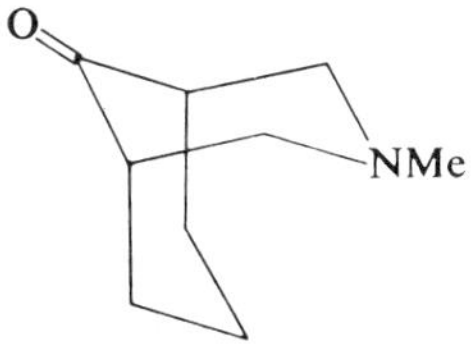   **(21)** 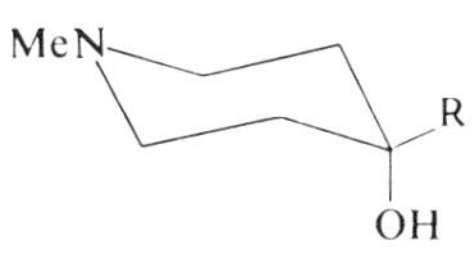   **(22)**

because of a suspicion that the presumed definitive lactone may in fact be a polymer.

E. Effect of Functional Groups on the Steric Course of Quaternization of Piperidines

The work described in the last section and other related results have incidentally served to demonstrate that fairly remote oxygen functions, for example, the hydroxyl groups required for lactone formation, have little effect on the steric course of quaternization.

1,2,3,6-Tetrahydropyridines (Δ_3-piperideines) appear to behave very similarly to piperidines as regards the stereoselectivity and stereospecificity of quaternizations (the less stable Δ_1- and Δ_2-piperideines have not been examined). The reactions $>$NR + MeX and $>$NMe + RX have been examined (29) in the base system **23**, configurations being assigned by correlating nmr spectra and (effectively) by use of the product ratio method (Sect. V-A-1) described above. The first type of quaternization gave the higher degree of axial product, as with the saturated piperidines discussed. Similarly, the double bond in the base **24** has little effect, and that in **25** no perceptible effect on the steric course of quaternization with trideuteriomethyl iodide (38). A higher stereoselectivity was observed (39) for the reaction $>$NMe + RX than for $>$NR + MeX in the tetrahydroisoquinoline

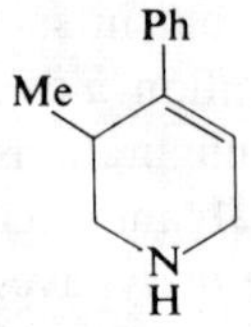

(23) 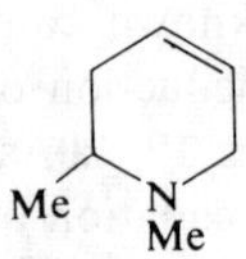 (24) 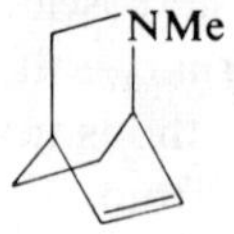 (25) 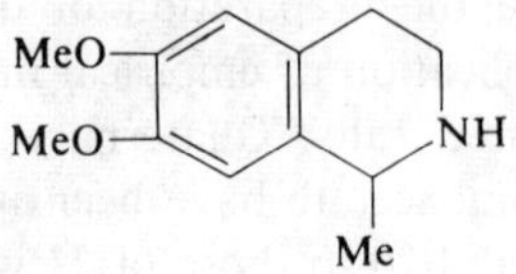 (26)

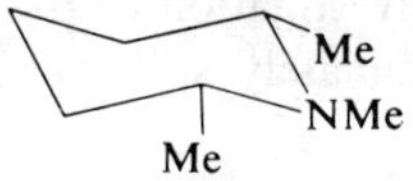

(27)

system **26**, a result suggesting preferred equatorial quaternization in reactions of marked stereoselectivity; isomerization experiments also appear to indicate that the more stable N-methyl-N-benzyl salt is formed preferentially by benzylation of the N-methyl base. A satisfactory comparison with saturated analogs is not presently possible, and the specific effect (if any) of the benzene ring cannot be assessed. The equatorial preference apparently exhibited in the quaternizations may be at least partly associated with the presence of the methyl group adjacent to nitrogen (cf. the summary given in the following section). The quaternizations were also studied for bases with other C-alkyl groups which gave more complex stereoselectivity patterns.

One can therefore say, from the relatively little that is known about the effects of functional substituents on the preferred steric course of quaternization, that remote oxygen functions (—OH and =O) are unlikely to have much influence, nor will remote (Δ_3) double bonds. There are hints of a more complex general pattern, however, and this is an area where much further work is desirable, particularly to discover specific effects associated with functional groups as distinct from any steric compressions they may exercise (as do nonfunctional alkyl groups) in competitive transition states.

F. Summary of Conclusions on Stereospecificities of Quaternizations of Piperidines

It is necessary to extrapolate and perhaps oversimplify to some extent in providing a general summary, but it is hoped the following will be useful for qualitative predictions and that it will be accurate in many cases. Some modifications may naturally be required as a result of further work.

1. The reaction $>$N-Alk + MeX (or $>$NMe + CD$_3$X) will usually show a preference for axial quaternization, or *boat-axial* (i.e., equatorial, configurationally) in the few cases where the piperidine ring in the quaternary salts has a boat conformation.

2. The reaction ⟩N-Me + Alk-X will show a reduced preference for axial quaternization, near-zero stereoselectivity, or some preference for equatorial quaternization. Since one is usually dealing with more nearly energetically equivalent competitive transition states here than in reaction (1), the qualitative prediction is more difficult, but there is often likely to be a much stronger tendency towards equatorial isopropylation than alkylation with a primary alkyl group. See also under *4*, below.

3. Variation among the commoner leaving groups X (e.g., halide, tosylate) is not too important a factor in determining stereospecificity, but this will not necessarily be true for more extreme variations, for example, ethylation with triethyloxonium fluoroborate (see previous comments, p. 290). Likewise, solvent variation among those commonly used for quaternizations (e.g., acetone, alcohol, acetonitrile, or even benzene) is not an important factor. In qualitative predictions we are not concerned with small variations in product ratios of the type illustrated in Table II.

4. The balance in the preference for axial versus equatorial quaternization depends quite markedly on the base system employed, and further work is certainly needed to clarify the general pattern in this connection. However, it is evident that strong *syn*-axial interactions with the *N*-axial substituent in the quaternary salt or associated transition states do not necessarily result in an increased tendency for equatorial alkylation (cf. the marked axial stereospecificities for the slow, sterically retarded quaternizations in the camphidine system). Of course, there is absolutely no *a priori* reason to expect otherwise, since the notion of preferred "approach (of reagent) from the less hindered side" (of the substrate) cannot legitimately be applied to examples where there may be a marked conformational change in the substrate during attainment of the transition state, as is the case with the quaternizations.

One factor which does seem to result in a reduced tendency for axial and/or an increased tendency for equatorial quaternization is equatorial α-alkylation. The results given above for the 2-methylpiperidine, *trans*-decahydroquinoline, and 4-aza-5α-cholestane systems and the tetrahydroisoquinoline **26** bear this out, as do other recent examples; for example, reaction of the piperidine **27** with ethyl iodide in cold acetone gave a product ratio of ~8:1 favoring equatorial ethylation, as compared with an equatorial preference for <1.5:1 with 1,2-dimethylpiperidine (**2b**). However, equatorial α-substitution or (particularly) α,α'-disubstitution, is well known (40) to alter or even invert the order of steric compressions at axial and epimeric equatorial positions, so that the result with base **27** may indeed be quite "normal," corresponding to preferred axial ethylation of, say, N-methylcamphidine; the attack is on the more hindered nitrogen face in each case.

The preferred steric course of quaternization of tropines is an important specific problem, an unambiguous solution of which is awaited with interest. The bridge position of the nitrogen atom in these bases, between two rings, makes difficult both qualitative predictions of stereospecificity and qualitative interpretations of the rather high stereoselectivities observed in the quaternizations (see Table I).

5. Remote hydroxyl or keto groups or double bonds are likely to have little effect on the steric course of the quaternization. Nothing can usefully be said presently about the effects of these functional groups when they are nearer the displacement center, or the effects of other functional groups.

VI. DEGRADATION OF DIASTEREOMERIC QUATERNARY SALTS

From studies (41,42) of the removal of N-methyl, N-ethyl, and N-benzyl groups from quaternary salts by nucleophilic displacement or (for N-ethyl groups) by Hofmann elimination, an interesting parallel has been found between preferred direction of quaternization (i.e., usually axial in the examples studied) and preferred stereochemistry for removal of the quaternizing groups (again usually axial). The initial work (41) in this area dealt with the removal of N-methyl and N-ethyl groups from diastereomeric pairs of quaternary salts by reaction with lithium aluminum hydride or (in more quantitative studies) with sodium thiophenate. In more recent work (42), which is somewhat easier to summarize, the substrates were N-methyl methiodides, N-ethyl ethiodides, and N-benzyl benziodides in which the exocyclic alkyl groups were partly isotopically labeled with ^{14}C (methyl and ethyl groups) or deuterium (—CD_2Ph in place of —CH_2Ph).

Some of the results are collected in Table III. If we consider the second column in this table (for reaction A) we note that irrespective of the distribution of the ^{14}C-label in the reactant methiodide, if the nucleophile SPh^- attacked axial and equatorial N-methyl groups with equal ease, then (neglecting isotope effects) one would get 50% of the ^{14}C activity in the product thioanisole in each case. More than 50% activity in the thioanisole samples indicates that the group more easily inserted by quaternization (the axial methyl rather than the equatorial, except perhaps in the nortropane system) is also being more readily removed by nucleophilic displacement. By combining the figures in the second column (Table III) with the observed product ratios from the N-methyl bases and trideuteriomethyl

TABLE III

Degradation of Isotopically Labeled Quaternary Ammonium Salts (42)

Reaction A: Degradation of quaternary salt mixtures from N-methyl bases and [^{14}C] MeI with NaSPh in glycol to give PhSMe + N-methyl bases (^{14}C-label distributed between products).

Reaction B: Degradation of quaternary salt mixtures from N-ethyl bases and [^{14}C] EtI with KOH in glycol to give ethylene + N-ethyl bases + H_2O (^{14}C-label distributed between organic products).

Reaction C: Degradation of quaternary salt mixtures from bases $\rangle NCD_2Ph$ and $PhCH_2I$ with NaSPh in glycol to give $PhSCH_2Ph + PhSCD_2Ph + \rangle NCD_2Ph + \rangle NCH_2Ph$.

Base system	Reaction A: % ^{14}C-label in recovered PhSMe ($\pm 3\%$)	Reaction B: % ^{14}C-label in recovered N-ethyl base ($\pm 2\%$)	Reaction C: % deuterium in benzyl methylene group of recovered N-benzyl base ($\pm 3\%$)
2-Methylpiperidine	50	—	—
4-Phenylpiperidine	60	48	67
trans-Decahydroquinoline	52	(54)	58
Camphidine	73	37	—
Nortropane	67	30	—

iodide and again neglecting isotope effects, the relative rates of nucleophilic attack on axial and equatorial N-methyl may be calculated. As the figures in the second column themselves suggest, the calculated displacement rate constant ratios are higher (~ 5) for the last two bases than for the others; it is interesting to note that quaternization stereoselectivities are also higher in the camphidine and tropane systems than in the others (cf. results given in Table I).

A similar trend is evident in the figures in the third and fourth columns (Table III). Those in the third column would in this case be expected to be less than 50% if an axial ethyl group (usually) were more readily eliminated as ethylene than an equatorial group. This is indeed so, except in the *trans*-decahydroquinoline system (bracketed figure), and again marked stereoselectivity is observed in the camphidine and nortropane systems. For benzyl group removal with sodium thiophenoxide in appropriately labeled *N*-benzyl benziodides, the two figures in the fourth column (each >50%) again indicate preferred nucleophilic displacement at the axial exocyclic group. The results are summarized in Figure 4.

The demonstrated parallelism between stereochemical preference for quaternization and for dequaternization is theoretically very satisfying (2a,42); in most cases an exocyclic group attached to nitrogen by a partial bond (either a partly *attached* or a partly *detached* group) is effectively less space demanding than the same group when fully attached (Fig. 5). As we implied earlier, however (cf. the discussion of Fig. 1), no satisfactory quantitative calculations of the energies of the diastereomeric transition states represented in Figure 5 can presently be made, and the corresponding observed experimental results could certainly not have been predicted. By way of qualitative comment (hardly interpretation) one might say that of the two obvious ways in which the space demands of a partly attached alkyl group should differ from those of the same fully attached group, i.e., because of the longer partial bond (making the partly attached group *less* space demanding) and because of the bipyramidal geometry at the displacement center (making the partly attached group *more* space demanding), the former appears to be the more important.

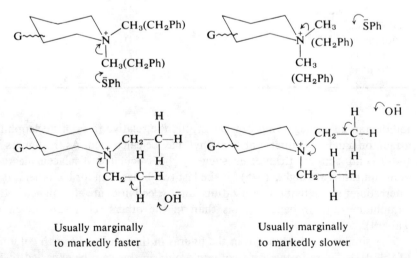

Fig. 4. Degradation of diastereomeric quaternary ammonium salts. G = reference or biasing group or fused ring.

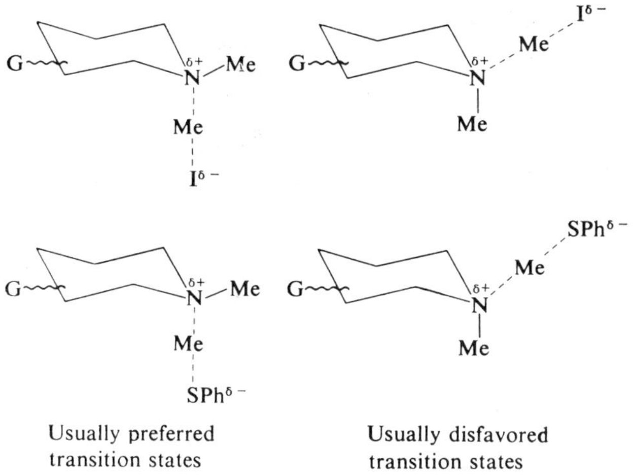

Usually preferred
transition states

Usually disfavored
transition states

Fig. 5. Correspondence in stereospecificity between quaternizations and quaternary-salt degradations. G = reference or biasing group or fused ring. Similar comparative diagrams may be shown for displacement of epimeric benzyl groups or Hofmann elimination with epimeric ethyl groups.

VII. RELATED WORK

The oxidation of certain tropine bases (43) and simple α-substituted piperidines (2b) to the amine oxides with hydrogen peroxide or peracids was thought to follow preferred axial stereospecificity, analogous to quaternizations at least in the latter systems, as seems reasonable; configurations were assigned mainly by empirical nmr correlations.

A series of pairs of diastereomeric N-methylquinolizidinium salts (**28, 29**, and C-methyl derivatives) has been prepared and examined (44). In a sense quinolizidines are piperidines, just as are, for example, decahydroquinolines, but in the case of the nitrogen-bridgehead compounds formal interconversion between diastereomeric quaternary salts is associated with a *cis–trans* change in ring fusion. This factor introduces an additional complexity not encountered in the other systems considered in this review, and it does not seem possible to interrelate the two types adequately on presently available information. It is interesting to note, however, that on comparative nmr and other evidence the *trans* isomer (**28**) appears to be formed predominantly by direct methylation of quinolizidine, while the *cis* isomer (**29**) is the chief product from the cyclization of 2-4′-iodobutyl-1-methylpiperidine, as indicated. These results correspond to a preference for axial quaternization in each case.

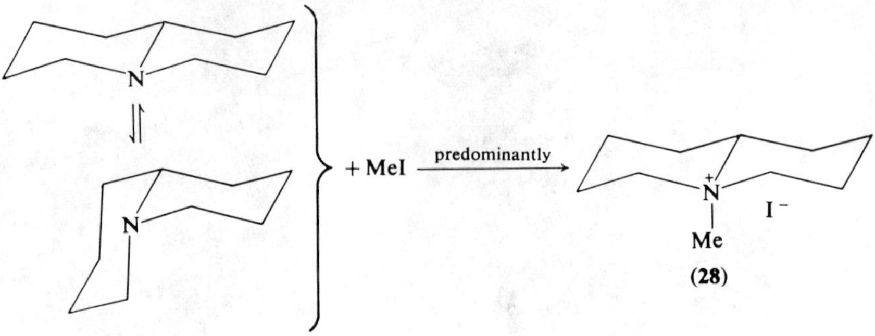

(28)

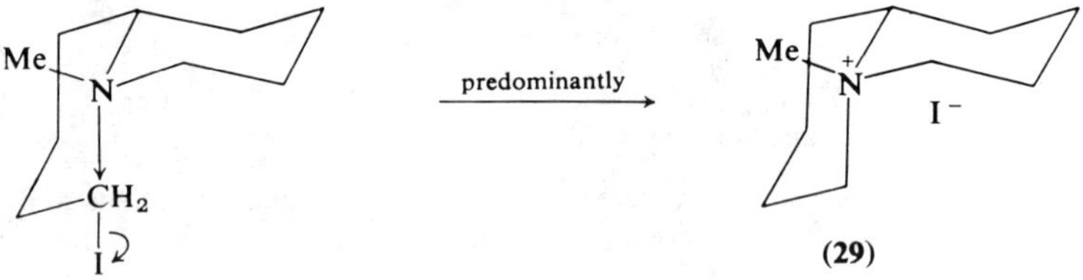

(29)

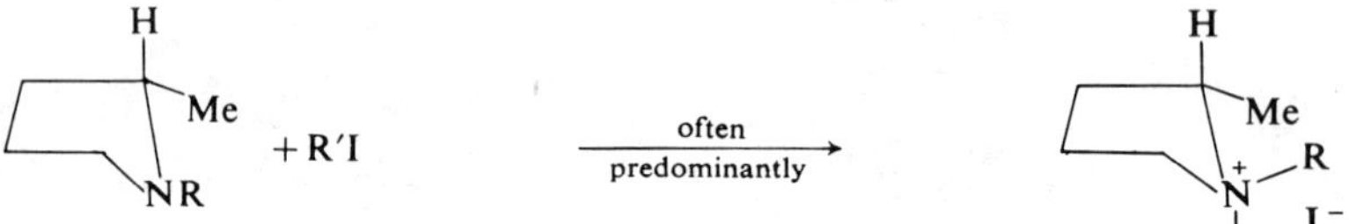

(30)

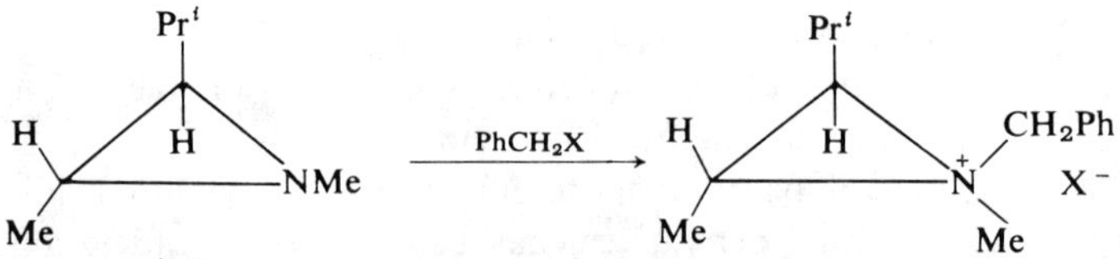

(31)

(weakly preferred diastereomer)

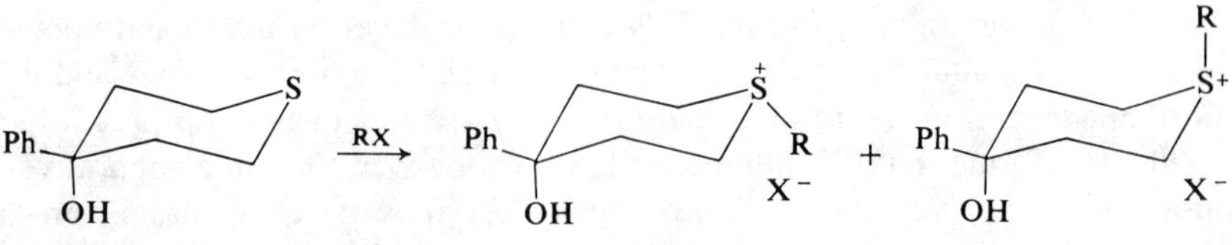

major product minor product

(32) (33) (34)

Chart 6. Related work on reaction stereospecificities.

2-Methylpyrrolidines (11) (30) and certain tertiary aziridines (22), for example 31, often show some tendency to undergo quaternization in the more hindered exocyclic quadrant at the nitrogen atom, *cis* to the largest adjacent *C*-substituent, as represented in Chart 6. This again corresponds to the preference for axial quaternization of (many) tertiary piperidines. The configurations of the appropriate pyrrolidinium and aziridinium salts were established by methods identical or analogous to those used for the piperidinium salts.

The preferred steric course of ternerization of 4-hydroxy-4-phenylthiacyclohexane (32) with methyl iodide, ethyl bromide, and methyl bromoacetate has recently been examined (45). A preference for equatorial ternerization was demonstrated, but the more stable product (33) appeared to be predominantly formed in each case by kinetic control; the equilibrium mixture (34 ⇌ 33) for the *S*-methyl derivatives had an equilibrium constant lower than the product ratio for methylation, indicating that a partly attached alkylating group in this thiacyclohexane is effectively *more* space demanding than is the same fully attached group. This result is evidently in contrast to that usually observed for quaternizations of tertiary piperidines; the different behaviors may be related to differences in significant bond lengths and angles between the two heterocyclic systems. Configurations for the sulfonium salts were established by the equilibration evidence, by what is equivalent to the product ratio method (Sect. V-A-1) discussed above (stereoselectivity lower for methylation; higher for the other alkylations) and by application of the lactonization procedure to the salt obtained with methyl bromoacetate.

References

1. D. Y. Curtin, *Record Chem. Progr.*, **15**, 111 (1954).
2. (a) J. McKenna, J. M. McKenna, and J. White, *J. Chem. Soc.*, **1965**, 1733; (b) Y. Kawazoe and M. Tsuda, *Chem. Pharm. Bull. Japan*, **15**, 1405 (1967); (c) A. T. Bottini and R. L. VanEtten, *J. Org. Chem.*, **30**, 575 (1965); (d) cf. G. L. Closs, *J. Amer. Chem. Soc.*, **81**, 5456 (1959).
3. For reviews see G. Fodor, *Experientia*, **11**, 129 (1955); *Tetrahedron*, **1**, 87 (1957); G. Fodor, K. Koczka, and J. Lestyán, *J. Chem. Soc.*, **1956**, 1411.
4. J-L. Imbach, A. R. Katritzky, and R. A. Kolinski, *J. Chem. Soc., B*, **1966**, 556.
5. J. McKenna, International Symposium on Conformational Analysis, Brussels, September, 1969.
6. cf. A. T. Bottini and M. K. O'Rell, *Tetrahedron Letters*, **1967**, 423.
7. See E. L. Eliel, N. L. Allinger, S. J. Angyal, and G. A. Morrison, *Conformational Analysis*, Wiley, New York, 1965, pp. 47–50.
8. J. McKenna and J. M. McKenna, *J. Chem. Soc., B*, **1969**, 644.
9. H. Booth, *Chem. Commun.*, **1968**, 802.
10. cf. M. J. T. Robinson, Plenary Lecture to International Symposium on Conformational Analysis, Brussels, September 1969.
11. J. McKenna, J. M. McKenna, A. Tulley, and J. White, *J. Chem. Soc.*, **1965**, 1711.

12. D. R. Brown, R. Lygo, J. McKenna, and J. M. McKenna, *J. Chem. Soc., B*, **1967**, 1184.
13. S. P. Findlay, *J. Amer. Chem. Soc.*, **75**, 3204 (1953).
14. J. McKenna, J. M. McKenna, and A. Tulley, *J. Chem. Soc.*, **1965**, 5439.
15. B. G. Hutley, J. McKenna, and J. M. Stuart, unpublished work.
16. R. Lygo, J. McKenna, and I. O. Sutherland, *Chem. Commun.*, **1965**, 356.
17. H. O. House and B. A. Tefertiller, *J. Org. Chem.*, **31**, 1068 (1966).
18. C. H. McGillavry and G. Fodor, *J. Chem. Soc.*, **1964**, 597.
19. C. C. Thut and A. T. Bottini, *J. Amer. Chem. Soc.*, **90**, 4752 (1968).
20. (a) G. Fodor, J. D. Medina, and N. Mandava, *Chem. Commun.*, **1968**, 581; (b) G. Fodor, International Symposium on Conformational Analysis, Brussels, September, 1969; (c) G. Fodor, personal communication, November 1969.
21. D. R. Brown, J. McKenna, and J. M. McKenna, *J. Chem. Soc., B*, **1969**, 567.
22. A. T. Bottini, B. F. Dowden, and R. L. VanEtten, *J. Amer. Chem. Soc.*, **87**, 3250 (1965).
23. H. O. House, B. A. Tefertiller, and C. G. Pitt, *J. Org. Chem.*, **31**, 1073 (1966).
24. G. S. Hammond, *J. Amer. Chem. Soc.*, **77**, 334 (1955).
25. C. G. Swain and E. R. Thornton, *J. Amer. Chem. Soc.*, **84**, 817 (1962).
26. E. R. Thornton, *J. Amer. Chem. Soc.*, **89**, 2915 (1967).
27. D. R. Brown, J. McKenna, and J. M. McKenna, *Chem. Commun.*, **1969**, 186.
28. D. R. Brown, B. G. Hutley, J. McKenna, and J. M. McKenna, *Chem. Commun.*, **1966**, 719.
29. A. F. Casy, A. H. Beckett, and M. A. Iorio, *Tetrahedron*, **22**, 2751 (1966).
30. R. Brettle, D. R. Brown, J. McKenna, and J. M. McKenna, *Chem. Commun.*, **1969**, 696.
31. J. K. Becconsall and R. A. Y. Jones, *Tetrahedron Letters*, **1962**, 1103; J. K. Becconsall, R. A. Y. Jones, and J. McKenna, *J. Chem. Soc.*, **1965**, 1726.
32. A. T. Bottini and M. K. O'Rell, *Tetrahedron Letters*, **1967**, 429.
33. H. O. House and C. G. Pitt, *J. Org. Chem.*, **31**, 1062 (1966).
34. D. R. Brown, J. McKenna, and J. M. McKenna, *J. Chem. Soc., B*, **1967**, 1195.
35. K. Zeile and W. Schulz, *Chem. Ber.*, **88**, 1078 (1955); J. Trojánek, H. Komrsová, J. Pospíšek, and Z. J. Čekan, *Collection Czech. Chem. Commun.*, **26**, 2921 (1961).
36. R. Brettle, D. R. Brown, J. McKenna, and R. Mason, *Chem. Commun.*, **1969**, 339.
37. H. Dorn, A. R. Katritzky, and M. R. Nesbit, *J. Chem. Soc., B*, **1967**, 501.
38. D. R. Brown and J. McKenna, *J. Chem. Soc., B*, **1969**, 570.
39. G. Bernáth, J. Kóbor, K. Koczka, L. Radics, and M. Kajtar, *Tetrahedron Letters*, **1968**, 225; *Acta Chim. Acad. Sci. Hung.*, **55**, 331 (1968), cf. *Chem. Abstr.*, **69**, 59076g (1968).
40. Cf. E. L. Eliel, N. L. Allinger, S. J. Angyal, and G. A. Morrison, *Conformational Analysis*, Wiley, New York, 1965, pp. 267–268.
41. J. McKenna, B. G. Hutley, and J. White, *J. Chem. Soc.*, **1965**, 1729.
42. B. G. Hutley, J. McKenna, and J. M. Stuart, *J. Chem. Soc., B*, **1967**, 1199; T. James, J. McKenna, and J. M. Stuart, unpublished work.
43. N. Mandava and G. Fodor, *Can. J. Chem.*, **46**, 2761 (1968).
44. T. M. Moynehan, K. Schofield, R. A. Y. Jones, and A. R. Katritzky, *J. Chem. Soc.*, **1962**, 2637; K. Schofield and R. J. Wells, *Chem. Ind. (London)*, **1963**, 572; C. D. Johnson, R. A. Y. Jones, A. R. Katritzky, C. R. Palmer, K. Schofield, and R. J. Wells, *J. Chem. Soc.*, **1965**, 6797.
45. M. J. Cook, H. Dorn, and A. R. Katritzky, *J. Chem. Soc., B*, **1968**, 1467.

AUTHOR INDEX

Numbers in parentheses are reference numbers and show that an author's work is referred to although his name is not mentioned in the text. Numbers in *italics* indicate the pages on which the full references appear.

Abe, K., 263(107), *274*
Abraham, R. J., 170(2), 173(2), 174(2), *200*, 207(7d), 240(51), 260(7d, 104, 105), *271–273*
Ackerman, Th., 136(267), *164*
Adler, A. J., 115(209), *163*
Agosta, W., 36(22), *66*
Akimoto, H., 50(58, 59), 54(59), 58(58), *67*
Alger, T. D., 233(38), *272*
Allen, G., 268(118), *274*
Allen, R. G., 200(99), *203*
Allinger, N. L., *28, 30,* 65–67, 188(56), *202,* 207(3), *271,* 280(7), 290(7), 301(40), *307, 308*
Ambrose, E. J., 98(138), *161*
Amdur, I., 146(293, 294), *165*
Anderson, J. E., 207(6), 250(78), 251(78), *271, 273*
Andrae, J. H., 219(17), *271*
Andreeva, N. S., 110(184), *162*
Angyal, S. J., 207(3), *271,* 280(7), 290(7), 301(40), *307, 308*
Anthonissen, I. H., 8(31b), *28*
Appelbaum, J., 11(43), *28*
Applegate, K., 134(264, 265), *164*
Applequist, J., 112(191), 135(268), 154(191), *162, 164*
Arakawa, S., 90(107, 108), 154(107), *160*
Araki, T., 186(46), *201*
Arbuzov, B. A., 244(57), *272*
Aritomi, J., 90(105), *160*
Arnott, S., 71(9, 10), *158*
Asadourian, A., 125(243), *164*
Asai, M., 119(227), *163*
Astbury, W. T., 116(211, 212), 119(217), *163*
Auer, H. E., 112(194, 195), *162*
Avram, M., 26(76), *29*

Bach, R. D., 38(32), 59, *66*
Badger, G. M., 49(56, 57), 54(56), *67*

Bailey, J., 246(64), 247(64), 248(67), *272*
Bair, H. E., 136(269, 274), *164*
Balasubramanian, D., 75(39), 89(39, 88), 97(135), *158, 160, 161*
Ballard, D. G. H., 90(95), *160*
Bamford, C. H., 71(4, 15), 90(95), 118(215), 151(215), *158, 160, 163*
Barfield, M., 188(55), *201*
Barnes, D. G., 73(29), 76(29, 4), *158*
Barnes, E. E., 73(25), *158*
Barnett, E., 33(4), *65*
Barsukov, L. I., 2(13), *27*
Bartell, L. S., 172(11a, 11b), 173(11), 174(11), 181(24), 184(24), *200, 201*
Basch, H., 73(28), 76(28), *158*
Bayley, K., 119(217), *163*
Bazhulin, P. A., 200(95), *203*
Beaudet, R. A., 191(68, 70–72), 200(70), *202*
Becconsall, J. K., 293(31), 294(31), *308*
Beckett, A. H., 291(29), 299(29), *308*
Bein, K., 53(71), 57, *68*
Bellamy, L. J., 173(16), 183(31), 184(39), *201*
Benedetti, E., 112(190, 200, 201), *162, 163*
Bentley, J. B., 254(85), *273*
Bergelson, L. D., 2(11–13), 20(12), *27*
Berger, A., 89(90), 105(155), 106(165, 168), 147(309a), 153(315), *160–162, 165, 166*
Bergmann, E. D., 11(43), *28*
Bernath, G., 299(39), *308*
Berson, A. J., 34
Berson, J., 34(11), *66*
Berthier, G., 74(32), *158*
Bestmann, H., 45(45), *67*
Bestmann, H. J., 10(40), 16(56), 22(70), *28, 29*
Beychok, S., 85(79), 88(79), 111(188), 115(79), *159, 162*
Bijvoet, J. M., 58(83), *68*
Binsch, G., 207(7a), 245(7a), 260(7a), *271*
Biquard, P., 245(60), *272*
Birum, G. H., 3(16), *28*
Bissing, D. E., 5(25), 7(32), *28*

309

Black, D. K., *66*
Blade-Font, A., 4, 5(21), *28*
Blake, C. C. F., 116(213), *163*
Bloom, S. M., 153(317), *166*
Blout, E. R., 82(51, 54, 57), 83(63, 64), 87(86), 90(63, 64, 97), 98(139), 106(161, 164, 167, 170, 172), 114(206), 115(206), 123(239), 124(241), 125(243), 153(316, 317), *159–164, 166*
Boardman, F., 81(47), 90(99), 154(318), *159, 160, 166*
Boden, N., 192(78), 193(78), *202*
Bodenheimer, E., 83(67), 154(67), *159*
Bohak, Z., 119(221), *163*
Bohlmann, F., 2(5), *27*
Bolt, R. H., 208(10), 209(10), *271*
Bommel, A. J. v., 58(83), *68*
Bonavent, G., 26(85), *30*
Bonham, R. A., 200(94), *202*
Boni, R., 136(270), *164*
Booth, H., 28(9), *307*
Borden, W. T., 56(75), *68*
Boskin, M. J., 4(18), *28*
Bothner-By, A. A., 187(49–53), 188(49–53), 189(52), 199(53), *201*
Bottini, A. T., 278(2c), 280(6), 286(19), 288(6, 22), 289(6), 290(6, 19), 291(6), 293(32), 299(19), 307(22), *307, 308*
Bovey, F. A., 75(38), 89(38), 93(112, 113), 97(112), 106(174, 175), 107(176), *158, 160, 162*
Boyd, D. R., 44(40), *67*
Bradbury, E. M., 72(16), 90(92–94, 103, 104), 95(118), 97(118), 119(218, 220, 225), *158, 160, 163*
Bradbury, J. H., 97(130, 131), 124(241), *161, 164*
Bradley, D. F., 90(98), 147(298–301), *160, 165*
Bragg, J., 125(254), 127–129, 135(254), *164*
Brahms, J., 90(91), *160*
Brant, D. A., 141(277), 144(277), 145(277), 146(277, 295), 147(277, 296, 297), 148(295), 152(295), *165*
Brause, A. R., 83(69, 70), 90(69, 70), 154(69), *159*
Brettle, R., 29(30), 297(36), *308*
Brevdo, V. I., 260(103), *273*
Brewster, J. H., 38(31), 41(35a), 43(37), 51(35a), 55(31), 56(31, 35a, 37), 59, *66, 67*

Brier, P. N., 268(118), *274*
Brooks, T. W., 2(10), 19(10), *27*
Brown, D. R., 284(12), 285(12), 287(12), 288(21), 290(12, 27, 28), 291(12, 30), 293(12), 294(28, 34), 295(28, 34), 296(12), 297(12, 36), 299(38), *308*
Brown, F. B., 266(111), 268(120), *274*
Brown, H. C., 243(56), *272*
Brown, L., 71(4, 7), 72(16), 90(93), 118(215), 119(218), 151(215), *158, 160, 163*
Brown, T. L., 185(45), 186, *201*
Buchanan, G. W., 173(15), *200*
Buchwald, M., 83(66), *159*
Bunnenberg, E., 53(67, 68), *68*
Burgada, R., 4(20), *28*
Butcher, S. S., 172(10), 174(10), *200*
Butenandt, A., 2(8), *27*
Butler, C. B., 2(10), 19(10), *27*
Butler, J. N., 200(98), *203*
Butsugan, Y., 11(46), *29*

Cahn, R. S., 31–33(1), 32(2), 33(2), 43(1), 48(1), 51(2), 59(1), 62(2), *65*
Caldin, E. F., 133(260), 134(260), *164*
Canceill, J., 26(81), *30*
Caraculacu, A., 235(42), 258(42), 261(42), *272*
Carlson, G. L., 260(88), 262(99), 270(132), *273, 274*
Carnevale, E. A., 242(54), *272*
Carpenter, B. G., 90(103, 104), *160*
Carroll, B. L., 172(11a), 174(11a), *200*
Carter, O. L., 48(51), *67*
Carver, J. P., 82(57), 106(161, 167), 114(206), 115(206), *159, 162, 163*
Caserio, M. C., 38(30), 46(30), 59, *66*
Cassim, J. Y., 83(62), *159*
Castellano, S., 187(52, 53), 188(52, 53), 189(52), 199(53), *201*
Casy, A. F., 291(29), 299(29), *308*
Cavalli, L., 260(104), *273*
Cekan, Z. J., 296(35), *308*
Chandrasekharan, R., 108(181), *162*
Chen, J. H., 232(28), 238(28), *272*
Cheney, B. V., 192(77), *202*
Chignell, D. A., 84(72), *159*
Choi, N. S., 141(286), *165*
Chopard, P. A., 5(26), *28*
Christmann, K. F., 9(39), 10(39, 53), 14(51, 53), 17(53, 58–60), 19(43, 61), 22(71, 72), 26(91), *28–30*

AUTHOR INDEX

Clague, A. D. H., 266(111), 268(120), *274*
Clark, A. E., 232(30), *272*
Closs, G. L., 26(78, 79), *29, 30,* 190(63), *202*
Cochran, W., 71(5), 151(5), *158,* 278(2d), *307*
Codding, E. G., 190(61), *202*
Conti, F., 95(120), *160*
Cook, B. R., 182(27–29), *201*
Cook, M. J., 307(45), *308*
Cooper, M. A., 240(51), *272*
Cope, A. C., 44(39), 57(39), *67*
Corey, E. J., 20(62), *29,* 56(75), *68*
Corey, R. B., 71(1, 2), 116(214), 141, 142(1), 151(2, 214, 313), *157, 163, 166*
Cosani, A., 97(136), 112(192, 196, 197), *161–163*
Costain, C. C., 180(19), *201*
Cowan, P. M., 106(158), 107, 108(183), 151(158), *161, 162*
Cowburn, D. A., 84(71), 87(71), *159*
Coyle, J. J., 26(78), *29*
Crabbé, P., 33(6, 8), 53(6), *66,* 85(81), 88(81), *159*
Craig, N. C., 200(97), *203*
Crane-Robinson, C., 95(118), 97(118), *160*
Creutzberg, J. E. G., 180(20), *201*
Crick, F. H. C., 71(5), 108(182), 15(5, 312), *158, 162, 166*
Criegee, R., 26(76), *29*
Crombie, L., 8(37), 21(65, 66), *28, 29,* 37(25), *66*
Crook, K. R., 232(32), 240(52), 268(121), *272, 274*
Crowder, G. A., 182(27–29), 185(41), 190(66), *201, 202,* 262(98), 269(123), *273, 274*
Cunlife, A., 270(131), *274*
Curl, R. F., Jr., 249(68), *272*
Curtin, D. Y., 278(1), 301(1), *307*

Daasch, L. W., 182(30), *201*
Danti, A., 266(111), 268(120), *274*
Dauben, W. G., 191(67), *202*
Davidson, B., 119(231, 235), *163, 164*
Davis, G. W., 112(190), *162*
Deber, C. M., 83(68–70), 90(68–70), 158(69), *159*
Dehmlow, E. V., 26(77), *29*
Del Pra, A., 110(187), 112(196), *162*
De Maeyer, L., 208(8), 226(8), *271*
De Mare, G. R., 189(60), *202*
Denney, D. B., 4(18), *28*

Deno, N. C., 196(85), *202*
Denzel, T., 16(56), *29*
DeSantis, P., 145(290), 148(290), 154(290, 322), 156(290, 322), *165, 166*
Dickerson, R. E., 71(6), *158*
Dickinson, S., 119(217), *163*
DiMarzio, E. A., 125(249, 250), *164*
Dinulescu, I. G., 26(76), *29*
Diorio, A. F., 106(171), *162*
Djerassi, C., 53(65–68, 70), *67, 68*
Donohue, J., 151(310), *165*
Dorn, H., 298(37), 307(45), *308*
Doty, P., 76(42), 78(45, 46), 82(50), 85(78), 88(78), 112(149, 195), 115(78), 119(234), 124(241), 125(242, 244), 136(271), *158, 159, 162, 164*
Dover, S. D., 71(10), *158*
Dowden, B. F., 288(22), 307(22), *308*
Downie, A. R., 72(16), 90(92–94), 106(166), 119(218, 225), *158, 160, 162, 163*
Drehfahl, G., 8(31a), 21(67), *28, 29*
Drewer, R. J., 49(56), 54(56), *67*
Dubois, J. E., 26(83), *30*
Dubois, M., 26(83), *30*

Ebersole, S. J., 187(53), 188(53), 199(53), *201*
Eccleston, G., 249(75, 76), 255(86), 270(129), *273, 274*
Edsall, J. T., 140(276), *165*
Eichler, S., 6, 7(28), *28*
Eigen, M., 133(259), 164, 208(8, 9), 226(8), 237(43), *272*
Eiter, K., 2(9), 8(36), *27, 28*
Eliel, E. L., 26(89), *28, 30,* 32(3), 33(3), *65–67,* 207(3), 250(77), 251(77), *271, 273,* 280(7), 290(7), 301(40), *307, 308*
Elliott, A., 71(4, 8, 13, 15), 72(16), 90(92–95), 98(138), 118(215), 119(218, 220, 223, 225), 123(240), 151(215), *158, 160, 161, 163, 164*
Emsley, J. W., 192(78), 193(78), *202*
Engel, J., 137(275), *165*
Entemann, A. E., 200(97), *203*
Epand, R. F., 147(302), 154(302), *165*
Esipova, N. G., 110(184), *162*
Evans, R. J. D., 36(18, 21), 46(21), *66*
Everard, K. B., 254(85), *273*
Eyring, H., 211(14), *271*

Fabbri, S., 33(7), *66*
Falcetta, J., 75(40), 89(40), *158*
Falk, H., 48(50), 59(85), *67, 68*
Falxa, M. L., 90(100), 93(100), 112(201), *160, 163*

Fasman, G. D., 83(64–67), 84(73), 89(87), 90(64), 106(169, 170, 172), 111(188), 112(193), 115(209), 116(73), 119(231, 233, 235), 121(237, 238), 153(317), 154(67), 156, *159, 160, 162–164, 166*
Fateley, W. G., 260(88), 262(99), 266(112), 270(132), *273, 274*
Feeney, J., 192(78), 193(78), *202*
Feinleib, S., 75(38), 89(38), *158*
Felix, A. M., 83(68–70), 90(68–70, 96, 98, 101), 154(69), *159, 160*
Fenn, M. D., 97(130, 131), *161*
Fenoglio, D. J., 26(90), *30,* 170(3, 4), 173(3, 4, 7), 174(3, 4, 7), 176(3, 4, 7), 177(3), 178(4), 179(3, 4), 180(22), 181(22), 182(4), 183(7), 186(4, 7), 187(4), *200, 201*
Ferretti, J. A., 95(119, 123, 124), 96(123, 124), 97, *160, 161*
Fessenden, R. W., 197(87), 198(87), *202*
Fetizon, M., 16(53a), *29*
Filipovich, G., 93(113), *160*
Findlay, S. P., 284(13), *308*
Fitts, D., 58
Fitts, D. D., 52(64), 54(64), 58(81), *67, 68*
Fitzpatrick, J. D., 26(76), *29*
Flitsch, W., 6(27), *28*
Flory, P. J., 135(266), 140(276), 141(277, 266), 144(277), 145(277), 146(277, 295), 147(277, 296, 297), 148(295), 149, 152(295), *164, 165*
Flygare, W. H., 190(64), 199(90), *202*
Fodor, G., 278(3), 281(3), 284(3), 286(18, 20a–c), 293(20a–c), 294(20a), 297(18, 20b), 293(3, 20a–c), 305(43),*307, 308*
Folt, V. L., 235(41), 258(41), *272*
Forchiassin, M., 26(86), *30*
Ford, J. A., 11(42), *28*
Fraisse, R., 26(85), *30*
Frank, G. A., 7, 8(31), *28*
Fraser, R. D. B., 72(16, 17), 119(218, 223, 224, 226), *158, 163*
Fruit, R. E., Jr., 196(85), *202*
Fujii, T., 104(142), 115(142), 123(142),*161*
Fukushima, K., 101(141), 104(142, 145), 115(142), 123(142), *161*

Gabard, J., 26(81), *30*
Gallagher, M. J., *28*
Garbers, C. F., 21(64), *29*
Gates, P. N., 235(40), 236(40), 258(40), *272*
Gebbie, H. A., 267(115), *274*

Geiduschek, E. P., 125(248), *164*
Gerlach, H., 40(34), 47(47), 50(34), 59, *66, 67*
Giacometti, G., 136(270), *164*
Gianni, M. H., 37(23), *66*
Gibbs, J. H., 125(249, 250), *164*
Giglio, E., 145(290), 146(290), 148(290), 154(290), 156(290), *165*
Glass, M. A. W., 35(16), 53(65, 66), 63(16), *66, 67*
Glick, R. E., 97(132, 133), *161*
Go, Y., 119(227), *163*
Goldman, H., 90(103), 95(118), 97(118), *160*
Goodman, M., 75(40), 81(47–49), 83(68–70), 89(40), 90(68–70, 96, 98–102), 93(100, 114), 105(154), 112(190, 198–201), 141(286), 154(69, 318, 319, 321), 156(321, 323), *158–163, 165, 166*
Gorbunoff, M. J., 119(232), *164*
Gornick, F., 106(171), *162*
Gottfried, N., 21(69), *29*
Gough, T. E., 181(23), *201*
Grant, D. M., 188(55), 192(77), *201, 202*
Grasemann, P. A., 35(14), *66*
Gratzer, W. B., 74(31), 77(31), 78(46), 79(31), 84(71, 72), 87(83, 71), 88(83), 106(169), 114(31), *158–160, 162*
Greenbaum, M. A., 34(11), *66*
Greenfield, N. J., 89(87), *160*
Grenville-Wells, H. J., 151(311), *166*
Grimshaw, D., 255(86), *273*
Grinter, R., 54(73), *68*
Groot, M. S. de, 239(45), 253–255(83), *272, 273*
Grosjean, M., 85(80), 88(80), *159*
Gross, J., 106(167), *162*
Guillory, J. P., 172(11a, 11b), 173(11), 174(11), 181(24), 184(24), *200, 201*
Gulati, A. S., 4(20), *28*
Günther, H., 187(50–53), 188(50–53), 189(52, 59), 199(53), *201, 202*
Gutowsky, H. S., 233(38), *272*
Guzhavina, L. M., 244(57), *272*
Gwinn, W. D., 267(113), *274*

Hagishita, S., 37(24), *66*
Hallam, H. E., 258(87), *273*
Haller, G., 48(51), *67*
Halpern, A., 73(24), 78(24), 79(24), *158*
Ham, J. S., 73(21), *158*
Hamblin, P. C., 239(48, 49), 249(72–76),250(73), 251(73, 74, 76), 253(74), *272, 273*

Hammond, G. S., 288(24), *308*
Hamori, E., 134(262), *164*
Hanack, M., 207(4), *271*
Hanby, W. E., 71(4, 15), 72(16), 90(92–94), 118(215), 119(218, 220, 225), 151(215), *158, 160, 163*
Hanlon, S., 97(125–128), *161*
Harkness, A. L., 146(293), *165*
Harrap, B. S., 119(224), *163*
Harrington, W. F., 87(85), 106(162, 163, 165, 173), *160, 162*
Harris, R. K., 252(81), *273*
Hartley, S. B., 6, 7(29), *28*
Hartman, K. O., 260(88), 262(99), *273*
Hashimoto, M., 90(105–108), 154(107), *160*
Hauser, C. F., 2(10), 19(10), *27*
Hayakawa, T., 119(227), *163*
Hayashi, H., 234(39), 261(95), 262(96), *272, 273*
Haylock, J. C., 95(121), *161*
Heasell, E. L., 242(53), 243, 244(53, 58), 245(58), *272*
Hecker, E., 2(8), *27*
Heinz, G., 14(52), 26(80), *29, 30*
Heitkamp, N. D., 266(111), 268(120), *274*
Heitman, H., 8(31b), *28*
Hemesley, P., 8(37), 21(65, 66), *28, 29*
Hendrickson, J. B., 146(291), *165,* 252(80), 260(80), *273*
Herbst, P., 2(5), *27*
Herschbach, D. R., 145(288), *165,* 187(47), *201*
Herzfeld, K. F., 208(11), 209(11), *271*
Hester, R. E., 269(125), *274*
Heublein, G., 16(55), *29*
Hill, R. K., 26(73), *29,* 51(61), 56(61), 57(61), 59, *67*
Hill, T. L., 125(252), *164*
Hippel, P. von, 106(163), *162*
Hirota, E., 187(48), 188(48), 199(91), *201, 202*
Hirsch, J. A., 158(56), *202*
Hirst, R. C., 188(55), *201*
Hoffman, R., 173(13), *200*
Hoffmann, H., 10(41), *28*
Holmes, W. S., 6, 7(29), *28*
Holtzer, A. M., 124(241), *164*
Holzwarth, G., 78(46), 85(78), 88(78), 115(78), *159*
Honig, B., 147(299), *165*
Hood, F. P., 106(174, 175), 107(176), *162*
Hopp, M., 2(8), *27*

Hopps, H. B., 35(16), 63(16), *66*
Horeau, A., 47(48), 48(48), *67*
Horner, L., 2(3), 3(15), 4(22, 23), 10(3, 41), 13(49), *27–29*
Horrocks, W. DeW., Jr., 59(84), *68*
House, H. O., 3(14), 5(14), 7, 8(31), *27, 28,* 286(17), 288(23), 289(23), 293(33), 298(17, 23), 299(17), *308*
Hoving, R., 115(209), *163*
Hsi, N., 171(6), 173–177(6), 180(22), 181(22), 193–195(83), 199(6), 200(83), *200–202*
Hudson, R. F., 5(26), *28*
Hueter, J. F., 208(10), 209(10), *271*
Huisman, H. O., 8(31b), *28*
Hunt, H. D., 73(22), *158*
Hutley, B. G., 285(15), 286(15), 290(28) 291(15), 293(15), 294(28), 295(28), 302(41, 42), 303(42), 304(42), *308*

Ichishima, I., 182(26), 183(37), 184(40), 186(37, 40, 46), *201*
Iitaka, Y., 50(58), 58(58), *67,* 119(219), *163*
Iizuka, E., 119(230), *163*
Ikeda, S., 121(238), *164*
Imahori, K., 78(43), 119(219), *158, 163*
Imbach, J-L., 278(4), 290(4), 293(4), *307*
Inamasu, S., 44(41), *67*
Ingold, C., 31–33(1), 32(2), 33(2), 43(1), 48(1), 51(2), 59(1), 62(2), *65*
Ingwall, R. T., 89(90), 147(309a), 153(315), *160, 165, 166*
Inhoffen, E., 2(5), *27*
Inouye, Y., 44(41), *67*
Iorio, M. A., 291(29), 299(29), *308*
Isemura, T., 108(177), *162*
Iso, K., 136(271), *164*
Itoh, K., 104(143, 150), *161*
Iwakura, Y., 104(150), *161*
Iwasaki, M., 263(109), *274*

Jacques, J., 26(81), *30*
Jacques, J. K., 6, 7(29), *28*
Jacquier, R., 26(85), *30*
Jambotkar, D., 8(34), *28*
James, T., 302–304(42), *202*
Jefferies, P. R., 49(57), *67*
Jenkins, I. D., *28*
Jenkins, P. A., 37(25), *66*
Johns, S. R., 34(11), *66*
Johnson, A. W., 2(1), 7(33), 8(35), *27, 28*
Johnson, C. D., 305(44), *308*

Johnson, C. K., 71(14), *158*
Jonathan, N., 269(128), *274*
Jones, E. R. H., 36(20a), *66*
Jones, G., 11(47), *29*
Jones, M. E., 8(38), 10(38), *28*
Jones, R. A. Y., 293(31), 294(31), 305(44), *308*
Jones, R. N., 180(21), 183(32), *201*
Jones, R. S., 43(37), 56(37), *67*
Jones, V. K., 7, 8(31), *28*
Jones, W. M., 38(26, 28, 29), 39, 47(29), *66*
Josephs, R., 87(85), *160*

Kagarise, R. E., 182(30), *201,* 263(110), *274*
Kajtar, M., 299(39), *308*
Kaneko, H., 13(48), *29*
Karabatsos, G. J., 26(90), *30,* 170(3, 4), 171(6), 173(4, 6, 7), 174(3, 4, 6, 7), 175(6), 176(3, 4, 6, 7), 177(3, 6), 178(4, 17), 179(3, 4), 180(22), 181(22), 182(4), 183(7), 186(4, 7), 187(4), 188(54), 192(76), 193(54, 76, 78-82), 194(54, 76, 80-83), 195(54, 76, 79-83), 199(6), 200(82, 83), *200-202*
Karasz, F. E., 136(269, 272-274), *164*
Karlson, R. H., 83(63, 64), 90(63, 64), *159*
Karplus, M., 169(1), *200*
Karpovich, J., 249(70), 250(70), *273*
Kartha, G., 108(179), *162*
Katchalski, E., 89(90), 105(155), 106(165, 168), 119(221), 147(309a), 153(315), *160-163*
Katritzky, A. R., 278(4), 290(4), 293(4), 298(37), 305(44), 307(45), *307, 308*
Kawazoe, Y., 278(2b), 284(2b), 301(2b), 305(2b), *307*
Kemp, C. M., 57(77, 78), *68*
Kendrew, J. C., 140(276), *165*
Ketcham, R., 8(34), *28*
Kilb, R. W., 172(8), *200*
Kilpatrick, J. E., 191(74), *202*
Kimber, R. W. L., 49(57), *67*
Kincaid, J. K., 211(14), *271*
Kirchoff, C., 209(12), *271*
Kirkwood, J., 58(81), *68*
Kistner, J. F., 56(35), *67*
Kittel, C., 211(15), *271*
Klaboe, D., 200(96), *203*
Klahre, G., 10(41), *28*
Klinger, H. B., 190(63), *202*
Klink, W., 13(49), *29*

Klotz, I. M., 97(125, 126, 128, 129), *161*
Knoeber, M. C., 250(77), 251(77), *273*
Kóbor, J., 299(39), *308*
Koch, W., 2(8), *27*
Kochi, J. K., 197(88), *202*
Koczka, K., 278(3), 281(3), 284(3), 299(3, 39), *307, 308*
Koenig, D. F., 116(213), *163*
Kolinski, R. A., 278(4), 290(4), 293(4), *307*
Komrsová, H., 296(35), *308*
Kossoy, A., 112(199, 201), *163*
Koster, D. F., 266(111), 268(120), *274*
Kratzer, O., 10(40), *28*
Krebs, K., 232(31), 243(31), 244(31), *272*
Krimm, S., 73(19), 113(202-204), 114(205), 115(202), *158, 163,* 235(41), 258(41), *272*
Krisher, L. C., 183(33, 35, 36), 185(44), 187(44, 47), *201*
Krow, G., 51(61), 56(61), 57(61), *67*
Krubiner, A. M., 21(68, 69), *29*
Krueger, P. J., 263(108b), *274*
Krumel, K. L., 193-195(81), *202*
Krusic, P. J., 197(88), *202*
Kubota, S., 119(222), *163*
Kuchitsu, K., 183(38), *201*
Kuebler, N. A., 73(28), 76(28), *158*
Kuhn, W., 53(71), 57, *68*
Kumosinski, T. F., 119(235), *164*
Kuratani, K., 182(26), 183(37), 186(37), *201*
Kurmeier, H. A., 26(77), *29*
Kurtz, J., 105(155), 106(168, 173), *161, 162*
Kwiatkowski, G. T., 20(62), *29*
Kyllingstad, V. L., 8(35), *28*

LaCount, R. B., 7(33), *28*
Lamb, J., 219(17), 224(18), 229(22), 232(31), 239(45), 242(53), 243(31, 53), 244(31, 53), 249(71), 250(71), 253-255(83), *271-273*
Lamberton, J. A., 34(11), *66*
Lande, S. S., 26(90), *30,* 171(7), 173(7), 174(7), 176(7), 183(7), 192-195(76), *200, 202*
Landor, P. D., 26(77), *29*
Landor, S. R., 26(77), *29,* 36(17, 18, 21), 46(21, 46a, 46b, 46c), *66, 67*
Landsberg, M., 83(66), *159*
Lane, G., 268(118), *274*
Laurie, V. W., 191(73), *202*
Lazarus, A. K., 34(12), 35(12), 52(12), 54(12, 55), 59(55), *66, 67*

AUTHOR INDEX

Leach, S. J., 141(280), 142(280), 150(314), 154(280), 156(280), *165, 166*
Lee, P. L., 182(25), *201*
Leemann, H. G., 33(7), *66*
Legare, R., 134(261), *164*
Legrand, M., 85(80, 82), 88(80, 82), 115(208), *159, 160, 163*
Leighly, E. M., 243(55), *272*
Lestyán, J., 278(3), 281(3), 284(3), 299(3), *307*
Lewis, G. E., 49(56), 54(56), *67*
Lifson, S., 125(255), 127–129(255, 256), 147(299), *164, 165*
Lignowski, J. S., 93(115), *160*
Lin, C. C., 172(8), 181, 182, *200*, 267(114), *274*
Lin, W. S., 181(23), *201*
Lincoln, D. N., 196(85), *202*
Lindblow, C., 83(67), 154(67), *159*
Linn, W. S., 38(30), 46(30), *66*
Linsley, S. G., 123(239), *164*
Lippincott, E. R., 147(303, 304), *165*
Liquori, A. M., 95(120),140(276), 141, 145(290), 146(290), 148(290), 149, 153, 154(290, 322), 156(290, 322),*160, 165, 166*
Listowsky, I., 81(47, 48), 90(99), *159, 160*
Litman, B. J., 73(30), 75(30), 89(30), *158*
Litovitz, T. A., 208(11), 209(11), 212(16), 232(30), 242(54), 246(65), 247(65), *271, 272*
Liu, J. S., 196(85), *202*
Liu, K. J., 93(115), *160*
Loder, J. D., 36(20a), *66*
Longworth, J. W., 75(38), 89(38), *158*
Lorenz, D., 8(31a), *28*
Lotan, N., 89(90), 147(309a), 153(315), *160, 165, 166*
Low, B. W., 151(311), *166*
Lowe, G., 55(74), *68*
Lowe, J. P., 207(7b), 260(7b), *271*
Lowe, J. T., 199(90), *202*
Lowry, T. M., 57(76), *68*
Loze, C. de, 72(18), 153(317), *158, 161*
Lumry, R., 134(261), *164*
Lundberg, R. D., 125(244), *164*
Lüttke, W., 189(58), 190(62), *202*
Luzzati, V., 72(18), *158*
Lygo, R., 284(12), 285(12, 16), 280(12, 16), 287(12), 290(12), 291(12), 293(12, 16), 296(12), 297(12), *308*
Lyle, G. G., 41(36), *67*

Maateescu, G., 26(76), *29*
McAlpine, R. D., 200(98), *203*
McClure, J. D., 16(54), *29*
McCoubrey, J. C., 6, 7(29), *28*
McCoy, L. L., 26(84), *30*
McCoy, W. H., 200(94), *202*
McDiarmid, R., 78(44, 45), *159*
McDonald, T. R. R., 90(93), 119(218),*160, 163*
McEwen, W. E., 4(19, 21), 5(21), *28*
McGavin, S., 106(158), 107, 108(183), 151(158), *161, 162*
McGillavry, C. H., 286(18), 297(18), *308*
McGinn, F. A., 35(13), 49(13), 54(55), 59(55), *66, 67*
Machleidt, H., 11(45), *28*
McKenna, J., 278(2a), 279(5), 281(5, 8, 11), 284(12), 285(2a, 12, 14–16), 286(12, 14–16), 287(12), 288(21), 290(12, 27, 28), 291(2a, 12, 14, 15, 30), 293(12, 15, 16, 31), 294(28, 31, 34), 295(28, 34), 296(11, 12), 297(11, 12, 14, 36), 299(38), 302(41, 42), 303(42), 304(2a, 42), 307(11),*307,308*
McKenna, J. M., 278(2a), 281(8, 11), 284(12), 285(2a, 14), 286(12, 14), 287(12), 288(21), 290(12, 27, 28), 291(2a, 12, 14, 30), 293(12), 294(28, 34), 295(28, 34), 296(11, 12),297(11, 12, 14), 304(2a), 307(11), *307, 308*
McKervey, M. A., 44(40), *67*
McPhail, A. T., 48(51), *67*
MacRae, T. P., 72(17), 119(223, 224, 226), *158, 163*
McSkimin, H. J., 226(19), *271*
Maercker, A., 2(2), *27*
Mahr, T. G., 112(191), 154(191), *162*
Mair, G. A., 116(213), *163*
Maisey, R. F., 11(47), *29*
Malcolm, B. R., 71(8), 123(240), *158, 164*
Malkus, H., 196(86), *202*
Mammi, M., 112(196), *162*
Mandava, N., 286(20a), 293(20a), 294(20a), 299(20a), 305(43), *308*
Mandelkern, L., 97(132, 133), 106(171),*161, 162*
Marica, E., 26(76), *29*
Mark, H., 118(216), *163*
Mark, J. E., 114(205), 154(321), 156(321, 323), *163, 166*
Marks, V., 4(20), *28*
Marlborough, D. I., 93(116, 117), 94(116), *160*
Marsden, R. J. B., 254(85), *273*
Marsh, R. E., 151(313), *166*
Martin, J. G., 26(73), *29*
Martin, J. S., 189(60), *202*

Martin, W., 26(75), *29*
Martinelli, L., 8(34), *28*
Mason, E. A., 146(294), *165*
Mason, R., 297(36), *308*
Mason, S. F., 38(27), 46(27), 54, (73), 55(27), 57(77, 78), *66, 68*
Masuda, Y., 81(47), 93(114), 101(141), 104(142, 144, 145), 105(153, 154), 115(142), 123(142), *159–161*
Matsuura, T., 11(46), *29*
Matthews, C. N., 3(16), *28*
Mechoulam, R., 4(17), *28*
Mechtler, H., 48(49), *67*
Medina, J. D., 286(20a), 293(20a), 294(20a), 299(20a), *308*
Medyantseva, E. A., 33(9), *66*
Mehta, A. S., 44(39), 57(39), *67*
Meijere, A. de, 189(58), 190(62), *202*
Melikhova, L. P., 261(91), 269(126), *273, 274*
Melillo, J. T., *67*
Merwe, J. P. van der, 21(64), *29*
Messick, J., 251(79), *273*
Mettee, H. D., 263(108b), *274*
Meyer, K. H., 116(210), 118(216), *163*
Miles, M. L., 2(10), 19(10), *27*
Miller, B. J., 46(46a, 46b), *67*
Miller, F. A., 266(112), *274*
Miller, M. A., 188(56), *202*
Miller, W., 134(261), 146(295), 148(295), 152(295), *164, 165*
Millionova, M. I., 110(184), *162*
Minkin, V. I., 33(9), *66*
Mislow, K., 33(5, 10), 34(12), 35(12–16), 43(38), 44(5, 43), 49(13, 14, 53), 52(12, 62, 64), 53(15, 65–68, 70), 54(12, 55, 64, 72), 57(38), 59(55), 63(16), *65–68*
Mitsui, Y., 105(152), *161*
Miyake, A., 184(40), *201*
Miyazawa, T., 98(139), 101(141), 104(142, 144–148, 151), 105(152–154), 115(142), 123(142), *161*, 182(26), 183(37), 184(40), 186(37, 40, 46), *201*, 263(107), *274*
Mizushima, S., 142(287), *165*, 182(26), 183(37), 184(40), 186(37, 40, 46), *201*, 207(1), 234(39), 240(1), 257(1), 258(1), 260(1), 263(107), *271, 272, 274*
Moffitt, W., 83(59–61), 84, *159*
Mohrman, D. W., 243(55), *272*
Mole, M. F., 6, 7(29), *28*
Moloney, M. J., 183(36), *201*
Momany, F. A., 200(94), *202*
Montagner, C., 97(137), *161*

Mooney, E. F., 235(40), 236(40), 258(40), *272*
Moore, W. R., 38(32), 59, *66*
Morino, Y., 183(38), *201, 260*
Morita, K., 87(86), *160*
Morozov, E. V., 190(65), *202*
Morrison, G. A., 207(3), *271*, 280(7), 290(7), 301(40), *307, 308*
Moscowitz, A., 43(38), 53(68), 57(38), 58(82), *67, 68*, 83(61), 84(74, 75), *159*
Moss, R., 26(78), *29*
Mousseron, M., 26(85), *30*
Moynehan, T. M., 305(44), *308*
Müller, G., 9, 10(39), 14(51), *28, 29*
Murai, M., 11(46), *29*
Murakami, M., 262(97), *273*
Murata, K., 261(95), 262(96, 97), *273*
Musso, N., 53(69), *68*
Myer, Y. P., 115(207), *163*

Naar-Colin, C., 187(49, 50), 188(49, 50), *201*
Naddaka, V. I., 33(9), *66*
Nagai, K., 125(253), *164*
Nagakura, S., 73(23), *158*
Nagase, T., 26(77), *29*
Nagashima, N., 105(152), *161*
Nakagawa, I., 182(26), 183(37), 184(40), 186(37, 40, 46), *201*
Nakagawa, M., 37(24), *66*
Nakahara, T., 104(143, 150), *161*
Nakamura, K., 90(107), 154(107), *160*, 234(39), *272*
Nash, H. A., 147(300, 301), *165*
Nelson, R., 180(18), *201*
Nemethy, G., 140(276), 141(279, 280), 142(279, 280), 150(280), 154(280), 156(280), *165*
Nenitzescu, C. D., 26(76), *29*
Nerdel, F., 7(30), *28*
Nesbit, M. R., 298(37), *308*
Neumann, E., 136(267), *164*
Newman, P., 34(12), 35(12), 52(12), 54(12), *66*
Newmark, R. A., 233(36, 37), *272*
Nielsen, E. B., 75(34, 35), 89(34, 35), *158*
Nielsen, J. R., 200(96), *203*
Noack, K., 180(21), *201*
Noguchi, J., 119(222, 227), *163*
Norland, K. S., 83(64), 90(64), *159*
North, A. C. T., 108(183), 116(213), *162, 163*
North, A. M., 246(64), 247(64), 248(67), *272*
Northam, F., 185(41), 190(66), *201, 202*

AUTHOR INDEX

O'Brien, R. E., 53(65, 66), *67*
Oedinger, H., 8(36), *28*
Ohno, M., 44(41), *67*
Okabayashi, H., 108(177), *162*
Okazaki, M., 13(48), *29*
Olah, G. A., 196(85), *202*
Oliver, J., 190(66), *202*
Oliveto, E. P., 21(68, 69), *29*
Ooi, T., 112(189), 141(189), 147(189, 302), 153(189), 154(189, 302), 156(189), *162, 165*
O'Reilly, J. M., 136(269, 272–274), *164*
O'Rell, M. K., 280(6), 288–291(6), 293(32), *307, 308*
Orrell, K. G., 93(116), 94(116), *160*
Orville-Thomas, W. J., 230(23), 232(29), 238(29, 44), 239(44, 47), 241(29), 244(47), 245(59), 261(92–94), 263(108a), *271–274*
Osborne, C. E., 193(79), 195(79), *202*
Oshima, T., 261(95), 262(96, 97), *273*
Osipova, L. P., 200(95), *203*
Owen, N. L., 269(125), *274*
Oya, M., 104(143, 150), *161*

Pachler, K. G. R., 260(104, 105), *273*
Padmanabhan, R. A., 232(26), 238(26), 244(58), 245(58), *272*
Palmer, C. R., 305(44), *308*
Pandit, U. K., 8(31b), *28*
Panse, P., 6, 7(28), *28*
Paolillo, L., 95(124), 96(124), 97, *161*
Park, P. J. D., 231(24), 232(24, 32), 233(24, 33), 238(24), 259(24), 262(100, 101), 268(119, 121), *271–274*
Parker, R. C., 134(264, 265), *164*
Parks, A. T., 181(24), 184(24), *201*
Parry, K., 207(7d), 260(7d), *271*
Pattenden, G., 8(37), 20(63), 21(65, 66), *28, 29*
Pauling, L., 71(1, 2), 116(214), 141, 142(1), 151(2, 214, 313), *157, 163, 166*
Peerdeman, A. F., 58(83), *68*
Peggion, E., 97(136), 112(192, 196–198), *161–163*
Peller, L., 125(247), *164*
Pelosi, E. T., 41(36), *67*
Pentin, Y. A., 190(65), *202*
Pentin, Yu. A., 261(89, 91), 269(126, 127), *273, 274*
Perkin, W. H., 40(33), 50(60), *66, 67*
Perlmutter, H. D., 33(10), *66*

Perutz, M. F., 71(3), *157*
Peters, H., 6(27), *28*
Peterson, D. L., 73(26), *158*
Pethrick, R. A., 207(7c), 226(20), 233(34, 35), 239(35, 48), 249(72, 74, 76), 251(74, 76), 253(74, 84), 254(20, 84), 255(84), 260(7c), 268(119), *271–274*
Petrauskas, A. A., 232(28), 238(28), *272*
Pettit, R., 26(76), *29*
Phillips, D. C., 116(213), *163*
Pierce, L., 180(18), 183(33), *201*
Piercy, J. E., 228(21), 232(25, 27), 239(21, 27, 46), 246(62, 63), 247(62, 63), 249(21), 250(21), 251(21, 82), 270(25), *271–273*
Pignolet, L. H., 59(84), *68*
Pilot, J. F., 3(16), *28*
Pinkerton, J. M. M., 247(66), *272*
Piskala, A., 13, 14(50), 26(91), *29, 30*
Pitt, C. G., 288(23), 293(33), 298(23), *308*
Pittman, C. U., Jr., 196(85), *202*
Pitzer, K. S., 146(292), *165*, 191(67, 74), *202*, 267(113), *274*
Platt, J. R., 73(21), *158*
Poland, D. C., 147(305–309), 154(305), *165*
Pommer, H., 2(7), 21(7), *27*
Ponsold, K., 21(67), *29*
Pope, W. J., 40(33), 50(60), *66, 67*
Pople, J. A., 170(2), 173(2), 174(2), 199(89), *200, 202*
Pospíšek, J., 296(35), *308*
Potter, J., 115(209), 119(233), 121(237), *163, 164*
Pozdyshev, V. A., 269(127), *274*
Pracejus, H., 44(42), *67*
Prelog, V., 31–33(1), 32(2), 43(1), 44(43, 44), 48(1), 51(2), 59(1), 62(2), *65, 67*
Prichard, F. E., 245(59), *272*
Prince, F. R., Jr., 90(102), *160*
Privett, J. E., 41(35a), 51(35a), 56(35), *67*
Pullman, B., 74(32), *158*
Punja, N., 46(46c), *67*
Pysh, E., 119(229), *163*

Quadrifoglio, F., 89(89), 97(134), 120(236), *160, 161, 164*

Radcliffe, K., 189(57), 199(57), *202*, 268(122), 269(124), *274*
Radics, L., 299(39), *308*
Ramachandran, G. N., 73(19, 20), 108(179–181), 140(276), 141(20, 281–283), 142(282, 283), 148, 151(281), 154(281–283), 156(281–283), *158, 162, 165*

Ramakrishnan, C., 73(20),141(281−283), 142(282, 283), 151(281), 154(281−283), 156(281−283), *158, 165*
Ramey, K. C., 251(79), *273*
Ramirez, F., 3(16), 4(20), *28*
Randall, A., 106(166), *162*
Rasmusson, G. R., 3(14), 5(14), *27*
Rassing, J., 270(130), *274*
Rattle, H. W. E., 95(118), 97(118), *160*
Ray, T. C., 258(87), *273*
Raymond, M. A., 2(10), 19(10), *27*
Records, R., 53(67), *68,* 90(98), *160*
Regan, J. P., 36(21), 46(21, 46a), *66, 67*
Rerick, M. N., 26(89), *30*
Rhodes, W., 73(29), 76(29, 41), 106(169), *158, 162*
Ricci, J. E., 54(55), 59(55), *67*
Rice, S. A., 125(248, 251), *164*
Rich, A., 108(182), 151(312), *162, 166*
Riddell, F. G., 250(78), 251(78), *273*
Ripamonti, A., 145(290), 146(290), 148(290), 154(290), 156(290), *165*
Risaliti, A., 26(86), *30*
Riveros, J. M., 249(69), *273*
Roberts, D. E., 106(171), *162*
Robin, M. B., 73(28), 76(28), *158*
Robinson, M. J. T., 250(78), 251(78), *273,* 281(10), *307*
Rogowski, R. S., 195(84), *202*
Rogulenkova, V. N., 110(184), *162*
Roig, A., 125(255), 127−129(255), *164*
Romers, C., 180(20), *201*
Rometsch, R., 53(71), *68*
Rosen, I., 8(49), *159*
Rosenheck, K., 76(42), 119(228), *158, 163*
Roux, A., 26(82), *30*
Rüchardt, C., 6, 7(28), *28*
Russell, G. A., 196(86), *202*
Russo, S. F., 97(125, 128), *161*
Ruterjans, H., 136(267), *164*
Rutkin, P., 34(12), 35(12), 52(12), 53(65, 66), 54(12), *66, 67*
Rydon, H. N., 93(116, 117), 94(116), 95(121, 122), *160, 161*

Saegebarth, E., 170(5), 184(5), *200*
Sage, A. J., 112(193), *162*
Sakakibara, S., 108(177), *162*
Saludjian, P., 72(18), *158*
Sarachman, T. N., 191(75), 199(92), *202*
Sarkar, P. K., 119(234), *164*

Sarma, V. R., 116(213), *163*
Sasisekharan, V., 106(159, 160), 108(180), 141(281, 282), 142(282), 151(159, 160, 281), 154(281, 282), 156(281, 282), *162, 165*
Sauer, J., 26(74), *29*
Scatturin, A., 110(187), *162*
Schaufele, R. F., 267(116), *274*
Schechter, E., 82(54, 57), 106(164),114(206), 115(206), *159, 162, 163*
Schellman, C., 87(84), 145(84), *160*
Schellman, J. A., 73(30), 75(30, 34, 35), 84(77), 87(84), 89(30, 34, 35), 125(245, 246), 145(84), 150, *158−160, 164*
Scheraga, H. A., 89(90), 91(110, 111), 112(189), 134(262), 140(276), 141(189, 278, 279, 280), 142(279, 280), 142(279, 280), 145(278, 289), 146(289), 147(189, 278, 302, 306−309, 309a), 148(278), 149, 150(280, 314), 152, 153(189), 154(189, 280, 302, 305), 156(189, 280), *160, 162, 164−166*
Scherrer, H., 44(43), *67*
Schick, H., 21(67), *29*
Schlögl, K., 48(49−52), 49, 59, *67*
Schlosser, M., 9(39), 10(39, 53), 13(50), 14(50−53), 17(53, 58−60), 19(53, 61), 22(71, 72), 26(80, 91), *28−30*
Schmitt, E. E., 154(319), *166*
Schmueli, V., 106(156, 157), *161*
Schneider, B., 235(42), 258(42), 261(42),*272*
Schneider, D. F., 21(64), *29*
Schneider, W. P., 26(87), *30*
Schnitt, G., 8(31a), *28*
Schofield, K., 305(44), *308*
Schöllkopf, U., 2(4), 8(4), *27*
Schröder, G., 26(74), *29*
Schroeder, R., 147(303, 304), *165*
Schroeter, S. H., 26(89), *30*
Schuler, R. H., 197(87), 198(87), *202*
Schulz, W., 296(35), *308*
Schupp, O. E., 11(42), *28*
Schwartz, G., 83(69), 90(69, 96), 132(257, 258), 133(258), 134(257, 263), 154(69),*159, 160, 164*
Schweitzer, E. E., 16(57), *29*
Schwendeman, R. N., 172(12), 173(14), 174(12), 175(14), 182(25), 184(12), 190(61),195(84), *200−202*
Scoffone, E., 110(187), 112(192), *162*
Scott, D. W., 262(98), *273*
Scott, R. A., 112(189), 141(278), 145(278, 289), 146(289), 147(189, 278, 302), 148(278),

AUTHOR INDEX

150(314), 153(189), 154(189, 302), 156(189), *162, 165, 166*
Searle, R. J. G., 5(26), *28*
Searles, S., 243(55), *272*
Sederholm, C. H., 233(36, 37), *272*
Seelig, J., 132(257), 134(257), *164*
Segal, D. M., 87(85), 108(178), 110(178), *160, 162*
Sela, M., 106(162, 165), *162*
Seshagiri, Rao, M. G., 232(27), 239(27), *272*
Seus, E. J., 11(42), *28*
Seyden-Penne, J., 26(82), *30*
Shahak, I., 11(43), *28*
Shemyakin, M. M., 2(11–13), 20(12), *27*
Sheppard, N., 200(93), *202,* 207(2), 240(2), 257(2), 258(2), 260(2), 261(2, 90), 264(2), *271, 273*
Sherwood, J., 249(71), 250(71), *273*
Shibnev, V. A., 110(184), *162*
Shido, N., 182(26), *201*
Shimanouchi, T., 104(143, 150), 142(287), *161, 165,* 182(26), 183(37), 184(40), 186(37, 40, 46), *201,* 234(39), 263(107), 267(116), *272, 274*
Shingu, K., 37(24), *66*
Shiori, T., 50(58), 58(58), *67*
Shipman, J. J., 235(41), 258(41), *272*
Shiro, Y., 261(95), 262(96, 97), *273*
Siegel, M., 52(62, 64), 54(55, 64), 59(55, *67*
Siegel, S., 191(69), *202*
Sim, G. A., 48(51), *67*
Simon, W., 35(16), 63(16), *66*
Simons, E. R., 87(86), *160*
Simpson, W. T., 73(22, 24–26), 78(24), 79(24), *158*
Sinnott, K. M., 183(34), *201*
Siomis, A. A., 34(11), *66*
Skell, P. S., 200(99), *203*
Sliam, E., 26(76), *29*
Slie, W. M., 246(65), 247(65), *272*
Slocum, D. W., 54(72), *68*
Slutsky, L. J., 134(264, 265), *164*
Smith, C. P., 3(16), 4(20), *28*
Sommer, B., 119(228), *163*
Sondheimer, F., 4(17), *28*
Sonnichsen, G. C., 178(17), 180(22), 181(22), *201*
Spach, G., 90(91), *160*
Speziale, A. J., 5(25), *28*

Spinner, E., 183(32), *201*
Spragg, R. A., 252(81), *273*
Staab, H. A., 26(77), *29*
Stake, M. A., 97(128, 129), *161*
Stapleton, I. W., 72(17), 119(223), *158, 163*
Steckelberg, W., 53(69), *68*
Steigman, J., 97(136, 137), *161*
Steinberg, D., 53(65166), *67*
Steinberg, I. Z., 106(165), *162*
Stephens, R. M., 90(104), *160*
Stevens, L., 121(237), *164*
Stewart, F. H. C., 119(224, 226), *163*
Stewart, W. E., 97(132, 133), *161*
Stokr, J., 235(42), 258(42), 261(42), *272*
Stothers, J. B., 173(15), *200*
Strasorier, L., 97(137), *161*
Street, A., 116(211), *163*
Stuart, J. M., 285(15), 286(15), 291(15), 293(15), 302–304(42), *308*
Suard, M., see Suard-Sender, M.
Suard-Sender, M., 74(32, 33), *158*
Suares, H., 34(11), *66*
Subrahmanyam, S. V., 239(46), 246(62, 63), 247(62, 63), 251(82), *272, 273*
Sugai, S., 119(222), *163*
Sugano, S., 104(145), *161*
Sugita, T., 44(41), *67*
Sugiura, M., 183(38), *201*
Susi, H., 104(149), *161*
Sutcliffe, L. H., 192(78), 193(78), *202*
Sutherland, I. O., 285(16), 286(16), 293(16), *308*
Sutton, L. E., 254(85), *273*
Suzuki, E., 119(223, 224, 226), *163*
Swain, C. G., 288(25), 289(25), *308*
Swalen, J. D., 180(19), *201,* 267(114), *274*
Szasz, G. J., 263(106), *274*

Tabor, W. J., 185(42), *201*
Tabuchi, D., 246(61), *272*
Taller, R. A., 188(54), 193–195(54, 80, 82), 200(82), *201, 202*
Tamburro, A. M., 110(187), *162*
Tamm, K., 237(43), *272*
Tamres, M., 243(55), *272*
Tanaka, J., 78(43), *158*
Tatchell, A. R., 46(46a, 46b), *67*
Tatevskii, V. M., 261(89), 269(127), *273, 274*
Taylor, E. W., 83(62), *159*
Taylor, R. P., 59(84), *68*

Taylor-Smith, R., 36(17), *66*
Tefertiller, B. A., 286(17), 288(23), 289(23), 298(17, 23), 299(17), *308*
Terbojevich, M., 112(197), *163*
Thomas, L. C., 183(31), *201*
Thomas, T. H., 238(44), 239(44, 47), 244(47), 261(92), *272, 273*
Thompson, D. S., 233(36), *272*
Thompson, J. G., 16(57), *29*
Thornton, E. R., 288(25, 26), 289(25), *308*
Thut, C. C., 286(19), 290(19), 299(19), *308*
Tiers, G. V. D., 93(113), *160*
Tiffany, M. L., 113(202–204), 115(202), *163*
Timasheff, S. N., 119(232, 235), 121(237), *164*
Tinoco, I., 58(79), *68*, 73(24), 78(24), 79(24), 84(76), 147(298), *158, 159, 165*
Tömösközi, I., 4(24), *28*, 45(45), *67*
Toniolo, C., 75(40), 89(40), 90(100), 93(100), *158, 160*
Tooney, N., 119(231), *163*
Townend, R., 119(235), 121(237), *164*
Traub, W., 71(12), 106(156, 157), 108(178), 110(178, 185, 186), *158, 161, 162*
Trippett, S., 8(38), 10(38), *28*
Trojanek, J., 296(35), *308*
Trotter, I. F., 71(4, 7), 118(215), 151(215), *158, 163*
Trouard, P. A., 58(80), *68*
Truesdell, C., 210(13), *271*
Truscheit, E., 2(9), *27*
Tsuboi, M., 99(140), 105(152), 119(219), 124(140), *161, 163*
Tsuchiya, S., 234(39), *272*
Tsuda, M., 278(2b), 284(2b), 301(2b), 305(2b), *307*
Tulley, A., 281(11), 285(14), 286(14), 291(14), 296(11), 297(11, 14), 307(11), *307, 308*
Turner, J. A., 58(80), *68*
Turner, J. J., 261(90), *273*
Turner, J. O., 196(85), *202*
Turolla, A., 136(270), *164*
Tutweiler, F. B., 38(26), *66*
Twiss, R. Q., 267(115), *274*
Tyminski, I. J., 188(56), *202*

Ullman, R., 93(115), *160*
Ulrich, T. A., 16(57), *29*
Ul'yanova, O. D., 261(91), 269(126), *273, 274*

Unland, M. L., 190(64), *202*
Uno, K., 104(150), *161*
Urnes, P., 82(50), *159*
Urry, D. W., 75(37), 89(37, 89), 91(109), 97(134), 120(236), *158, 160, 161, 164*

Valentin, E., 26(86), *30*
Vand, V., 71(5), 151(5), *158*
Vanderkooi, G., 91(110), 112(189), 141(189), 147(189, 302), 153(189), 154(189, 302), 156(189), *160, 162, 165*
Van der Werf, C. A., 4, 5(21), *28*
Vane, G. W., 38(27), 46(27), 55(27), *66*
Van Etten, R. L., 278(2c), 288(22), 307(22), *307, 308*
Vaver, V. A., 2(13), *27*
Velluz, L., 85(80, 82), 88(80, 82), *159, 160*
Venkatachalam, C. M., 73(19), *158*
Verdier, P. H., 172(9), *200*
Verdini, A. S., 97(137), 112(192, 196, 197), *161–163*
Viennet, R., 115(208), *163*
Vill, J. J., 4(18), *28*
Vogel, E., 26(75), *29*
Vold, R. L., 233(38), *272*
Volkenshtein, M. V., 207(5), 260(103), 261(5), *271, 273*
Volltrauer, H. N., 172(12), 173(14), 174(12), 175(14), 184(12), *200*
Vournakis, J. N., 91(111), *160*

Wada, A., 105(152), 125(248, 251), 141(285), 154(320), *161, 164–166*, 260(102), *273*
Wadsworth, D. H., 11(42), *28*
Wahl, G. H., Jr., 35(16), 63(16), *66*
Wailes, P. C., 2(6), *27*
Walborsky, H. M., 44(41), *67*
Walbrick, J. M., 38(29), 39, 47(29), *66*
Walker, S., 248(67), *272*
Wallach, O., 40(33), *66*
Waters, W. L., 38(30), 46(30), *66*
Watts, L., 26(76), *29*
Weedon, B. C. L., 20(63), *29*
Weigang, O. E., 58(80), *68*
Weinlich, J., 26(76), *29*
Weiss, J., 53(66), *67*
Weiss, V., 190(64), 199(90), *202*
Wellman, K., 53(67), *68*
Wells, M., 115(209), *163*
Wells, R. J., 305(44), *308*
Wendisch, D., 189(59), *202*
Wessels, P. L., 260(105), *273*

AUTHOR INDEX

Wessendorf, R., 11(45), *28*
Westen, H., *59*
Wetlaufer, D. B., 75(39), 89(39, 88), *158, 160*
Weyerstahl, P., 7(30), *28*
White, J., 278(2a), 281(11), 285(2a), 291(2a), 296(11, 12), 297(11, 12), 302(41), 304(2a), 307(11), *307, 308*
White, J. W., 267(117), *274*
White, R. F. M., 239(48, 49), 249(72–76), 250(73), 251(73, 74, 76), 253(74), *272, 273*
Whiting, M. C., 36(20a), *66*
Williams, E. J., 239(47), 244(47), *272*
Williams, R. L., 173(16), 183(31), 184(39), *201*
Willis, H. A., 235(40), 236(40), 258(40), *272*
Wilson, E. B., Jr., 170(5), 172(8–10), 173(10), 174(10), 184(5), 185(43, 44), 187(44), 191(68), *200–202,* 249(69), *273*
Wilson, J. W., Jr., 38(26, 28), 39, *66*
Winkler, H., 4(22), *28*
Wippel, H., 10(41), *28*
Witkowski, R. E., 260(88), 262(99), 266(112), *273, 274*
Wittig, G., 2(4), 8(4), 26(76), *27, 29,* 49(54), *67*
Wolf, A. P., 4(19), *28*
Wolf, H., 75(36), 89(36), *158*
Wonacott, A. J., 71(9), *158*
Wood, J. L., 189(57), 199(57), *202,* 268(122), 269(124), *274*
Woods, H. J., 116(212), *163*

Woodward, A. J., 269(128), *274*
Woody, R. W., 58(79), *68,* 147(298), *165*
Woolford, R. G., 181(23), *201*
Wu, S., 173(15), *200*
Wyn-Jones, E., 207(7c), 226(20), 230(23), 231(24), 232(24, 29, 32), 233(24, 33–35), 238(24, 29, 44), 239(35, 44, 47–49), 240(52), 241(29), 244(47), 249(72–76), 250(73), 251(73, 74–76), 253(74, 84), 255(84, 86), 259(24), 260(7c), 261(92–94), 262(100, 101), 263(108a), 268(119, 121), 270(129, 130), *271–274*

Yahara, I., 119(219), *163*
Yakel, H. L., 71(11), *158*
Yamada, S., 50(58, 59), 54(59), 58(58), *67*
Yamaoka, K., 82(53), *159*
Yan, J. F., 91(110, 111), *160*
Yang, J. T., 82(52, 55, 56, 58), 83(60), 84(52), 87(52), 115(52), 119(230), 125(242), *159, 163, 164*
Yasumi, M., 263(107), *274*
Yomosa, S., 73(27), *158*
Yonath, A., 108(178), 110(178, 185, 186), *162*
Yphantis, D. A., 154(319), *166*

Zeile, K., 296(35), *308*
Zhdanov, Yu. A., 33(9), *66*
Zimm, B. H., 125(254), 127–129, 135(254), 136(271), *164*
Zimmerman, H., 49(54), *67*

SUBJECT INDEX

Absolute configuration, determination of,
    by physical methods, 52
    by theoretical methods, 53
    of dissymmetric molecules, 31
    X-ray determination of, 58, 59
Absorption coefficient, of sound, 209, 226
Acetaldehyde, 172, 199
Acetaldoxime, 200
    O-methyl ether of, 200
Acetic acid, 199
Acetone, 180
Acetophenone, 19
Acetophenones, substituted, 183
Acetylenic alcohols, chiral, 36
Acetyl halides, 183
Acoustic enthalpies, 240
Acoustic frequency of, 214
Acoustic spectrum, 208
Acrolein, 173, 254, 255
Acrylonitrile, 16
Acrylyl chloride, 269
Acrylyl fluoride, 269
Activation energy, 230
Activation enthalpy, 230
Acyl halides, conformations of, 183
Acylurea formation, stereochemical model
    for, 48
Adamantanes, chiral, 33
Adiabatic compressibility, 220
L-Alanylglycyl anhydride, 89
Aldehydes, aliphatic, conformational populations of, 173
    Wittig reaction with triphenylphosphonium alkylids, 15
N-Alkylaldimines, 193
Alkylation of β-oxido phosphorus ylids, 23
Alkylidenecycloalkanes, absolute configuration of, 40, 50, 56
    by stereoselective decarboxylative debromination, 50
    chiral, 32
    sequence rule method, application to, 62

Allenes, absolute configuration of, 36, 45, 55
    using ORD, 37
    asymmetric synthesis of, 45
    chiral, 32
    circular dichroism of, 55
    configuration of, by rearrangement of vinyl propargyl ethers, 36
    by stereospecific rearrangements, 36
    by S_Ni' rearrangement, 36
    semihydrogenation, stereoselective, 37
Allenic carboxylic acids, chiral, 37
Aluminum complex, chiral, 46
Amide A band, 98
Amide B band, 98
Amide I band, 98
Amide II band, 98
Amide III band, 99
Amide IV band, 100
Amide V band, 100, 101
    of *meso* polypeptides, 105
    of racemic polypeptides, 104
Amide VI band, 100, 101
Amide VII band, 104
Amides, 73
    electronic spectra of, 74, 75
    dimers, 77
    energy levels of, 74
    molecular orbitals of, 75
    rotational barriers in, 245
Amine oxides, from piperidines, 305
Amines, conformational equilibria in, table of, 244
    primary, ultrasonic relaxation of, 243
    rotational barriers, table of, 244
    secondary, ultrasonic relaxation of, 243
    tertiary, ultrasonic relaxation of, 242
L-3-Aminopyrrolidin-2-one, 89
t-Amyl radical, 197
Anamolous X-ray dispersion, for configurational assignments, 54
Angelic acid ester, 3
Angle deformation, in polypeptides, 141

SUBJECT INDEX

Anisole, 269
Aporphines, 53
Arrhenius rate equation, 230, 234, 257
Asymmetric atrolactic acid synthesis, 44
Asymmetric hydride reduction, empirical rule for, 46
Asymmetric hydroboration, empirical rule for, 47
Asymmetric reduction, allene configurations from, 46
Asymmetric synthesis, of allenes, 45
  conservative, 44
  self-immolative, 33, 44
Atropisomers, circular dichroism curves of, 53
  optical rotatory dispersion of, 53
Attenuation measurements, 226
Axially dissymmetric molecules, absolute configuration of, 31
3-Azabicyclo[3.2.1]octanes, 283, 286, 293
3-Azabicyclo[3.3.1]nonanes, 283, 286, 291, 293
4-Aza-5α-cholestanes, 283, 286, 301
Azimuthal angle, 206, 265
Aziridines, tertiary, 307
Aziridinium salts, 307
Azopolypeptides, 112

$\pi \to \pi^*$ Band, of helical polypeptides, 78
Band splitting, of amide chromophore, 77
Beckman rearrangement, stereospecific, 41
Benzaldehyde, 5, 6, 7, 13, 17, 173
3,4-Benzophenanthrenes, chirality of, 57
Benzoyl chloride, 24
1-Benzylidene-4-methylcyclohexane, 41
1-Benzyl-2-methylpiperidine benziodide, 294
N-Benzyl-N-methylpiperidinium salts, equilibration of, 291
1-Benzyl-4-phenylpiperidine, 297
1-Benzyl-4-phenylpiperidine methobromide, X-ray analysis of, 297
O-Benzyl L-serine, 119
Betaine-lithium salt complex, twist-boat conformation of, 27
Betaines, 14
  decomposition of, reversible, 13
  diastereoisomeric, equilibration of, 17
  diastereomers of, 17
  erythro, 26
  erythro-threo ratios, 25, 26
  formation of, 26
  reversibility of, 5
  threo, formation of, 26
Betaine ylids, 17, 22
Bianthracyls, absolute configuration of, 49, 50
Bianthryls, electronic absorption spectra of, 54
  circular dichroism spectra of, 54
Biaryls, absolute configuration of, 49, 53
  applications of sequence rule to, 62
  chiral, 32
Bimer, 77
  absolute configuration of, by X-ray, 58
  bridged, 35, 44
  chiral, 35, 44
  configuration of, chemical correlation of, 49, 50
Biopolymers, 70
Biphenyls, absolute configuration of, 44, 49
  bridged, 35
  chiral, 34, 35
  configuration of, chemical correlation of, 49, 50
  doubly bridged, 35
Bromine, anomalous dispersion effect of, 58
Bromoacetaldehyde, 174, 176, 178, 179
Bromoacetone, 182
m-Bromoanisole, 269
o-Bromoanisole, 269
p-Bromoanisole, 269
2-Bromo-3-chloropropene, 269
Bromocyclohexane, 251, 270
2-Bromoethyl mercaptan, 262
2-Bromohexane, 238, 261, 264
3-Bromohexane, 238, 261, 264
Bromolactone formation, from allenic acids, stereospecific, 37
2-Bromo-2-methylbutane, 232, 235, 262
2-Bromo-2-methylpropionyl bromide, 184, 185
2-Bromopentane, 238, 261, 264
3-Bromopentane, 238, 261, 264
1-Bromopropane, 200
1-Bromopropene, 200
3-Bromopropene, 189
Buckingham potential, 145
n-Butane, 232, 263
1-Butene, 188
cis-2-Butene, 191
trans-2-Butene, 191
Butylacetaldehyde, 174
t-Butylacetaldehyde, 174, 177
t-Butyl alkyl ketones, 181

SUBJECT INDEX 325

sec-Butyl bromide, 238
sec-Butyl chloride, 238
t-Butyl diethylcarbinyl ketone, 181
sec-Butyl iodide, 238
3-t-Butylpropene, 189
2-Butyl radical, 197
t-Butyl radical, 197
Butyraldehyde, 174

Calorimetric measurements, of helix-coil transitions, 135
Calvet differential microcalorimeter, 136
Camphanic acid chloride, 48
Camphidines, 282, 286, 291, 301, 303, 304
Carbalkoxymethylids, 5
Carbodiimides, absolute configuration of, 48
Carbonium ions, conformations of, 195
Carbonyl compounds, see Aldehydes; Ketones; etc.
Carbonyl olefination, procedures for, 20
  cis-stereoselective, 21
  of steroidal ketones, 11, 21, 22
  see also Wittig reaction
N-Carboxyanhydrides, "random" copolymerization of, 108
Characteristic dimensionless ratio, 147
Chemical methods, correlation of configuration by, 33
Chemical-shift increments, for assignment of configuration of piperidinium quaternary salts, 295
Chemical shifts, 168
Chirality rule, 60, 61
Chiral space-model, 65
Chiral structures, secondary, 65
Chiral symbol, 65
Chloroacetaldehyde, 173, 174, 176, 178, 179
Chloroacetone, 182
α-Chloroacetophenone, 183
m-Chloroanisole, 269
o-Chloroanisole, 255, 269
p-Chloroanisole, 269
m-Chlorobenzaldehyde, 5, 13
1-Chloro-2-bromopropane, 238
1-Chlorobutane, 200
2-Chlorobutane, 261
Chlorocyclohexane, 251, 270
2-Chloroethyl mercaptan, 262
3-Chlorohexane, 237

2-Chloro-2-methylbutane, 232, 235, 262
  ultrasonic absorption data for, 227
4-Chloromethyl-1, 3-dioxolan-2-one, 253
2-Chloropentane, 238, 261, 264
3-Chloropentane, 238, 261, 264, 270
m-Chlorophenylpropene, 13
1-Chloropropane, 199
1-Chloropropene, 191, 200
2-Chloropropene, 190
3-Chloropropene, 189
m-Chlorostyrene, 13
Cinnamaldehyde, 254
Cinnamic acid esters, 7, 8
Circular dichroism (CD), 49, 82, 84
  associated idealized absorption band, 86
  of helical structures, 84
  of polypeptides, 73
  of right-handed helical polypeptides, 88
Cis addition of hydrogen, stereoselective, 40
Citral, 254
Claisen rearrangement, stereospecific, 36, 46
Cobalt(salicyladehyde)$_2$-(+)-2,2$'$-diamino-6,6$'$-dimethylbiphenyl, absolute configuration of, 59
Coefficient of thermal expansion, 220
Coil conformation, 125
Coil nucleation, 132
Coil-promoting solvents, 147
Collagen, 108
  optical properties of, 106
  polyhexapeptide models of, 110
  polytripeptide models of, 110
Collagen II, 144
Compressibility, adiabatic, 220
  isothermal, 220
Configurational energy, of polypeptides, 130
Configurational entropy, of polypeptides, 152
Configuration by direct chemical correlation, 49
Conformation(s), bisecting, 170
  enthalpy differences between, 176, 177
  table of, 241
  of homopolypeptides, 148
  as related to planar and axial chirality, 65
  see also specific compound or structure; for example, Acyl halides, conformations of
Conformational analysis of ketones, 179
Conformational chirality, 61
Conformational entropy change, in polypeptides, 135
Conformational helix, 61
Conformational populations, of aliphatic aldehydes, 173

calculated from coupling constants and chemical shifts, 168
  dependence on dielectric constant, 176
  effect of dipole moments on, 175, 184, 186
  effect of non-bonded interactions on, 182, 184, 193
  enthalpy of, 186
Conformational selection rules, 61
β-Conformation of peptides, optical activity of, 119
β-Conformations, of polypeptides, 76
Conformer stabilities, 175
  of polymer chain, 125
Conservative asymmetric syntheses, 44
Cotton effect, 85
  of isolated absorption band, 86
  of polypeptides, 83
Coupled oscillator model, for configurational assignments, 54
Coupling constants, 168
Cowan–McGavin helix, 106, 107
Cram-Prelog type rule, 45
Cross-β-structure, 118
Crotonaldehyde, 254
Crotonitrile, 200
Curtin-Hammett principle, 278
1-Cyano-2-bromoethane, 261
1-Cyano-2-chloroethane, 261
cis-1-Cyanopropene, 191
3-Cyanopropene, 188
Cyclic peptide molecules, 89
trans-Cycloalkenes, applications of sequence rule to, 64
  chiral, 32
  configuration of, 57, 43
Cyclobutylcarboxaldehyde, 174
Cyclohexane, 251
Cyclohexanecarboxaldehyde, 173, 174
Cyclohexane derivatives, ring inversion in, 249
Cyclooctatetraenes, chiral, 33
trans-Cyclooctene, 38, 44, 64
  configuration of, 43, 57
Cyclopentadiene, 36
Cyclopentanecarboxaldehyde, 174, 176
Cyclopropane-allene conversion, stereospecific, 38
Cyclopropanecarboxaldehyde, 172, 173, 174, 176
Cyclopropanecarboxylic acid, 184

Cyclopropane forming ylid reactions, 4
Cyclopropyl semidiones, 196

trans-Decahydroquinolines, 282, 285, 286, 295, 297, 301, 303, 304, 305
Degradation, of diastereomeric quaternary salts, 302
Deuteration of β-oxido phosphorus ylids, 24
Diacetylenic allenes, naturally occurring, 56
Diastereomeric piperidinium salts, nmr spectra of, 291
2, 7-Diazaspiro[4.4]nonane, correlation with 2-methyl-2-ethylsuccinic acid, 51
Dibromoacetaldehyde, 175, 176, 178
1, 2-Dibromobutane, 238
dl-2, 3-Dibromobutane, 238, 262
meso-2, 3-Dibromobutane, 233, 235, 262
Dibromocarbene, stereospecific addition of, 38
1, 2-Dibromo-1, 1-dichloroethane, 233, 268
1, 2-Dibromo-1, 1-difluoroethane, 233, 268
2, 3-Dibromo-2, 3-dimethylbutane, 233, 264
1, 2-Dibromoethane, 232, 235, 260, 261, 263
1, 1-Dibromo-3-fluorobutadiene, 262
1, 2-Dibromo-2-methylpropane, 232, 235, 241, 264
1, 2-Dibromopentane, 238
1, 4-Dibromopentane, 238
1, 2-Dibromopropane, 238
2, 3-Dibromopropene, 269
1, 2-Dibromo-1, 1, 2, 2-tetrafluoroethane, 232, 241
1, 2-Dibromotetrafluoroethane, 268
1, 2-Dibromotrifluoroethane, 239, 268
Di-t-butylacetaldehyde, 174, 177
Dichloroacetaldehyde, 171, 175, 176, 178
sym-Dichloroacetone, 182
dl-2, 3-Dichlorobutane, 237, 238, 262
meso-2, 3-Dichlorobutane, 233, 235, 262
  ultrasonic absorption data for, 227
2, 2'-Dichloro-6, 6'-dimethyl-4,4'-diaminobiphenyl, 53
2, 3-Dichloro-2, 3-dimethylbutane, 264
6, 6'-Dichloro-2, 2'-diphenic acid, 52
1, 2-Dichloroethane, 230, 232, 235, 257, 258, 260, 261, 263, 265, 268, 270
  internal rotation, 206
  potential energy diagram, 206
1, 1-Dichloro-3-fluorobutadiene, 262
1, 2-Dichloro-2-methylpropane, 232, 235, 241, 264

1, 2-Dichloropropane, 238, 264, 270
2, 3-Dichloropropene, 269
1, 2-Dichlorotetrafluoroethane, 268
Dielectric relaxation, 134
Diethylacetaldehyde, 174, 177
Diethyl adipate, 248
4-N-Diethylaminobutane-2-one, 244
1-Diethylamino-2-chloropropane, 244
3-N-Diethylaminoethylamine, 244
3-N-Diethylaminopropiononitrile, 244, 245
3-N-Diethylaminopropylamine, 244
Diethyl carbethoxymethylphosphonate, 12
N,N-Diethylcyclohexylamine, 244
Diethyl ketone, 180
Diethyl malonate, 248
Diethyl oxalate, 248
Diethyl succinate, 248
5, 5-Diethyltrimethylene sulfite, 251
N, N'-Diferrocenylcarbodiimide, 48
Differential microcalorimetry, 136
sym-Difluoroacetone, 182
1, 2-Difluoroethane, 200
1, 2-Difluoroethylene, 200
1, 2-Difluoro-1, 1, 2, 2-tetrachloroethane, 235, 268
1, 1-Difluoro-1, 2, 2-trichloroethane, 268
1, 2-Dihaloethanes, 263
Dihydrotestosterol, 12
(+)-Diisopinocamphenylborane, 46
1, 2-Dimethoxycyclooctane, 44
3, 3-Dimethoxypropene, 188, 189
α, β-Dimethylacrolein, 254
Dimethyl adipate, 248
1, 3-Dimethylallene, 38
  hydroboration of, 46
3, 3-Dimethylbutanal, 174
2, 3-Dimethylbutane, 232, 241
2, 3-Dimethylcyclopropanecarboxylic acid, 38
Dimethylcyclopropylcarbinyl cation, 196
2, 2-Dimethyl-1, 3-dioxane, 255
  boat-chair equilibrium, 270
6, 6$'$-Dimethyl-2, 2$'$-diphenic acid, 52
6, 6$'$-Dimethyl-2, 2$'$-diphenyldiamine, 53
Dimethyl glutarate, 248
Dimethyl malonate, 248
Dimethyl oxalate, 248
1, 2-Dimethylpiperidine, 301
Dimethyl succinate, 248
1, 2-Dimethylthioethane, 262

4, 6-Dimethyltrimethylene sulfite, 251
5, 5-Dimethyltrimethylene sulfite, 251
6, 6$'$-Dinitro-2, 2$'$-diphenic acid, 52, 53
1, 3-Dioxane, 251, 252
  boat-chair equilibria, 270
  ring inversion in, 250
1, 3-Dioxane-spiro-2-cyclopentane, boat-chair equilibrium, 270
Dipeptide model, spatial description of, 140
Diphenic acids, correlation of, 52
  meta-bridged, chiral, 33
1, 3-Diphenylallene, 38, 46, 55
2, 3-Diphenylcyclopropanecarboxylic acid, 38
Dipole-dipole interactions, in polypeptides, 141, 146
Dipoles, orientation in polypeptide helix, 78
3-N-Dipropylaminopropiononitrile, 244
Disordered polypeptides, 113
cis-3, 3-Ditrifluoromethyl-1, 3-difluoropropene, 192
Drude equation, 82

Electronic spectra, of ordered systems, 81
Electronic transition, $n_O \rightarrow \pi^*$, of amides, 74
Electrostatic interactions, 141, 147
End effect factor, C_r, dependence on ln w (statistical factor), 131
Energy barriers, in ethane derivatives, 231, 232
Energy contour maps, 141
Enthalpy, acoustic, 240
  of helix-coil transition, 135
Enthalpy differences, between conformations in substituted propenes, table of, 188
  from infrared measurements, table of, 261
  interactions affecting, 263
  of phase change, 260
  from sound measurements, 228
  from vibrational spectroscopy, 258
Entropy differences, from sound measurements, 228
  from vibrational spectroscopy, 260
Entropy of activation, 230
Epoxide forming ylid reactions, 4
Equilibration, erythro to threo, 27
Esters, conformational equilibria in, table of, 245, 247, 248
  rotational barriers in, table of, 245, 247, 248
Ethane derivatives, A factors in, 235

energy barriers in, 231, 232
torsional frequencies in, 235
Ethane-1, 2-dithiol, 261
Ethanes, fluorinated, 263
  halogenated, 263
  substituted, rotational barriers in, 230
Ethylacetaldehyde, 174, 177
Ethyl acetate, 247
Ethyl bromoacetate, 186
2-Ethylbutanal, 174
N-Ethylcamphidine methiodide, X-ray analysis of, 297
Ethyl chloroacetate, 186
Ethyl dichloroacetate, 186
Ethyl difluoroacetate, 186
Ethylene carbonate, 253
Ethylene sulfite, 252
Ethyl fluoroacetate, 186
Ethyl formate, 246, 247, 249
γ-Ethyl L-glutamate, copolymers of, CD spectra of, 112, 113
N-Ethylnortropine methobromide, X-ray analysis of, 297
1-Ethyl-4-phenylpiperidine, 290
1-Ethylpiperidines, 4-substituted, table of quaternizations, 289
Ethyl propionate, 247
2-Ethyl-3-propylacrolein, 254
Ethyl radical, 197
Exciton resonance coupling, 77
Exciton splitting, of $\pi \rightarrow \pi^*$ transition, 77, 107
6-Exponential function, 145
Eyring-Jones model of optical activity, 43, 56
Eyring rate equation, 230, 233

Factorization rule, 32, 60
Farnesol, 22
Fibrous structural proteins, conformations of, 116
Fluoroacetone, 182
m-Fluoroanisole, 269
o-Fluoroanisole, 269
p-Fluoroanisole, 269
1-Fluoro-2-chloroethane, 200
2-Fluoroethanol, 263
1-Fluoropropane, 199
cis-1-Fluoropropene, 191
3-Fluoropropene, 187, 188
1-Fluoro-1, 1, 2, 2-tetrachloroethane, 233, 235, 268

Formaldehyde, 199
Formamide, 76
Free radicals, conformations of, 195
Furacrolein, 254
Furan-2-aldehyde, 206, 254, 257, 265
  internal rotation in, potential energy diagram, 207
3(2-Furyl)acrylophenone, 254
2-Furylideneacetone, 254

Geraniol, 22
Glutinic acid, 36
Glycidaldehyde, 173, 177
Glycyl-L-alanine, 142, 143
Glycyl-glycine, 142, 150
Glycyl-L-isoleucine, 143
Glycyl-L-valine, 143

α-Haloacetaldehyde, rotamers of, 26
α-Haloacetates, 185
Haloacetyl halides, 183, 184
2-Haloethanols, 263
Halogenation of β-oxido phosphorus ylids, 24
3-Halopentane, conformations of, 236
1-Halopropenes, microwave spectra of, 192
Hammond postulate, 288
Helical conformation, 125
Helical polypeptides, 70, 79
  orientation of dipoles in, 78
Helical sequences, 133
  number-average length of, 130
Helicenes, application of the helicity rule to, 64
Helicity, criteria of, 157
Helicity rule, 60
  application of, to helicenes, 64
  use of, 61
α-Helix, 71, 72, 138, 149, 150, 153, 154, 155
  left-handed, 150
ω-Helix, 72, 73, 150
3_{10} Helix, 150
3.7_{13} Helix, 71
Helix, left-handed, 81, 154
  right-handed, 83
Helix-breaking solvents, 93
Helix-coil transitions, 70, 124, 126, 134
  application of relaxation methods to, 133
  heat capacity of, 135
  kinetics of, 131
  by nmr, 93, 95
  thermodynamics of, 134
  van't Hoff heat of, 135

SUBJECT INDEX

Helix conductor model, of optical activity, 43, 56
Helix content, by nmr, 95
Helix-helix transitions, 124, 137
Helix nucleation, 132
Hexahelicene, 64
  chiral, 32
  configuration of, 58
Hexanal, 174
n-Hexane, 239
α, n-Hexylcinnamaldehyde, 254
Horeau's method, 48, 49
Hydride reduction, stereochemical model of, 46
Hydride transfer, asymmetric, stereochemical model for, 35
Hydroboration, stereoselective, 46
Hydrogen bonding, intramolecular, 132
  in polypeptides, 141, 147
Hydrogen bonds, fraction of intersegment, 127
Hydrophobic bonding, in random coil, 147
(β-Hydroxy-α, β-diphenylethyl)-diphenylphosphine oxide, 13
4-Hydroxy-4-phenylthiacyclohexane, 307
β-Hydroxyphosphonamides, 20
Hypochromism, of helical polypeptides, 70, 79
  of α-helix, 77

Imino compounds, table of enthalpy differences between rotamers, 194
Insulin, 70
Intensity, of vibrational bands, 259
Interchain distance, of β proteins, 116
Interferometry, Michelson, 267
Internal rotation, mechanism of, 233
Iodoacetone, 182
2, 4'-Iodobutyl-1-methylpiperidine, 305
2-Iodoethanol,
2-Iodo-2-methylbutane, 232, 235, 262
3-Iodopropene, 189
Isoborneol, 40
2-^2H-Isobornyloxymagnesium bromide, asymmetric synthesis, using, 40
Isobutyl bromide, 232, 241, 261, 264
Isobutyl chloride, 232, 241, 261, 264
Isobutyl iodide, 232
Isobutyl radical, 197
Isobutyraldehyde, 172, 173, 174
Isopropylacetaldehyde, 174

Isopropyl formate, 247
Isopropyl mercaptan, 262
Isothermal compressibility, 220

cis-Jasmone, 21
Johnson-Bovey method, 294
pH Jump method, 134

Karplus equation, 169
β-Keratin, 124
  conformation of, 116, 118
Ketones, conformational analysis of, 179
  β, γ-unsaturated, chirality of, 53
  see also Unsaturated ketones
β-Ketophosphonamides, 20
Kinetic resolution, 34, 48
  with α-phenethylamine, 48
Kirkwood method, for assigning absolute configurations, 54, 58
Kronig-Kramer dispersion relations, 87

Lactonization method, for piperidinium salt configurations, 298
Lennard-Jones potential, 146
Lifson and Roig model, 127
Ligancy complementation, 60
Lindlar hydrogenation, 21
Liquids, structure of, 212
Lithium aluminum hydride sugar complex, asymmetric reduction with, 46
α-Lithium benzyldiphenylphosphine oxide, 13
Lithium dimethoxyaluminum hydride, asymmetric reduction with, 46
Lithium halides, effect of on Wittig reaction, 26
Lithium salts, effects of on Wittig reaction, 7, 16
London dispersion force, 146
Lowe's rule, 43, 47, 48, 51, 55, 56
Lysozyme, 116

M, configurational symbol, 61, 65
Magnetic non-equivalence, in quaternary ammonium salts, 294
Marasine, 46, 47
Mean square end-to-end distance, 147
Meerwein-Ponndorf reduction, 35
Menshutkin reaction, 276
Meridional reflection, 71
Mesityl oxide, 255

Metallocenes, centrally chiral, absolute configuration of, 48
Methacrolein, 254, 255
Methallyl chloride, 190
Methallyl iodide, 190
Methoxyacetaldehyde, 174, 176, 179
3-Methoxypropene, 188, 189
N-Methylacetamide, 255
  infrared spectrum of, 98
Methyl acetate, 247
N-Methyl-N-alkylcamphidinium salts, 281
N-Methyl-N-alkylpiperidinium salts, 293
β-Methyl L-aspartate, electronic spectrum of, 82
  oligomer, 81
N-Methyl-N-benzylcamphidinium salts, 295
3-Methylbutanal, 174
2-Methylbutane, 231, 232, 241
3-Methyl-1-butene, 188
3-Methyl-3-t-butyl-1-chloroallene, 36
N-Methylcamphidine, 302
N-Methylchloroacetamide, 186
Methyl m-chlorocinnamate, 4
Methyl chloroformate, 24
α-Methylcinnamaldehyde, 254
Methylcyclohexane, 250, 251
4-Methylcyclohexylideneacetic acid, 40
  trans addition of bromine to, 50
Methylcyclopropyl ketone, 181
4-Methyl-1, 3-dioxane, 250, 251
Methyldiphenylphosphonium benzylid, 8
1-Methyl-2, 6-diphenyl-4-piperidone, 41
2-Methyl-2-ethyl succinic acid, 1
5-Methyl-5-ethyltrimethylene sulfite, 251
Methylferrocene-α-carboxylic acid, 48
Methyl formate, 247, 249
γ-Methyl L-glutamate, electronic spectrum of, 82
  oligomer of, 81
N-Methyl-N-isopropyl salts, steric interactions in, 286
Methylmarasine, 46, 47
Methylmercaptoacetaldehyde, 175, 177, 179
2-Methylpentane, 238
3-Methylpentane, 238
N-Methylphthalimide, 6
2-Methylpiperidines, 282, 285, 286, 288, 297, 301, 303
3-Methylpiperidine, 295
N-Methylpiperidine, 29
N-Methylpropionaldimine, 193

Methyl propionate, 247
2-Methylpyrrolidines, 307
N-Methylquinolizidinium salts, 305
β-Methylstyrene, 13
4-Methyltrimethylene sulfate, 249, 250, 251, 252
4-Methyltrimethylene sulfite, 251
Methyl vinyl ether, 255
Minimum potential energy analysis, 91
Moffitt-Yang equation, 82
Molecular-beam scattering, 146
Molecular chirality, specification of, 59
Molecular orbitals of the amide group, 74
Monopole approximation, 147
Myoglobin, 70
Myristamide, 75

Neopentyl sulfite, 251, 252
N-mer, 78
NMR spectra, of diastereomeric piperidinium salts, 291
  of polypeptides, 94
  see also specific compound or class; for example, Diastereomeric piperidinium salts, nmr spectra of
Nomenclature of Cahn, Ingold, and Prelog, see Sequence rule
Nomenclature problems, of chirality, 65
Non-bonded interactions, effect on conformer populations, 182
Non-2-en-4, 6, 8-triyn-1-ol, 46
Norcamphor, 36
Nortropanes, 303, 304
Nucleation processes, 132
Nucleotides, minimum potential energy calculations, 147

2-Octanol, 34, 35, 45
Olefin formation, stereospecific, 41
Oligomers, 79
Optical displacement rules, 52, 54
Optical rotation, 84
  physical basis of helical structures, 84
Optical rotatory dispersion (ORD), 82, 84
  and idealized absorption band, 86
  of α-helical polypeptides, 87
  of polypeptides, 73
  of β-I-polypeptides, 119
  of β-II-polypeptides, 119
  of poly-L-proline, 106

SUBJECT INDEX

see also specific compound; for example, Polypeptides, optical rotatory dispersion, spectra of
Osmium tetroxide, *cis* stereospecific addition of, 44
Oxaphosphetane, 3, 6, 7
Oxidation, with ruthenium tetroxide, 299
β-Oxido phosphorus ylids, 17, 22
　alkylation of, 23
　deuteration of, 24
　halogenation of, 24
β-Oxido ylids, 21
　addition of electrophilic agents to, 22
Oxime, chiral, absolute configuration of, 41

P, configurational symbol, 61, 63, 65
Parachor, 244
[2.2]-Paracyclophanecarboxylic acid, 48
Paracyclophanes, absolute configuration of, 48
　applications of sequence rule to, 63
　chiral, 32
b_0 Parameter, 83
Parasantonide, 53
Partition function, for polypeptides, 126, 128
Pentanal, 174
n-Pentane, 239
3-Pentanone, 2-substituted, 181
Peptide bonds, *cis* and *trans*, 105
Peptide oligomers, ultraviolet spectra of, 81
Peptides, infrared spectra of, 99
Phananthrenes, chiral, 32
　configuration of, 57
　Cotton effects of, 57
α-Phenethylamine, kinetic resolution with, 48
Phenoxyacetaldehyde, 174, 176, 179
Phenylacetaldehyde, 174, 176
L-Phenylalanine, CD and ORD studies on block and random copolymers of, 112
2-Phenylbutene, 19
2-Phenylbutyric anhydride, 47
Phenylcyclopropane, 190
α-Phenyldihydrothebaine, 34
δ-Phenyldihydrothebaine, 34
4-Phenyl-1, 3-dioxane, 250, 251, 252
Phenylglycidic ester, 5
Phenyllithium, 13
Phenylmercuric bromide, 4
4-Phenylpiperidines, 282, 285, 288, 291, 295, 303

Phosphine oxides, carbanions of, 9
Phosphonate, α, α-disubstituted, 11
Phosphonic acid amides, 20
Phosphonic esters, 9
Phosphorus betaines, conformations of, 25
Phosphorus ylids, 2
　"reactive," 13
　resonance-stabilized, 5
Pilot atom, 63
Piperdine, boat conformations of, 286, 293
　protonations of, 280
　stereoselectivity of quaternizations of, 281
　stereospecificity of oxidation of tertiary, 305
　stereospecificity of quaternizations of, 284
Piperidine quaternizations, *syn*-axial interactions in, 301
　cross-products in, 279
　Curtin-Hammett analysis of, 278
　effective sizes of alkyl groups in, 285, 287
　energy schematics of, 277
　kinetics-based studies, 285
　leaving group variation in, 301
　product ratios of, 278, 284
　reaction coordinate in, 289
　solvent variation in, 30
　table of, 289
　thermodynamics based studies, 291
　with trideuteriomethyl iodide, 286
Δ_3-Piperidines, 299
Piperidinium quaternary salts, configurations of, 292
　degradation of diastereomeric, 305
　diastereomeric from α, ω-dihalides, 288
　infrared spectra of epimeric *N*-methyl-*N*-alkyl, 296
Piperidinium salts, degradation of diastereomeric, 304
　infrared spectra of diastereomeric, 296
　lactones from, 298
　spectroscopic analysis of, 292
　X-ray crystallographic analysis of, 297
Pivaladehyde, 181
Planar dissymmetric molecules, absolute configuration of, 31
Pleated sheet, antiparallel-chain, 117
　parallel-chain, 117
β-Pleated sheet structure, 118
PO-Activated carbonyl olefination, 11
Point-dipole approximation, 294
Polarizability, of substituents, 51, 55, 56

theory, for assigning absolute configurations, 54, 56
Polarized absorption spectra, 79
Polarized light, influence of an optically active sample on, 85
Polarized ultraviolet absorption spectrum, 78
Polar solvent, effect of on the Wittig reaction, 19
Poly-L-acetoxyproline, 107
Poly-O-acetylallothreonine, 119
Poly-O-acetyl L-serine, 119
Poly-O-acetylthreonine, 119
Poly-DL-alanine, infrared spectroscopy, 115
Poly-L-alanine, 89, 96, 98, 99, 152
   CD of, 90
   conformations of, 153
   far infrared spectrum of, 104
   helical conformation of, 152, 153
   infrared spectrum of, 103
   ultraviolet spectrum of, 91
   X-ray diffraction of, 71
β-Poly-L-alanine, 118
Poly-β-alkyl L-aspartates, 95
Poly-γ-alkyl L-glutamates, 95
   infrared spectra of, 101
Poly-α-amino acids, 157
Poly-L-α-amino-n-butyric acid, 104
Poly-L-p-aminophenylalanine, 112
Poly-L-aspartate esters, 91, 92
Poly-L-aspartic acid, 90, 105
   helix sense of, 92
Poly-β-benzyl D-aspartate, N-deuterated, infrared spectra of, 101
   infrared spectra of, 101
Poly-β-benzyl L-aspartate, 72, 73, 83, 119
   infrared spectra of, 100
Poly-S-benzyl L-cysteine, 119
Poly-γ-benzyl L-glutamate, 83, 93, 94, 119, 134, 155
   N-deuterated infrared spectra of, 101
   heats of solution of, 136
   infrared spectra of, 94, 99, 100
   nmr spectra of, 101
   opitcal rotations of, 136
   X-ray diffraction of, 71
Poly-O-benzyl L-serine, 119
Poly-S-benzylthio L-cysteine, 72, 73
Poly-O-t-butoxy L-serine, 119
Poly-S-carbobenzoxy L-cysteine, 119

Poly-S-carboxymethyl L-cysteine, 119
   CD spectrum of, 121
Poly-β-ethyl L-aspartate, 91
Poly-γ-ethyl L-glutamate, 97
   heats of solution of, 137
   optical rotations of, 137
Poly-L-glutamate, N-deuterated, infrared spectra of, 116
   infrared spectra of, 116
   sodium, 115
Poly-α-L-glutamic acid, CD of, 88
Poly-L-glutamic acid, ORD of, 87
Polyglycine, 148, 150
   conformations of, 149
   infrared of, 104
Polyglycine I, 123
Poly-L-histidine, 83
Poly-L-p-(p'-hydroxyphenylazo)phenylalanine, 113
Poly-L-hydroxyproline, 106, 108
   X-ray diffraction of, 108
Poly-L-leucine, 95, 104
Poly-L-lysine, 88, 134
   CD spectrum of, 89, 119, 121
   ORD spectra of the antiparallel β-form, 119
   rotatory dispersion of, 120
Poly-L-lysine hydrochloride, 76, 115
   ultraviolet absorption spectra of, 76
Poly-L-methionine, 95
   nmr spectra of, 96
Poly-N-methyl-L-alanine, 156
Poly-β-methyl-L-aspartate, 90, 154
Poly-S-methyl-L-cysteine, 119
Poly-γ-methyl-D,L-glutamate, 79
   infrared spectra of, 104, 105
Poly-γ-methyl-L-glutamate, 78, 154
   N-deuterated infrared spectra of, 103, 124
   infrared spectra of, 102, 124
   infrared spectrum of β-form, 123
   ultraviolet absorption spectrum of, 79
   X-ray diffraction of, 71
Poly-L-norleucine, 104
Polypeptides, antiparallel-chain β-form, infrared spectra of, 100
   with aromatic side chains, conformation of, 111
   bond angles in, 140
   bond lengths in, 140
   centrosymmetra, 150

SUBJECT INDEX

circular dichroism spectrum of, 109, 122
   of charged, 115
   of films of, 121
   of random coil, 115
composed of sequential alanyl and glycyl residues, 119
conformations of, 139, 151
   restrictions of randomness of, 114
   steric restrictions on, 144
   theoretical treatment of, 140
disordered form, infrared spectrum of, 100
helical form, infrared spectrum of, 100
infrared dichroism of, 119
infrared spectra of, 97, 98, 99, 102, 103, 124
   of random coil, 115
infrared spectra of, amide V bands of, 115
minimum potential energy calculations, 139
molar residue absorptivities for, 80
nmr spectra of, 94
optical rotatory dispersion, spectra of, 119
   spectrum of random coil, 115
parallel-chain β-form, infrared spectrum of, 100
potential functions describing, 141
rotational angles, notations for, 156
rotatory dispersion curves of, 82
sequential, 95
stereochemistry, 70
β-structures of, 116
synthetic, 70
ultraviolet spectra of, 80
β-Polypeptides, circular dichroism spectrum of, 119
   infrared spectra of, 123
   of amide I band of, 123
   of amide V bands of, 123
meso-Polypeptides, 105
Poly-L-phenylalanine, 83, 95, 111
Poly-L-*p*-(phenylazo)phenylalanine, 112
Poly-L-proline, 105
   circular dichroism spectrum of, 106, 107
   circular dichroism and ultraviolet absorption spectra of, 108
   conformations of, 156
   infrared spectra of, 107
   mutarotation of, 106, 107
   optical rotatory dispersion of, 106

   relaxation times for, 137
   rotatory dispersion of, 107
Poly-L-proline I, 106
Poly-L-proline II, 106, 138
Polyproline I, 138
Polyproline I - polyproline II transitions, 137
Polyproline II-type helix, 115
Poly-β-*n*-propyl L-aspartate, 119
Poly-L-serine, 115, 119
   circular dichroism spectrum of, 122
   ultraviolet spectrum of, 123
β-Poly-L-serine, circular dichroism spectrum of, 120
Poly-L-tryptophan, 83, 111
   circular dichroism spectra of, 112
   X-ray diffraction of, 112
Poly-L-tyrosine, 83, 111, 134
   circular dichroism spectrum of, 110, 111
   conformations of, 154
   optical rotatory dispersion spectrum of, 154
Poly-L-tyrosine helix, screw sense of, 112
Poly-L-valine, 119
   conformations of, 153
Porcupine quill, conformations of, 116
Potential barrier, 230
6-12 Potential function, 146
Potential functions, describing polypeptides, 141
PO ylids, 9, 11, 13, 20
Prelog's rule, 44
L-Proline, 105
Propanal, 14
Propene, 187, 199
   3-substituted, table of enthalpy differences between rotamers, 188
Propionaldehyde, 172, 173, 174, 177, 255
Propionaldehyde hydrazones, 193
Propionaldoxime, 193
Propylacetaldehyde, 174
n-Propyl bromide, 232, 241, 268
n-Propyl chloride, 268
   conformations of, 235
Propylene carbonate, 253
n-Propyl fluoride, 268
n-Propyl formate, 247
n-Propyl halide, 263
n-Propylid, 17
n-Propyl iodide, 268
1-Propyl radical, 197
Proteins, denaturation of, 124

SUBJECT INDEX

globular, 157
Pulse apparatus, block diagram of, 225
Pulse technique, for ultrasonic measurements, 225
Pyrethrins, 21
cis-Pyrethrolone, 21
Pyrethrones, 21
Pyrrolidinium salts, 307

Quasi-racemic compounds, 52
Quaternary ammonium salt, chiral, 43
  dealkylation of, 302
Quaternization, of piperidines, kinetic analysis of, 290
  with trideuteriomethyl iodide, 299
  table of, 282
Quinolizidine, 305

R, configurational symbol, 65
Racemic polypeptides, 104
Random coil conformation, 76, 119
Random coil polypeptides, 113, 115
Random copolymers, 83
Randomness of polypeptide conformations, restrictions on, 114
Random polypeptide chain, 114
Random polypeptides, ultraviolet absorption spectrum of, 114
Reduction with aluminum chloride, lithium aliminum hydride, stereospecific, 41
Relaxational response, for a periodic forcing function, 218
Relaxation frequency, 214
  for conformational equilibria, 239
Relaxation methods, periodic disturbance, 133
  single displacement, 133
Relaxation time, τ^*, 133, 139
Residue translation, 71
Reversibility factor, 9
Ring inversion, table of energy parameters for, 251
  theories of, 251
Rotational barriers, 206
  in amides, 245
  in amines, table of, 244
  of C-N bond, 145
  in esters, table of, 247
  by infrared spectroscopy, 269
    table of, 268

  by molecular orbital calculations, 173
  theory of, 265
  two-fold, 172
Rotational symbol, 65
Rotor, potential energy diagram for, 265
Ruthenium tetroxide, oxidation with, 299
Rydberg orbitals, in polypeptide spectra, 76

S, configurational symbol, 65
Santonide, 53
Schiff bases, chiral, 32, 33
Secondary structures, of polypeptides, 71
Selection rule, for axial chirality, 62
  for planar chirality, 63
Sequence rule, 59, 60, 61
  applications to biaryls, 62
  applications to trans-cycloalkenes, 63
  applications to paracyclophanes, 63
  applications to spirans, 62
Sheppard's Rules, 264
  table of results, 264
Silk, conformations of, 116
Slater-Kirkwood equation, 146
Solvent effects, on polypeptide conformations, 147
  on rates and stereochemistry of the Wittig reaction, 7
Solvent-polymer interactions, in polypeptides, 141
Solvent-solute interactions, 157
Sound absorption, 134
Sound absorption coefficient, 210
Sound absorption of, per unit wavelength, 214
Sound velocity, dispersion in, 215, 228
  measurements of, 226
  temperature dependence on, 211
Sound wave, pressure variation of, 217
  propagation of, 208
  velocity of, 211
Specific heat, 220
Spectroscopy, far infrared, 267
  laser Raman, 267
  neutron scattering, 267
Spectrum, acoustic, 208; see also Acoustic spectrum
Spiranes, absolute configuration of, 43, 47, 51, 56
  application of sequence-rule method, 62
  axially dissymmetric, 32, 33
Spiro[4.4]nonane-1, 6-diol, 48

Spiro[4.4]nonane-1,6-dione, 47
Squalane derivatives, 22
Stefan's Law, 209
Stereoselective synthesis, of simple olefins, 17
cis-Stereoselectivity, of Wittig reaction, 2, 14
syn-Stereoselectivity, 26
trans-Stereoselectivity, 21
  of Wittig reaction, 3, 5, 11
Stereoselectivity, of carbon-carbon linking reactions, 26
  due to electronic effects, 45
  due to steric effects, 35, 38, 45, 48
Stereospecificity, of quaternizations of piperidines, 300
  due to ring strain, 34
  due to steric effects, 39, 44
Steric map, 141
Stevens' rearrangement, stereospecific, 43
Stilbene, 12, 13, 21
  from Wittig reaction, stereochemistry of, 10
Stokes-Navier equation, 209
β-Structures (parallel and antiparallel), 116
Styrene, 13
  chiral, 33
N-Substituted amides, ultraviolet spectra of, 76
Sulfites, cyclic, ring inversion in, 250
Sulfonium salts, lactones from tertiary, 307
Swain-Thornton postulate, 288

Temperature jump method, 134
1,1,2,2-Tetrabromoethane, 232, 241
1,1,2,2-Tetrachloroethane, 268
Tetrahydroisoquinolines, 299, 301
1,2,3,6-Tetrahydropyridines, 299
Tetramethylene sulfite, 253
Thebaine, 34
Thermal analysis, 52
Thermal expansion, coefficient of, 220
Thermodynamic properties, from sound measurements, 228
Thiophene-2-aldehyde, 254
Tiglic acid ester, 3
Torsional energy, 266
Torsional frequencies, of ethane derivatives, 235
Torsional potential, 141, 144

$n \rightarrow \pi^*$ Transition, of amides, 70
  of helical polypeptides, 79
$\pi_1 \rightarrow \pi^*$ Transition, of amides, 74
$\pi \rightarrow \pi^*$ Transition, of peptides, 70
Transition probability, 230, 233
Transition temperature, 135
Traub helix, 106
Trialkylamines, possible conformations of, 243
Tri-n-alkylamines, 254
Trialkylphosphonium benzylids, 10
Trialkylphosphonium carbalkoxymethylids, 7
Triallylamine, 244
1,1,2-Tribromoethane, 232, 241
Tributylamine, 244
Trichloroacetyl chloride, 184
1,1,2-Trichloroethane, 232, 241, 264, 268
  conformations of, 231
1,2,2-Trichloropropane, 264
1,1,2-Trichlorotrifluoroethane, 268
Tricyclohexylphosphonium carbethoxymethylid, 8
Trideuteriomethyl iodide, 282, 299
  piperidine quaternizations with, 286
Triethylamine, 244, 245
Triethyloxonium fluoroborate, 290, 301
Trihexylamine, 244
Triisopentylamine, 244
Trimer, 78
Trimethallylamine, 244
Trimethylamine, 242, 243
Trimethylchlorosilane, 24
Trimethylene sulfite, 251, 252
Trimethyl phosphate, 255
4,5,6-Trimethyltrimethylene sulfite, 251
Tripentylamine, 244
Triphenylphosphine, 5
  Michael-type addition of, 16
Triphenylphosphonium acetylmethylid, 5
Triphenylphosphonium alkylids, 2, 13, 14, 15, 19, 21, 25
Triphenylphosphonium allylid, 8
Triphenylphosphonium benzylid, 8
Triphenylphosphonium carbalkoxymethylid, 5, 6, 7
Triphenylphosphonium carbamidomethylid, 5
Triphenylphosphonium carbomethoxyethylid, 3
Triphenylphosphonium carbomethoxymethylid, 5, 6, 7, 8

Triphenylphosphonium cyclohexylmethylid, 16
Triphenylphosphonium ethylid, 13, 14, 17, 19
Triphenylphosphonium methylid, 13, 23
Triphenylphosphonium phenacetylid, 5
Triphenylphosphonium propinylid, cis-stereoselectivity of, 8
Triphenylphosphonium salts, 9
Tripropylamine, 244
Tropanes, 283, 284, 286, 291, 294, 295, 297, 298, 301
Tropines, 278, 286, 289, 294, 302, 305
L-Tryptophan, copolymers of, circular dichroism spectra of, 112, 113
Tussah silk, conformation of, 118
Twist-boat conformation, of betaine-lithium salt complex, 27

Ultrasonic absorption, 134, 208
Ultrasonic absorption data, see specific compound
Ultrasonic parameters, variations with frequency, 215
Ultrasonic relaxation, 208, 212
Ultraviolet absorption (UV), of polypeptides, 73; see also Poly-L-proline, circular dichrosim and ultraviolet absorption spectra of
Ultraviolet spectra, of polypeptides, 80
of N-substituted amides, 76
α, β-Unsaturated aldehydes, internal rotation in, 253
table of, 254
α, β-Unsaturated carbonyl compounds, from Wittig reaction, 10
Unsaturated fatty acids, 20
Unsaturated ketones, energy parameters for internal rotation, table, of, 254
α, β-Unsaturated ketones, 21

internal rotation in, 253
from Wittig reaction, 1

Van der Waals contact distances, 142
Van der Waals interactions, 141, 145
Van't Hoff isochore, 259
Velocity dispersion, 134
Vibrational spectroscopy, for conformational analysis, 257
Vicinal coupling constants, 168
Vinylcyclopropanes, 189
Viscosity, shear, 210, 214, 224
volume, or compressional, 210, 224

Winstein-Eliel method, 280
Wittig carbonyl olefination, 2; see also Wittig reaction
Wittig reaction, 2
effect of lithium halides on, 26
mechanism of, 3, 25
effect of polar solvent on, 19
stereochemistry of, effects on by protic solvents, 8
effects on by soluble lithium salts, 8
stereoselective, 21
cis-stereoselective, 20, 26
trans-stereoselective, 19

X-ray determination, of absolute configuration, 58, 59
ortho-Xylene, 192

Ylids, moderated, 8
reactions of, 13
reactive, stereoselectivity in, stable, 4

Zimm and Bragg model, 125

CUMULATIVE INDEX, VOLUMES 1–5

	VOL.	PAGE
Absolute Configuration of Planar and Axially Dissymmetric Molecules *(Krow)*	5	31
Acetylenes, Stereochemistry of Electrophilic Additions *(Fahey)*	3	237
Analogy Model, Stereochemical *(Ugi and Ruch)*	4	99
Axially and Planar Dissymmetric Molecules, Absolute Configuration of *(Krow)*	5	31
Carbene Additions to Olefins, Stereochemistry of *(Closs)*	3	193
Carbenes, Structure of *(Closs)*	3	193
sp^2-sp^3 Carbon-Carbon Single Bonds, Rotational Isomerism about *(Karabatsos and Fenoglio)*	5	167
Chirality Due to the Presence of Hydrogen Isotopes at Noncyclic Positions *(Arigoni and Eliel)*	4	127
Conformational Changes, Determination of Associated Energy by Ultrasonic Absorption and Vibrational Spectroscopy *(Wyn-Jones and Pethrick)*	5	205
Conformational Changes by Rotation about sp^2-sp^3 Carbon-Carbon Single Bonds *(Karabatsos and Fenoglio)*	5	167
Conformational Energies, Table of *(Hirsch)*	1	199
Conjugated Cyclohexenones, Kinetic 1, 2 Addition of Anions to, Steric Course of *(Toromanoff)*	2	157
Cyclohexyl Radicals, and Vinylic, The Stereochemistry of *(Simamura)*	4	1
Electrophilic Additions to Olefins and Acetylenes, Stereochemistry of *(Fahey)*	3	237
Enzymatic Reactions, Stereochemistry of, by Use of Hydrogen Isotopes *(Arigoni and Eliel)*	4	127
Geometry and Conformational Properties of Some Five- and Six-Membered Heterocyclic Compounds Containing Oxygen or Sulfur *(Romers, Altona, Buys, and Havinga)*	4	39
Helix Models, of Optical Activity *(Brewster)*	2	1
Heterocyclic Compounds, Five- and Six-Membered, Containing Oxygen or Sulfur, Geometry and Conformational Properties of *(Romers, Altona, Buys, and Havinga)*	4	39
Heterotopism *(Mislow and Raban)*	1	1
Hydrogen Isotopes at Noncyclic Positions, Chirality Due to the Presence of *(Arigoni and Eliel)*	4	127
Intramolecular Rate Processes *(Binsch)*	3	97
Metallocenes, Stereochemistry of *(Schlögl)*	1	39
Nuclear Magnetic Resonance, for Study of Intramolecular Rate Processes *(Binsch)*	3	97
Olefins, Stereochemistry of Carbene Additions to *(Closs)*	3	193

	VOL.	PAGE
Olefins, Stereochemistry of Electrophilic Additions to *(Fahey)*	3	237
Optical Activity, Helix Models of *(Brewster)*	2	1
Optical Circular Dichroism, Recent Applications in Organic Chemistry *(Crabbé)*	1	93
Optical Purity, Modern Methods for the Determination of *(Raban and Mislow)*	2	199
Optical Rotatory Dispersion, Recent Applications in Organic Chemistry *(Crabbé)*	1	93
Phosphorus Chemistry, Stereochemical Aspects of *(Gallagher and Jenkins)*	3	1
Piperidines, Quaternization, Stereochemistry of *(McKenna)*	5	275
Planar and Axially Dissymmetric Molecules, Absolute Configuration of *(Krow)*	5	31
Polymer Stereochemistry, Concepts of *(Goodman)*	2	73
Polypeptide Stereochemistry *(Goodman, Verdini, Choi and Masuda)*	5	69
Quaternization of Piperidines, Stereochemistry of *(McKenna)*	5	275
Radicals, Cyclohexyl and Vinylic, The Stereochemistry of *(Simamura)*	4	1
Rotational Isomerism about sp^2-sp^3 Carbon-Carbon Single Bonds *(Karabatsos and Fenoglio)*	5	167
Stereochemistry, Dynamic, A Mathematical Theory of *(Ugi and Ruch)*	4	99
Stereoisomeric Relationships, of Groups in Molecules *(Mislow and Raban)*	1	1
Ultrasonic Absorption and Vibrational Spectroscopy, Use of, to Determine the Energies Associated with Conformational Changes *(Wyn-Jones and Pethrick)*	5	205
Vibrational Spectroscopy and Ultrasonic Absorption, Use of, to Determine the Energies Associated with Conformational Changes *(Wyn-Jones and Pethrick)*	5	205
Vinylic Radicals, and Cyclohexyl, The Stereochemistry of *(Simamura)*	4	1
Wittig Reaction, Stereochemistry of *(Schlosser)*	5	1

THE LIBRARY